Methods in Marine Zooplankton Ecology

Makoto Omori
Tokyo University of Fisheries

Tsutomu Ikeda
Antarctic Division
Department of Science and Technology
Australia

A WILEY-INTERSCIENCE PUBLICATION
JOHN WILEY & SONS
New York · Chichester · Brisbane · Toronto · Singapore

English translation and revision of the author's previous book
Methods of Zooplankton Ecology, published in Japanese by
Kyoritsu Shuppan, Tokyo, 1976.

Library of Congress Cataloging in Publication Data:

Omori, Makoto, 1937–
    Methods in marine zooplankton ecology.

    Translation of: Dōbutu purankuton seitai kenkyūhō.
    "A Wiley-Interscience publication."
    Bibliography: p.
    Includes index.
    1. Marine zooplankton—Ecology—Technique. 2. Marine
ecology—Technique. I. Ikeda, Tsutomu, 1944–
II. Title.

QL123.04613   1984   591.52′636′028   83-26012
ISBN 0-471-80107-0

Printed in the United States of America

10 9 8 7 6 5 4 3 2 1

Preface

This book is based on the English translation of our previous book *Methods of Zooplankton Ecology*, which was published in Japanese in 1976 by Kyoritsu Shuppan, Tokyo. The entire text was revised and a number of new paragraphs were added in light of recent progress in marine planktology.

The prime motive for writing this English edition was that the senior author (M.O.), through visiting universities and institutions of marine sciences in many countries, became aware of the strong demand by teachers and young scholars for a book which could provide comprehensive, up-to-date knowledge of research methods. In this book we have intended the review of as many subjects as possible relating to zooplankton methodology.

Research on biological processes, such as productivity and food webs of the marine plankton, has been thought by many people to be the most appealing theme of the ecology of the sea. Furthermore, the plankton is investigated in many places in relation to the problem of environmental pollution. Needless to say, definition of objectives, screening of adequate methods, and careful planning are the first steps and are essential toward successful research achievement. Notwithstanding a long history of plankton research which extends more than a century, many methods routinely used in the field and laboratory are not necessarily well defined in terms of accuracy, limitation, and interpretation of results. It is particularly true in field observation and sampling. During the last decade extensive experimental works have been carried out in the laboratory to characterize various processes of marine zooplankton. We believe that refinement of these experimental approaches, combined with further development of field methods, are necessary so that they become more powerful tools in evaluating functional roles of zooplankton in various marine ecosystems.

There are approximately 3700 species of holoplanktonic animals in the marine environment. However, the number of animals on which ecological studies have been carried out is relatively small. Those animals which have been used in laboratory experiments are even fewer and include mostly crustaceans. Thus, many methods described in this book may be applicable to only a limited number of species. For the remaining organisms, methods must be developed which are appropriate to the behaviors, physiology, and life histories of individual species. Further examination and refinement of presently available methods are also required as many inconsistent results are derived from these methods. We hope that this book is useful to readers in identifying the problems inherent with presently available methods and in developing better new methods.

We would like to express our deep gratitude to Professors S. Motoda and T. Kawamura, who led us to the study of plankton and have given us warm encouragement and guidance at the Hokkaido University. We express sincere appreciation to Dr. D. M. Checkley, Jr., for his full assistance in translating and editing the text. Many scientists read part of our English draft and provided valuable suggestions and comments. Among them, our special thanks are extended to Drs. M. V. Angel, J. R. Beers, C. M. Lalli, M. R. Landry, M. Maeda, M. M. Mullin, G.-A. Paffenhöfer, and E. L. Venrick. Special acknowledgment is also made to Mrs. Fumiko Kawamura for assistance in English translation, Miss Masumi Shimizu for help in searching the literature and preparing the bibliography, and Miss Sachiko Kasuya for aid in preparing certain illustrations.

This work was supported in part by the Ministry of Education, Science and Culture of Japan with a Grant-in-Aid for Publications of Scientific Research Result in 1981.

The references to chemicals and equipments in this book do not imply official endorsements of these products, nor is any criticism of similar products intended.

MAKOTO OMORI
TSUTOMU IKEDA

Tokyo, Japan
Kingston, Tasmania, Australia
March 1984

Acknowledgments _____

We are most grateful to the Tanaka Sanjiro Co., Dr. P. B. Ortner, the Yanagimoto Mfg. Co., and the Nikkaki Co., who kindly granted the use of the photographs for Figures 2.9, 4.4, 5.9, and 7.4, respectively. Acknowledgments are also given to those who granted permission for use of their published materials: A de C. Baker, J. R. Beers, T. E. Bowman, M. A. Buzas, S. H. Chuang, J. M. Colebrook, G. W. Comita, M. Dagg, W. T. Edmondson, A. Fleminger, P. R. Flood, J. Flüchter, R. P. Gerber, R. R. L. Guillard, N. A. Holme, A. G. Humes, W. Ikematsu, K. Itoh, R. Z. Klekowski, M. Lloyd, J. A. McGowan, C. B. Miller, S. Motoda, J. A. Northby, G.-A. Paffenhöfer, M. R. Reeve, F. M. H. Reid, H. S. J. Roe, N. Taga, J. Throndsen, D. I. Williamson, American Society of Limnology and Oceanography, Marine Biological Association of U.K., Institute of Oceanographic Sciences, U.K., Academic Press (London), Blackwell Scientific Publications, Crane, Russak & Co., Elsevier Biomedical Press BV., and Springer-Verlag KG.

Contents_____

Methods in Marine Zooplankton Ecology

What Is Plankton?

1.1 Definition of Plankton

Plankton is a community including both plants and animals that consists of all those organisms whose powers of locomotion are insufficient to prevent them from being passively transported by currents. Except for such large planktonic organisms as Rhizostomeae, their length is usually between a few microns and 20 mm. Some taxa having considerable swimming capabilities, such as Euphausiacea and fish larvae, also belong to the plankton. Some species perform extensive vertical migrations. In practice the majority of these planktonic organisms are either collected with nets or filters of standard pore sizes, which naturally has led to them being classified by size. There is no strict border between plankton and nekton. Small squids, pelagic shrimps, and fishes such as myctophids and cyclothones whose body lengths of approximately 20–100 mm are called micronekton. They are treated as both plankton and nekton.

1.2 Size of Plankton

Plankton is classified into several groups by size. Although such a division is artificial, it is meaningful when the food web in a planktonic community is investigated. The size categories defined at the symposium on "Measurements of Primary Production in the Sea" held in Bergen in 1957 are as follows: ultraplankton (<5 μm), nanoplankton (5–60 μm), microplankton (60–500 μm), mesoplankton (0.5–1 mm), macroplankton (1–10 mm), and megaloplankton (>10 mm) (Cushing et al., 1958). This classification is still used in some textbooks, but it is subjective, and the size range of various groups of plankton is not always standardized. Therefore, these categories are not convenient when one wants to measure quantitative relationships among the different size groups of plankton. Dussart (1965) divided plankton into two major groups, nanoplankton, including organisms which pass through a very fine plankton net (20 μm), and net plankton, which are collected by this plankton net. He then divided plankton into the following five sizes according to an exponential function,

2×10^{n} μm ($n = 0, 1, 2, \ldots$): ultrananoplankton (<2 μm), nanoplankton (2–20 μm), microplankton (20–200 μm), mesoplankton (200–2000 μm), and megaloplankton (>2000 μm). In this definition, however, macroplankton and mesoplankton, both frequently used terms, are treated as synonyms. Also, the difference between megaloplankton and micronekton is unclear.

In this book we propose to divide plankton into seven categories as shown in Table 1.1. The smallest size of net plankton, namely those accurately sampled by nets, is 200 μm. Plankton smaller than that are difficult to sample quantitatively even in the oligotrophic open ocean because of clogging of nets (see Chapter 2, Section 2.3.2). We do not separate micronekton from megaloplankton by size; the former consists of organisms with backbones or exoskeletons, such as fish and crustaceans, whereas the latter includes gelatinous plankton, such as salps and medusae, which are generally fragile and difficult to capture with nets without damage.

Since zooplankton contains small protozoans, its size ranges extraordinarily from 2 μm to about 200 mm, including micronekton. However, the principal components of zooplankton consist of mesoplankton and macroplankton, and in this book methodology for these organisms is mainly described.

1.3 Ecological Classification of Zooplankton

Besides classification by size, plankton has been grouped in various ways based on habitat, depth distribution, length of planktonic life, and so on. The following classifications are often used.

A. Classification by habitat
1. Marine plankton (haliplankton)
    a. Oceanic plankton: Those inhabiting waters beyond continental shelves.
    b. Neritic plankton: Those inhabiting waters overlying continental shelves.
    c. Brackish water plankton
2. Freshwater plankton (limnoplankton)
B. Classification by depth distribution
1. Pleuston: Those living at the surface of the sea, part of whose bodies project into the air. They are often treated as a separate category from plankton because they are transported by wind rather than by current. Examples: *Physalia* and *Velella* (Hydrozoa).
2. Neuston: Those living in the uppermost few to tens of millimeters of the surface microlayer.

Table 1.1. Proposed grouping of plankton by size based on classification by Dussart (1965)

Group[a]	Size limits	Major organisms
1. Ultrananoplankton	<2 μm	Free bacteria
2. Nanoplankton	2–20 μm	Fungi, small flagellates, small diatoms
3. Microplankton	20–200 μm	Most phytoplankton species, foraminiferans, ciliates, rotifers, copepod nauplii
4. Mesoplankton	200 μm–2 mm	Cladocerans, copepods, larvaceans
5. Macroplankton	2–20 mm	Pteropods, copepods, euphausiids, chaetognaths
6. Micronekton	20–200 mm	Cephalopods, euphausiids, sergestids, myctophids
7. Megaloplankton (gelatinous plankton)	>20 mm	Scyphozoans, thaliaceans

[a] Groups 1 to 3 are water bottle plankton, 4 to 6 are net plankton.

3. Epipelagic plankton: Those living shallower than about 300 m in daytime.

4. Mesopelagic plankton: Those living between about 300 and 1000 m in daytime.

5. Bathypelagic plankton: Those living between 1000 and 3000–4000 m in daytime.

6. Abyssopelagic plankton: Those living in water deeper than 3000–4000 m.

7. Epibenthic plankton (demersal or bottom-living plankton): Those living close to the bottom or living temporarily in direct contact with the bottom.

C. Classification by length of planktonic life

1. Holoplankton (permanent plankton): Those living as plankton during their entire life.

2. Meroplankton (transitory or temporary plankton): Those living as plankton during only a part of their life, such as the egg and/or larval stage.

The definition of depth in epi-, meso-, and bathypelagic plankton varies depending on environmental characteristics of the water. The above-mentioned depths are derived from the opinions of many scientists. From results of a number of previous studies on zooplankton vertical distribution patterns, it is considered more meaningful to divide the epiplankton into upper epiplankton (0–150 m in daytime) and lower epiplankton (150–300 m in daytime) and the mesoplankton into upper mesoplankton (300–700 m in daytime) and lower mesoplankton (700–1000 m in daytime).

1.4 Species of Zooplankton

The zooplankton is rich in species. The typical holoplankton consists of many phyla and classes of invertebrates. The estimated number of species of major holoplanktonic animals, excluding most of Protozoa (more than 1000 species in Tintinnida alone) and some minor taxonomic groups, is shown in Table 1.2.[1] Since many teleosts (vertebrates) and benthic organisms (mostly invertebrates) spend their early life stage in the meroplankton, most phyla of the animal kingdom are represented in the zooplankton. A classification of marine zooplankton and list of representative taxa (but see the note on the members of Mastigophora) are given below. "Part" means that some species of a class, order, or genus are holoplanktonic. Diagrams of some taxa are shown in Figures 1.1–1.7.

A. Phylum Protozoa
        Class Mastigophora (Flagellata)
            Subclass Phytomastigophorea

Table 1.2. **Estimated number of species of major holoplanktonic animals and meroplanktonic coelenterates in the ocean**[a]

Taxonomic group			Number of species
Protozoa	Foraminifera		37
Cnidaria	Hydrozoa	Siphonophora	140
		others	600
	Scyphozoa		220
Ctenophora			14–21
Nemeritinea	Enopla		97
Mollusca	Heteropoda		18
	Pteropoda		97
Annelida	Errantia		110
Crustacea	Cladocera		8
	Ostracoda		130
	Copepoda	Calanoida	1850
		Cyclopoida	250
		Harpacticoida	20
		Monstrilloida	90
	Mysidacea		100
	Amphipoda	Hyperiidea	283–300
		Gammaridea	150
	Euphausiacea		85
	Decapoda	Sergestidae	85
		others	85
Chactonatha			70
Chordata	Appendiculata		57
	Thaliacea		65

[a] Most protozoa and some minor taxonomic groups are excluded.

We treat members of the following two orders as phytoplankton in this book because of their holophytic nutrition.

Order Chrysomonadina
　　Part. *Dictyocha, Dinobryon, Emeliania, Isochrysis, Syracosphaera*

Order Dinoflagellata
　　One of the most representative groups of plankton, which contributes significantly to marine food chains. Some species cause red tides. *Ceratium, Dinophysis, Gonyaulax, Gymnodinium, Noctiluca, Peridinium*

Subclass Zoomastigophorea
　　Order Choanoflagellida
　　　　Part. *Diaphanoeca, Monosiga, Stephanoeca*

Class Sarcodinea
　　Subclass Rhizopoda

Order Foraminifera
> Important as fossils; part. *Globigerina, Globorotalia*

Subclass Actinopoda

Order Radiolaria
> Part. *Acanthometron, Aulosphaera*

Class Ciliata[2]

Order Holotricha
> Some species cause red tides. *Mesodinium*

Order Spirotricha
> Part. Suborder Tintinnina is important group of micro-zooplankton. *Condonella, Favella, Parafavella, Tintinnopsis, Tintinnus*

B. Phylum Cnidaria (Coelenterata)

Class Hydrozoa

Order Hydroida

Suborder Athecata (Anthomedusae)
> Many are meroplanktonic. *Leuckartiara, Sarsia*

Suborder Thecata (Leptomedusae)
> *Aequorea, Obelia*

Order Limnomedusae
> Brackish water. *Craspedacusta*

Order Trachylina (Trachymedusae)
> *Aglantha, Geryonia, Rhopalonema*

Order Siphonophora

Suborder Calycophorae
> *Abyla, Muggiaea*

Suborder Physophorae
> *Agalma*

Suborder Rhizophysaliae (Cystonectae)
> *Physalia*

Suborder Chondrophorae
> *Porpita, Velella*

Class Scyphozoa

Order Stauromedusae
> *Haliclystus*

Order Cubomedusae
> *Carybdea, Tamoya*

Order Coronatae
> Deep water. *Atolla, Atrella*

Order Semaeostomeae
> *Aurelia, Dactylometra*

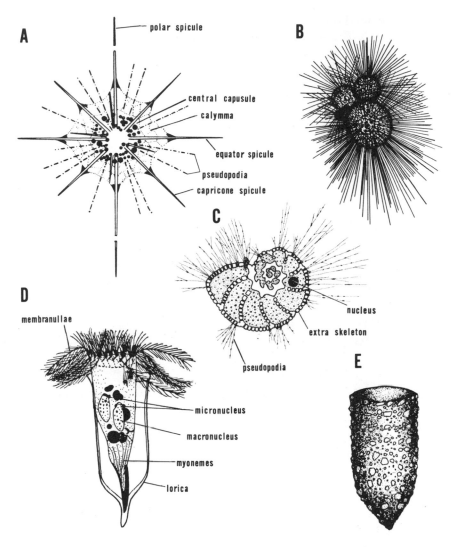

Figure 1.1. Diagrams of zooplankton. (A) Radiolaria, *Acanthometron*; (B) Foraminifera, *Globigerina*; (C) Foraminifera structure (in section); (D) Ciliata, *Favella*; (E) Ciliate, *Tintinnopsis* (lorica).

Order Rhizostomeae
> Some species in genera *Lobonema, Rhopilema,* and *Stomolophus* are unusually large plankton.

C. Phylum Ctenophora
> Class Tentaculata
>> Order Cydippida
>>> *Pleurobrachia, Hormiphora*

Order Lobata
Bolinopsis, Leucothea, Mnemiopsis
Order Cestida
Cestum
Class Atentaculata (Nuda)
Order Beroida
Beroe

D. Phylum Nemertinea
Class Enopla
Order Hoplonemertinea
Deep water. *Nectonemertes, Pelagonemertes*

E. Phylum Aschelminthes
Class Rotatoria
Order Monogononta
Brackish water. *Brachionus, Keratella, Notholca*

F. Phylum Mollusca
Class Gastropoda
Subclass Prosobranchia
Order Mesogastropoda
All of suborder Heteropoda: *Atlanta, Carinaria, Hydrobia, Janthina, Pterotrachea*
Subclass Opisthobranchia
Order Thecosomata (Pteropoda)
Cavolina, Clio, Creseis, Spiratella (Limacina)
Order Gymnosomata (Pteropoda)
Clione, Pneumoderma
Order Nudibranchia
Part. Warm water. *Glaucus*

G. Phylum Annelida
Class Polychaeta
Order Errantia
Part. *Aliciopa, Lepidametria, Poeobius, Sagitella, Tomopteris, Vanadis*

H. Phylum Arthropoda
Class Crustacea
Subclass Branchiopoda
Order Cladocera
Part. Coastal and brackish water. *Evadne, Penilia, Podon*
Subclass Ostracoda
Order Myodocopida
Part. *Archiconchoecia, Conchoecia, Gigantocypris*

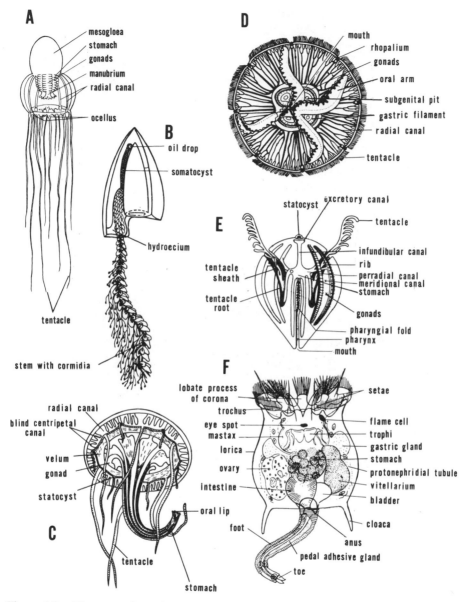

Figure 1.2. Diagrams of zooplankton. (A) Athecata, *Leuckartiara*; (B) Siphonophora, *Muggiaea*; (C) Trachylina, *Geryonia*; (D) Semaeostomae, *Aurelia*; (E) Ctenophora, *Hormiphora*: (F) Rotatoria.

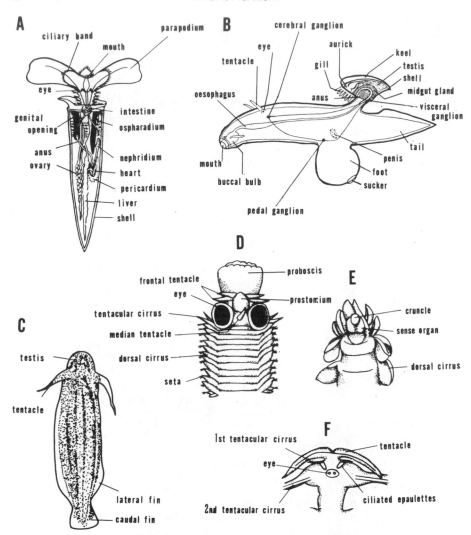

Figure 1.3. Diagrams of zooplankton. (A) Thecosomata; (B) Mesogastropoda, *Carinaria*; (C) Nemertinea, *Nectonemerites*; (D) Polychaeta, Alciopidae head; (E) Polychaeta, Typhloscolecidae head; (F) Polychaeta, Tomopteridae head.

Subclass Copepoda
    Order Calanoida
        One of the most important marine zooplankton taxa.
        Acartia, Calanus, Candacia, Centropages, Eucalanus, Eu-
        chaeta, Eurytemora, Haloptilus, Metridia, Paracalanus, Pseu-
        docalanus, Scolecithrix, Sinocalanus, Temora, Undinula
    Order Cyclopoida
        Part. *Corycaeus, Oithona, Oncaea, Sapphirina*

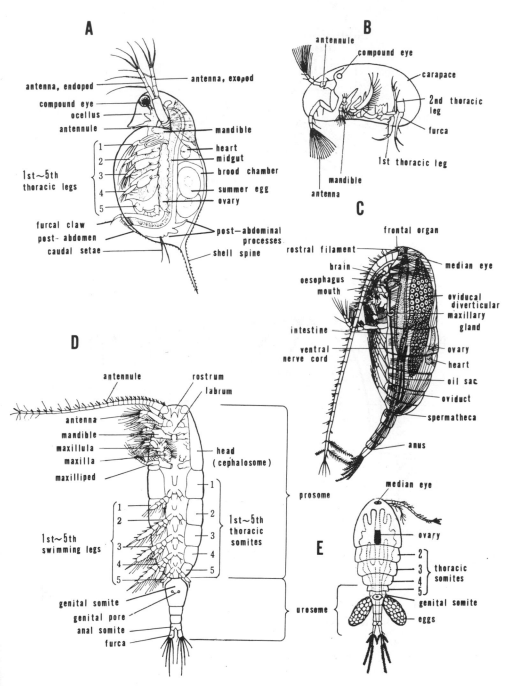

Figure 1.4. Diagrams of zooplankton. (A) Cladocera; (B) Ostracoda; (C) and (D) Copepoda, Calanoida female; (E) Copepoda, Cyclopoida female.

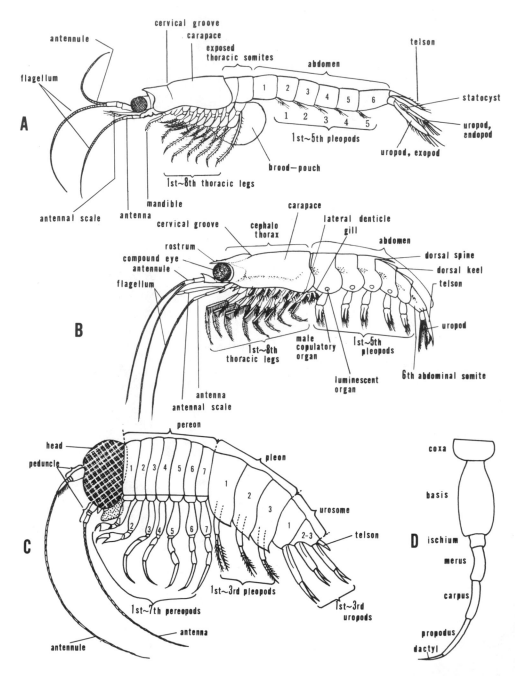

Figure 1.5. Diagrams of zooplankton. (A) Mysidacea; (B) Euphausiacea; (C) Amphipoda, Hyperiidea male (after Bowman and Gruner, 1973); (D) Pereopod (thoracic leg) of Amphipoda and Decapoda.

Order Harpacticoida
> Part. Majority are coastal or epibenthic. *Diosuccus, Euterpina, Harpacticus, Microsetella, Tigriopus, Tisbe*

Order Monstrilloida
> Part. *Monstrilla*

Subclass Malacostraca

Order Mysidacea
> Part. Many are coastal or brackish and epibenthic; some are meso- and bathypelagic. *Archiomysis, Holmesiella, Lophogaster, Mysis, Neomysis, Siriella*

Order Cumacea
> Part. Many are epibenthic. *Dimorphostylis*

Order Amphipoda
> Part of suborder Gammaridea and all Hyperiidea are planktonic or parasitic. Some scientists consider that hyperiids are not free-living amphipods but are all parasitoids, which develop obligatorily on gelatinous hosts (see Laval, 1980). *Cyphocaris, Hyperia, Parathemisto, Phronima, Themisto, Vibilia*

Order Euphausiacea
> Important taxa in epi- and mesopelagic plankton. *Euphausia, Meganyctiphanes, Nematoscelis, Thysanoëssa, Thysanopoda*

Order Decapoda

Suborder Dendrobrachiata
> *Bentheogennema, Gennadas, Acetes, Lucifer, Sergestes, Sergia*

Suborder Pleocyemata
> *Acanthephyra, Hymenodora*

I. Phylum Chaetognatha

Class Sagittoidea
> *Eukrohnia, Krohnitta, Pterosagitta, Sagitta*

J. Phylum Echinodermata

Class Holothuroidea
> Part. Deep water. *Enypniastes, Pelagothuria*

K. Phylum Chordata

Class Appendiculata (Larvacea)

Order Appendicularia (Copelata)
> *Fritillaria, Oikopleura*

Class Thaliacea

Order Pyrosomata
> *Pyrosoma*

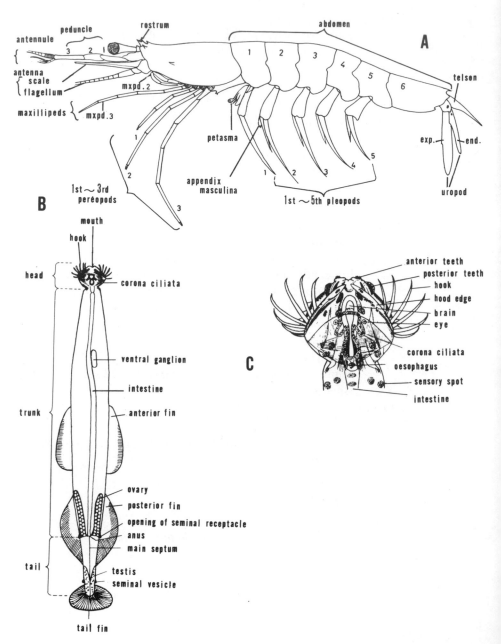

Figure 1.6. Diagrams of zooplankton. (A) Decapoda, *Acetes* male; (B) Chaetognatha; (C) Chaetognatha head.

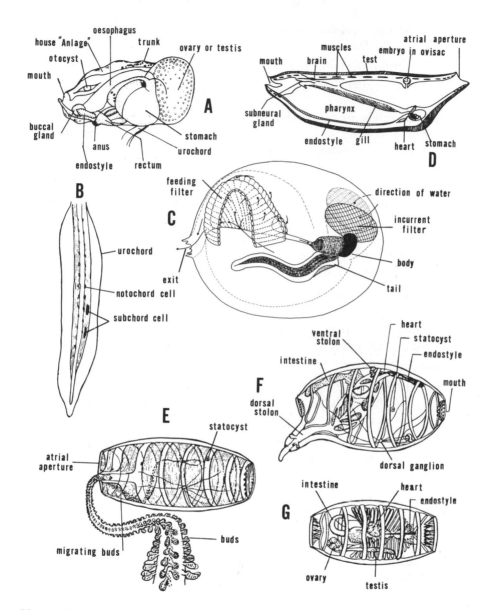

Figure 1.7. Diagrams of zooplankton. (A) Appendicularia body; (B) Appendicularia tail; (C) *Oikopleura* in house (after Flood, 1978); (D) Desmomyaria, *Salpa* structure (in section); (E) *Salpa*, asexual adult; (F) *Salpa*, the same (young); (G) *Salpa*, sexual adult.

Order Cyclomyaria (Doliolida)
Doliolum

Order Desmomyaria (Salpida)
Salpa, Thalia, Thetys

    Among invertebrate larvae that appear as plankton, some have specific
names. The following are representative groups of organisms and the
phyla to which they belong:

A. Phylum Porifera
    parenchymula, amphiblastula, olynthus
B. Phylum Coelenterata
    1. planula, actinula, . . . , larvae of Hydrozoa
    2. scyphistoma, strobila, ephyra, . . . , larvae of Scyphozoa
    3. diconula, conaria, rataria, . . . , larvae of Chondrophorae

C. Phylum Platyhelminthes
    Müller's larva, Götte's larva, . . . , larvae of Polycladida

D. Phylum Nemertinea
    Desor's larva, pilidium, . . . , larvae of Heteronemertea

E. Phylum Mollusca
    1. trochophore, veliger, . . . , larvae of Gastropoda and Bivalvia
    2. rhynchoteuthion, . . . , larvae of Cephalopoda Ommastrephidae

F. Phylum Annelida
    Loven's larva, trochophore

G. Phylum Arthropoda
    1. nauplius, metanauplius, . . . , larvae of Crustacea
    2. cypris, pupa, . . . , larvae of Cirripedia
    3. protozoea, zoea, metazoea, mysis, . . . , larvae of Malacostraca
    4. manca, . . . , larvae of Isopoda and Cumacea
    5. calyptopis, furcilia, . . . , larvae of Euphausiacea
    6. elaphocaris, acanthosoma, . . . , larvae of Sergestidae
    7. phyllosoma, puerulus, . . . , larvae of Palinuridae
    8. glaucothoë, . . . , larvae of Paguroidea
    9. megalopa, . . . , larvae of Brachyura
    10. erichthoidina, erichthus, alima, pseudozoea, . . . , larvae of
        Stomatopoda

H. Phylum Tentaculata
    1. actinotrocha, . . . , larvae of Phoronidea
    2. cyphonautes, . . . , larvae of Bryozoa

I. Phylum Echinodermata
    1. doliolaria, pentacrinoid, . . . , larvae of Crinoidea
    2. bipinnaria, brachiolaria, . . . , larvae of Asteroidea

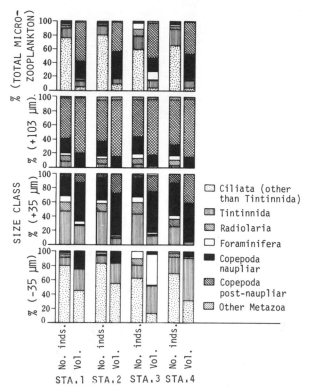

Figure 1.8. Microzooplankton assemblages in the upper 100 m, off San Diego, California. Percentages of the numerical abundance per m³ (left bar) and estimated volume per m³ (right bar). Station 1 was over the continental slope; stations 2, 3, and 4 were further offshore in oceanic water. (After Beers and Stewart, 1969.)

    3.   ophiopluteus, pluteus, . . . , larvae of Ophiuroidea
    4.   pluteus, . . . , larvae of Echinoidea
    5.   auricularia, doliolaria, . . . , larvae of Holothuroidea
J. Phylum Hemichordata
    tornaria, . . . , larvae of Enteropneusta

    Among microzooplankton the most important groups in marine food webs are dinoflagellates, ciliates, and copepod nauplii. Figure 1.8 shows the composition of the microzooplankton off San Diego, California, studied by Beers and Stewart (1969). They collected the microzooplankton, excluding dinoflagellates, that passed a 202-µm mesh and separated it into three size groups: <35 µm, 35–103 µm, and >103 µm. About 60% of total number of individuals in the upper 100 m were ciliates. They were particularly important in the size group smaller than 35 µm. Between 35 and 103 µm tintinnids and copepod nauplii were the most abundant,

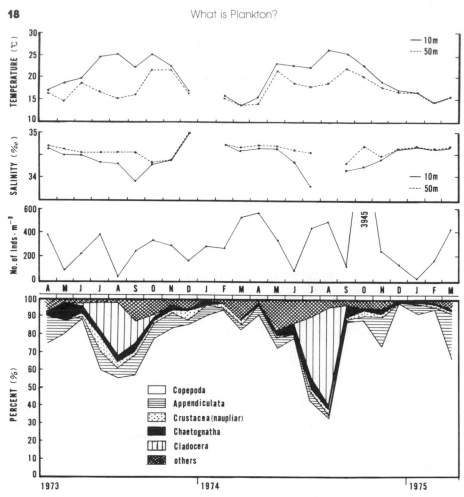

Figure 1.9. Seasonal variation of numbers of individuals and composition by major taxonomic groups of zooplankton at Station 10 (see Fig. 2.1) in Suruga Bay, Japan, from April 1973 to March 1975. Samplings were made with NORPAC net (330-μm mesh) vertical haul from a 150-m depth to the surface.

and then considerable foraminiferans and radiolarians, whereas postnaupliar copepods always comprised more than 50% of the total individuals in the group larger than 103 μm.

In net plankton copepods are generally the most abundant taxon throughout the year. Copepods usually comprise 80% or more of the total number of individuals in most waters, but in coastal waters and inlets cladocerans may undergo rapid increases in numbers by parthenogenetic reproduction; massive occurrences of decapod crustacean larvae may also be seen in particular seasons. Figure 1.9 shows the monthly occurrence of net plankton in the 0–150 m layer at a selected station in Suruga Bay,

Japan. Second to copepods, appendicularians and chaetognaths were most numerous throughout the year and cladocerans became important in summer. In offshore waters, besides copepods, several species of amphipods, euphausiids, and appendicularians generally constitute the main components of the zooplankton and become important food items of larger carnivores.

Notes

[1] It is a pleasure to acknowledge the assistance of the following scientists who have generously given information on the number of species in the indicated taxa: Dr. A. Alvariño (Coelenterata and Ctenophora), Dr. T. E. Bowman (Amphipoda), Dr. M. Imajima (Annelida), Dr. F. Iwata (Nemeritinea), Dr. M. Murano (Mysidacea), Dr. T. Okutani (Mollusca), Mr. N. Shiga (Chordata), and Dr. M. Terazaki (Chaetognatha).

[2] Corliss (1979, p. 189) reviewed major taxonomic controversies on ciliate classification and systematics and proposed his own scheme of classification. According to this scheme the systematic position of Tintinnina is as follows:

   Phylum Ciliophora
      Class Polyhymenophora
         Order Oligotrichida
            Suborder Tintinnida

CHAPTER 2 _____

Sampling

2.1 Choice of Methods

Zooplankton is distributed in the seas as well as in ponds, swamps, rivers, and lakes. One can collect numerous plankton simply by towing a net from the seashore. Even a glassful of water scooped from a tide pool contains many living organisms. But it is difficult to collect the specific plankton needed for research, whether it is a particular species or an assemblage of species. Even within a species many differences can be found between larvae and adults in their sizes, swimming capabilities, and habitats. Therefore, it is almost impossible to sample the entire zooplankton assemblage of particular interest from a large volume of water by using a single collection method. Many factors (e.g., sampling gear, frequency of sampling, degree of net avoidance, and horizontal/vertical migration of species) influence the perception of the true density and distribution of zooplankton. Under present sampling technique plankton samples some-times do not represent the true plankton community in the sea. Our observations by scuba diving reveal that when using a conventional plankton net or small pump it is difficult to collect adequately the organisms that can sense, swim, or tend to swarm/school. Thus, there is a great possibility that many previous "quantitative" studies of macroplankton and micronekton were based on unnatural (artificial) assemblages consisting of specimens captured by accident. In spite of this, scientists are sometimes not sufficiently careful in the choice of equipment and methods for sampling, simply willing to examine the plankton provided by inadequately designed means of collection.

The sampling design is as important as the analyses and techniques used in the laboratory because field sampling provides the most valuable information for investigating the lives of plankton in nature. The infor-mation content of samples is measured by the *accuracy* with which they reflect the true characteristics of the parent population. Well-designed sampling procedures increase the amount of information attainable. One may emphasize *precision* in sampling and analysis, but *precision* does not imply a true and accurate description of a population's characteristics.

The concepts of *accuracy* and *precision* are distinct and should not be confused.

Cochran (1963) and Venrick (1978a) discussed the important steps in a sampling program. They may be summarized as follows:

1. A rigorous statement of objectives including the following:

    a. The ultimate goal—the relationships to be examined, the hypotheses to be tested, the prediction to be made, and so on.
    b. The analytical methods to be employed.
    c. The accuracy and precision desired.

2. Definition of the target population including th following:

    a. Specification of the sampling unit.
    b. Setting of the sampling frame.
    c. Selection of random samples.

If one defines the target population to be the mesoplankton in a particular bay, and the sampling unit the mesoplankton collected with a NORPAC net, the sampling frame is all possible NORPAC net samples that potentially could be collected from the bay. Improper definition of the target population is a frequent source of inefficiency in field studies. Too small a target population may not be representative of the dynamics of the whole population. At the same time an unnecessarily large target population will cause inefficient expenditure of effort. Plankton populations are mobile and in most cases invisible from the surface. Therefore, we can rarely be sure of sampling the same population on subsequent attempts. Also, we cannot sample many locations simultaneously. To study a well-formulated hypothesis, equipment and locations must be carefully selected and each location must be represented with a few replicated samples.

2.1.1 *Quantitative Sampling*

Numerous variations of zooplankton occur over a wide spectrum of spatial and temporal scales. Some variations may be due to growth, reproduction, death, and migration in populations, and others may be due to random effects. These parameters may also vary due to the advection and patchy distribution (spatial heterogeneity in density and composition) of zooplankton. The influence of diel and tidal rhythms on the variance can be a significant problem when interpreting zooplankton data from a time series of samples made at regular intervals of a day or longer and a bias may result. For example, when tidal and/or lunar effects are important, regular sampling at intervals of one month may miss important biological events that occur with the tidal/lunar rhythms.

We must assume that a certain limited volume of water represents the whole volume under examination. With this assumption several samples are taken from the limited volume of water to estimate the abundance and composition of a particular species or assemblage of species in the entire volume of water being considered. In proceeding with the study of zooplankton in a certain area, the following steps should be taken in a typical investigation:

1. On the basis of preliminary knowledge of the area of interest, pilot sampling should be carried out at several locations and depths to obtain material for species composition and variability estimates. It is a good rule to use samples which are well complemented by simultaneous observations of physical and chemical variables.

2. Various methods of sampling with different types of equipment should be tested and compared to determine the best method of gaining samples representative of the zooplankton under investigation. The sampling unit is usually dictated by the available sampling gear. To tackle problems associated with the *behavior* of plankton in the sea, it is necessary to use more than one sampling method. This provides data on the differences in population-specific biology that alternative methods supply (see Omori and Hamner, 1982).

3. In order to examine temporal and spatial variability, a few replicate samples should be taken at both the same time and at different times at a few fixed geographic stations. This will leave less doubt as to the representativeness of the samples collected and will also provide a mean number for each, allowing greater accuracy in population estimates.

4. The entire area of investigation should be divided into several divisions or strata. The distribution and size of these divisions can be designed to distinguish the effects of suspected sources of variation. Ideally, the divisions should be internally homogeneous with maximum variability between one other. These divisions need not be any special shape, but for convenience they are usually square or rectangular[1] as shown in Figure 2.1. Representative samples are collected from each division or stratum.

5. When the area under investigation is divided into m divisions, the population of the entire area is expressed by the following equation:

$$N = \sum_{j=1}^{m} N_j w_j \qquad (2.1)$$

    where N is the abudance of zooplankton in the entire area of investigation, N_j is the average biomass or number of individuals per unit area or volume at a station in the jth division, and w_j is the area of the jth division.

6. The sampling strategy should be altered if necessary so that the investigator may attain his/her final objectives.

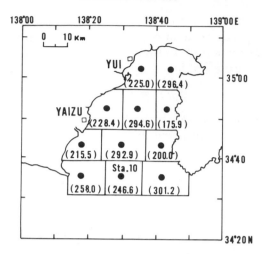

Figure 2.1. Sampling stations for *Sergia lucens* larvae in Suruga Bay, Japan, and area (km²) represented by each station.

Ideally, the variability, precision, and bias of samples should be determined by thorough statistical analysis. However, this is not always practical in a zooplankton study since the complexities of three- and four-dimensional distributions are far removed from the theoretical variations underlying parametric statistical procedures (Venrick, 1978a). Also, the scale on which variability is measured is often arbitrary. With so many variable factors the spread of observations (variance) about the mean is often very wide, even when sampling is repeated at fixed geographic locations, fixed temporal or spatial intervals, and/or fixed depths. According to Cushing (1962) the coefficient of variation for the population density, that is, $(s/\bar{x})$ × 100% (where s is the standard deviation and $\bar{x}$ is the mean), of three plankton species in net samples collected at one location at different times varied between 15 and 70% in calm seas, but reached 300% in storms. Cassie (1963) found similar coefficients of variation. In estuaries, where tidal influences significantly affect sample contents, the hourly fluctuations in zooplankton are often more important than the differences between stations separated by considerable distance or by long times periods (e.g., Sameoto, 1975). Gagnon and Lacroix (1981), in a study of the upper St. Lawrence estuary in Quebec, determined the 95% confidence limits of single zooplankton samples taken over short time intervals to be about 25–400% of the mean. It should be remembered that small, statistically untreated differences between samples collected at different times or locations do not necessarily deserve serious attention.

If the migrations of zooplankton result in a periodic pattern of catch numbers in samples taken at a particular location, this pattern may be represented by a simple mathematical equation. The standing stock of zooplankton can then be estimated by this equation. For instance, vertical migration often produces a sine curve of catch numbers at a particular location over a 24-hr period. In this case the standing stock of zooplankton can be expressed using the following equation:

$$\log_{10} Y = a + b \cos T \tag{2.2}$$

where Y is the standing stock of zooplankton at time T with midnight and noon being $0°$ and $180°$, respectively, and a and b are constants (King and Hida, 1954).

Figure 2.2 shows an estimate of zooplankton standing stock obtained by the California Cooperative Fisheries Investigation (CalCOFI). This figure was obtained by first calculating the median biomass of samples taken in each of three time periods, daytime (D), night (N), and twilight (T, from 1.5 hr before to 1.5 hr after local sunrise or local sunset). Day and night catch numbers were then multiplied by N/D and N/T, respectively, to compensate for day, twilight, and night effect on standing stock estimated to approximate equivalent night time catches. The standing stock values are grouped using an exponential function, 4^n ($n = -2, -1, 0, 1, 2, \ldots$).

It is difficult to make specific recommendations about the most appropriate intervals between samples in time and space. In principle, the most precise estimates for most parameters are obtained by taking more samples at the times or places when populations are most variable. Thus, sampling should be intensified in times and/or places of environmental transition. In the absence of knowledge about a population, the distance from shore or gradients of temperature and salinity may be useful to establish horizontal partitioning, while seasonal cycles of biomass or productivity may be used to categorize populations in time. In practice, with limited manpower and expenditure for sampling, an efficient allocation of effort may call for mixed sampling strategies. For example, to investigate the specific composition of a zooplankton assemblage during the seasons when water is mixed, it is better to sample more often; however, because there is less spatial heterogeneity in the populations, the number of stations can be reduced. When stratification develops, changes in time become less important but assemblage become more different from one place to another. In this case it is worth sampling a greater number of locations at the expense of reducing the frequency in sampling. Margalef (1978) suggested that the average distance between samples be chosen in relation to the scale of the phenomenon under study (Table 2.1). This suggestion was for phytoplankton studies but is also applicable to zooplankton investigations. One should refer to Platt et al. (1970), Cassie (1971), Platt (1975), Ibánez (1976), and Margalef (1978) for discussion of the criteria for selecting sampling stations in relation to the spatial and temporal distribution of the plankton.

Table 2.1. **Recommended average distances between samples in relation to the different scales of the phytoplankton phenomenon under study**[a]

	Approximate size	Separation between samples		
		Horizontal	Vertical	Time
Upwelling	100–1000 km	10–100 km	10–50 m	100 days
Coastal areas	10–100 km	1–10 km	1–10 m	10 days
Red tides, pollution	1–10 km	100 m–1 km	0.1–1 m	0.1–1 day
Microdistribution	100 m–1 km	10–100 m	0.01–0.1 m	0.01–0.1 day

[a] After Margalef (1978).

Lastly, the following suggestions concerning sampling duration are made. For the study of tidal variations, horizontal or vertical transport, or diel migration of the plankton, it is desirable to use frequently repeated samples at fixed locations and at a number of depths. This time series should extend for more than 24 hr. To study annual cycles, the continuation of a sampling program at fixed location over many years always reveals considerable differences between successive yearly cycles. To compare between years, samples should be collected from a population at least once a month over 3 or 4 years. If the objective is to measure the growth or mortality rate of a certain species, a number of samplings must be scheduled in relation to an estimate of the species' reproductive period and life span. These samplings must be spaced closely enough so as not to miss mechanisms of egg production and development of cohorts. Reproduction of cladocerans and copepods living in the coastal waters of Japan occurs from April to October, with the life span of each individual being 20 to 50 days. In order to quantitatively describe the fluctuations in abundance of these species, sampling should be conducted at least once or twice a week during those months.

2.1.2 *Qualitative Sampling*

In order to study species richness, distributions, or seasonal fluctuations qualitatively, preliminary sampling and observation is necessary to determine locations representative of the entire area. Normally, stations are selected where sampling may be repeated using the same method throughout the year and where physicochemical environmental factors are well known. When we want to collect as many species of plankton as possible from a certain region, it is necessary to use a variety of sampling gear, each with different capabilities. To collect organisms that tend to aggregate or occur rarely, it may be necessary to increase the volume of water filtered per one sampling. Wiebe (1972) reported that the towing distance is more

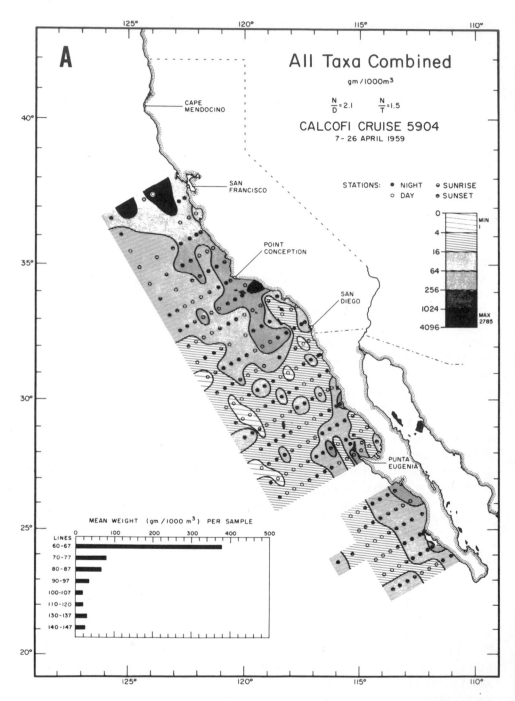

Figure 2.2. Examples to show abundance (biomass) of zooplankton off California. Samples were taken at different times of the 24-hr cycle. N/D means ratio of the biomass between night and day samples; N/T means that between night and twilight samples. (A) unadjusted

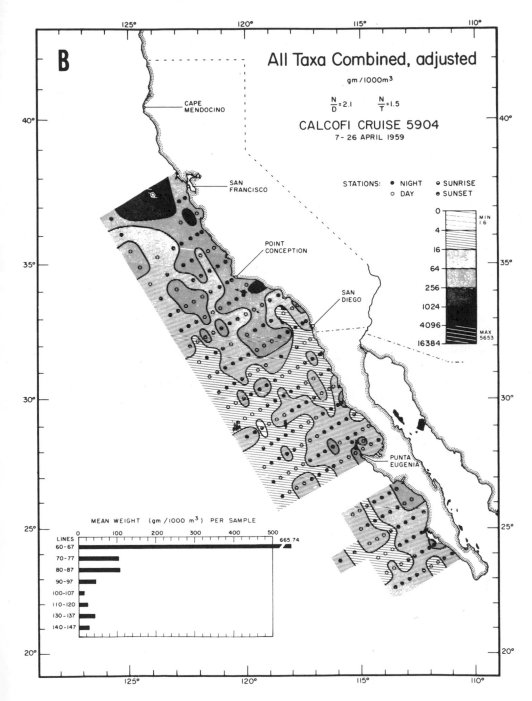

values; (B) day and twilight values adjusted by the ratio of the medians method to approximate median night values. The distribution of the biomass estimates is contoured in the usual manner. (After Isaacs et al., 1969.)

important than the size of the net in determining the accuracy of samples. The method for determining the sample size from species–individual abundance relationships will be discussed in Chapter 10, Section 10.3.4.

2.1.3 *Standard Sampling*

To investigate long-term variations in species composition and abundance of a plankton community in a particular region it is best to choose a particular piece of equipment and standard sampling procedure. Such a method must be capable of sampling a wide range of plankton, including species that are both quantitatively and qualitatively representative of that region. Uniformity in sampling methods also makes comparison of the abundance of certain groups of plankton or of total plankton biomass in wide geographical areas possible. Attempts at sampling standardization have been codified by an international group of scientists who recommended particular types of plankton nets for various sizes of zooplankton as well as methods of deployment. As a result, the NORPAC and IIOE nets were used as the standard samplers for the Cooperative Survey of the Kuroshio and the International Indian Ocean Expedition, respectively. The nets were towed vertically at a speed of 1 m·sec^{-1} from 150 m (NORPAC) or 200 m (IIOE) to the surface.

When a new type of sampler is to be adopted as a standard sampler, it is necessary to examine the new sampler relative to existing standard ones in order to compare their sampling capabilities.

2.2 Microzooplankton

2.2.1 *Sampling*

Sampling plankton with very fine mesh nets often results in clogging problems. Generally, a fixed amount of water is collected with either water bottles or a pump and the organisms are concentrated.

Sampling with Water Bottles

Surface water can be obtained by gently scooping water into a container of a suitable size from the leeward side of the ship. In inshore waters where plankton is very dense, a 100-ml water sample from surface water may be sufficient, but under the usual conditions of study it is advisable to sample 1–50 liters of water depending on the density of the plankton.

In shallow waters, particularly in inlets and creeks, one may collect specimens with a simple "snatch" sampler (Fig. 2.3).

The Van Dorn bottle (Van Dorn, 1957) is one of the most frequently used water samplers (Fig. 2.4). It is a cylinder with both ends open and is normally made of nonmetallic material, such as polyvinyl chloride or

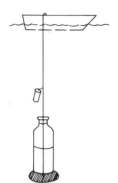

Figure 2.3. Simple water sampler.

polyethylene, and has a capacity of 5–20 liters. Its two ends are sealed by stretched rubber lids pulled tight by rubber tubing inside the cylinder. Prior to taking the sample, the rubber lids are hooked on a release attached to the sampler so that water can pass freely through the cylinder; the lids are released and will close when a messenger strikes the release at a particular depth. Usually 3–10 samplers are attached on a wire so that water from various depths can be collected at once.

If the wire from the ship is not vertical, the depth of the sampler is estimated by measuring the wire angle and the length of the wire paid out by means of an incrinometer and, if necessary, to let out additional wire to reach the desired depth. However, in sampling deep layers, it is

Figure 2.4. Van Dorn sampler.

desirable to measure the depth accurately by attaching an STD, CTD, or set of protected and unprotected reversing thermometers just below the deepest sampler.

Sampling with Pumps

Pumps have been used by many workers for sampling microzooplankton as well as net plankton in recent years (Aron, 1958; Beers et al., 1967; Lenz, 1972; Icanberry and Richardson, 1973; Mullin and Brooks, 1976; Miller and Judkins, 1981). Using a pump, it is possible to obtain a sample from either a point source or integrate over a defined layer with reliable measurement of the volume of water filtered. It can also collect physico-chemical parameters in exactly the same water at exactly the same time. However, pumping systems are never light or easy to handle. Due to the use of hose and the need for ship's power by most pumps, their sampling depth is practically limited to 100 m or less. Other disadvantages are:

1. Frictional resistance of water in the hose may cause turbulence and smearing of samples taken at various depths when collecting from different strata.
2. The amount of water filtered by pump sampling is smaller than net sampling and avoidance of the intake by highly motile organisms may occur, though the latter problem has not been studied in detail.
3. Specimens may be physically damaged and/or suffer from adverse physiological effects.

Firm-bodied zooplankton can be obtained in good condition, often alive, if a diaphragm pump or vortex pump[2] is used, but gelatinous animals are usually broken into unidentifiable pieces by the turbulent flow or by the pump itself (particularly centrifugal pumps). Therefore, pump sampling is perhaps the best method to investigate the small-scale distribution of microzooplankton and smaller firm-bodied net plankton in the shallow layers.

There are two types of pumping systems, a hose-on-suction-side system (pump on deck) and a hose-on-output-side system (submersible pump). Examples of the component selection process of the two pumping systems and design criteria are given by Miller and Judkins (1981) and shown in Figure 2.5.

The collection of samples on deck must take into account the time required for water to travel through the hose from the point of intake to discharge. To determine this "delay" time, it is best to introduce dye into the intake of the pumping system and measure the time required for it to reach the point of discharge. Ultimately, water delivered by hose to the deck of a ship undergoes mesh filtration.

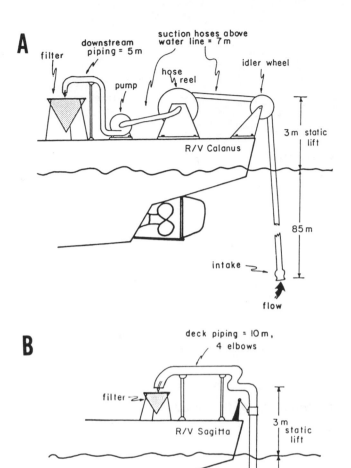

Figure 2.5. Example of component selection process for hose-on-suction-side (A) and hose-on-output-side (B) pump zooplankton samplers. Design criteria: (A) Minimum of 750 liters·min^{-1} from depths to 85 m. Typical discharge height will be 3 m above water line. Suction hose (10.2 cm in diameter) will be coiled on deck on a reel. Total above-water line piping about 7 m. Downstream piping about 5 m with three elbows. Independence from ship's power disirable so that it can be used easily on many ships; (B) Minimum of 750 liters·min^{-1} from depths to 90 m through straight hose. Configuration of ship determines that water must be released 3 m above water line. Additional hose and piping on deck will be 10 m with four wide elbows. (After Miller and Judkins, 1981.)

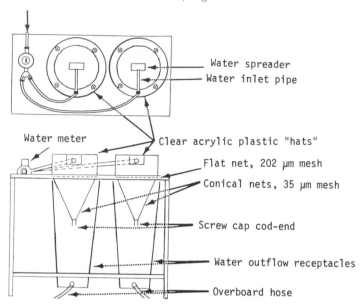

Figure 2.6. The deck-mounted microzooplankton collector. (Modified after Beers and Stewart, 1970, and Beers, personal communication.)

2.2.2 *Concentration of Samples*

Mesh Filtration

At first, a fixed volume of water is passed through a mesh with an opening of about 200 μm in order to exclude mesoplankton. Then the microzooplankton in this water is concentrated in a volume of less than 200 ml. Concentrating with a mesh is performed with small conical nets of fine gauze (20–35-μm mesh opening). Gravity and the flow of the stream carry the plankton into the cod-end where it is concentrated and thus easily removed. In order to protect fine mesh from wear, the net is better to be backed by an outer layer of coarse mesh. Also, it should have the filtering surface "underwater" so there is minimal damage to the microzooplankton. A concentrator used at the Scripps Institution of Oceanography is shown in Figure 2.6 (see also Beers et al., 1967). After use, the meshes must be cleaned with filtered seawater to prevent clogging.

Since many small flagellates and ciliates pass through the 20–35-μm mesh, a fraction of the total volume of water which has been passed through the fine mesh is retained, fixed, and then filtered through membrane filters having appropriate pore sizes or is concentrated either by centrifugal separation or by sedimentation.

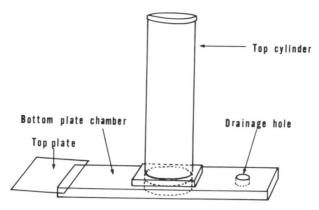

Figure 2.7.   Combined plate chamber.

Centrifugal Separation

Centrifugal separation of microzooplankton from seawater takes 5–10 min at 1000–2000 rpm. This method cannot be employed aboard a small boat due to vibration. It is also difficult to concentrate a large amount of water. Some zooplankton species remain suspended and cannot be separated by centrifugation because their specific gravity is similar to that of seawater. Continuous centrifugation is reported to be effective, but again there is a risk that part of the sample will be lost while draining.

Settling–Sedimentation

After fixation, the organisms in the sample are allowed to settle undisturbed to the bottom of sample containers in the laboratory on land for 1–2 days. Then, by use of a siphon, the surface water is gently removed. If necessary, this process is repeated until the sample in a cylinder is concentrated in 10–50 ml. The settling method is a simple process, although it is time consuming and there is risk of losing or damaging the sample when transferring from one cylinder to another.

The inverted-phase microscope method, or Utermöhl method, is the technique most commonly and successfully used for settling and counting phytoplankton and microzooplankton organisms (Utermöhl, 1931, 1958; Lund et al., 1958). Specially designed combined plate chamber[3] is used with an inverted-phase microscope. The chamber consists of a top (sedimentation) cylinder of 10-, 50-, or 100-ml capacity and a bottom plate chamber (Fig. 2.7).

The principle is to remove the top cylinder after sedimentation, leaving organisms in the bottom plate chamber, which has a height less than the working distance of the condenser of the inverted-phase microscope. The combined plate chamber is ready for use when a top cylinder of the desired capacity is placed on top of the bottom plate chamber. The water sample (often preconcentrated to known volume) is poured into the cylinder and a square plate is placed on the top to eliminate dust and evaporation and prevent the sample from draining out. Usually it takes 18–48 hr to settle. Then the top cylinder is pushed slowly to the side of the bottom plate chamber with the square plate until the cylinder reaches the small drainage hole at one end of the bottom plate. It is necessary to equalize the pressure such that water does not "escape' at the bottom of the cylinder. The top plate of the cylinder is then removed to drain supernatant water out of the cylinder through the hole (see Hasle, 1978, for details). The sample in the bottom plate chamber can be used for identification and counting immediately. Tangen (1976) constructed a sedimentation table which provides a firm base for a number of chambers during filling, settling, and draining of the supernatant water. The table has openings into which the chambers fit and smaller openings corresponding to the smaller drainage holes of the plate chambers. An apparent advantage of the Utermöhl method is that a quantitative sample of microplankton can be concentrated and directly examined without repeated transfer. One problem in the use of this method is that cells sometimes attach to the inside wall of the cylinder and do not settle onto the bottom plate during the sedimentation period.

Membrane Filtration

Membrane filters are usually used to retain particles in water samples. A filter is then cleared with a bleaching substance having the same refractive index as the filter itself and mounted on a glass slide for direct microscopic examination (see Chapter 5, Section 5.2.1). It is also possible to use membrane filters of different pore sizes to simply concentrate microzooplankton from a large volume of water, although in practice it is often difficult to quantitatively wash off specimens from the filters. Filtration is done at low pressure, about 250 mm Hg.

Among the commonly used membrane filters, Millipore and Gelman filters are made of cellulose ester, whereas Nuclepore filter is a polycarbonate membrane whose surface is smooth and can withstand a certain amount of pressure. Therefore reverse washing may be possible; pore sizes up to 12 μm are available.

2.3 Net Plankton

2.3.1 *Introduction to Sampling with Plankton Nets*

Middle-sized plankton are obtained by ordinary plankton net sampling. A net is towed vertically, horizontally, or obliquely. Shapes of nets commonly used are (1) conical, (2) conical–cylindrical (conical net with a

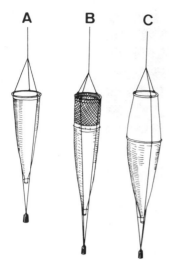

Figure 2.8. Three representative shapes of plankton net. (A) Conical; (B) Conical–cylindrical; (C) Conical with mouth-reducing cone. (After Motoda, 1974.)

collar), and (3) conical with a mouth-reducing cone (Fig. 2.8). Rectangular conically shaped nets have also been designed. The net is attached to the wire directly or, in most net designs, with a bridle. At the cod-end of a plankton net, a sampling bucket is attached.

In net sampling a comparatively large amount of water is filtered and, as a result, a representative sample of organisms in a given volume is obtained. But certain problems are associated with this method, such as loss of organisms through the meshes, net avoidance, and variation in filtration efficiency. Filtration efficiency differs for different types of nets, and the sampled organisms vary in relation to the mesh size and mouth area of the sampler as well as the towing speed. In the case of quantitative sampling, filtering a large amount of water may minimize the effect of rarity due to the patchiness of plankton but, consequently, precludes the collection of organisms from a particular small volume of water. In other words, by net sampling we can only measure the average density (often underestimated because of net avoidance or extrusion) of organisms integrated over a certain volume but cannot measure the density or distribution of organisms on small spatial scales.

2.3.2 *The Gauze of Plankton Nets*

Silk gauze has been used as plankton net for a long time. Although it has still a few advantages in dry sieving and filtering processes, silk gauze is never superior to nylon gauze when used in the sea. Silk gauze is expensive, abrades easily, holds water in the fibers, and shrinks. In comparison nylon gauze is durable, a few times stronger than the silk, and has better water permeability than the silk. Polyester fiber gauze is also durable but is often

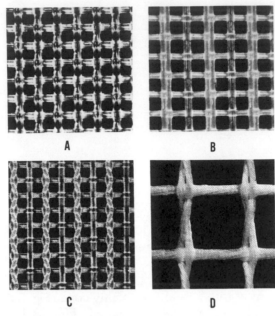

Figure 2.9. Major types of weaves used in plankton gauze. (A) Plain weave (heat-set); (B) Simple locked weave (or normal quality); (C) Semi-twist locked weave (or leno weave); (D) Twist locked weave.

too stiff and unsuitable for plankton net. Some fine meshes with polyester monofilament plain weave may be used for mesh filtration of microzoo-plankton.

Figure 2.9 shows major types of weave used in nylon gauze. Monofilament strand is stronger than multifilament strand under tow. Distortion resistance also depends on the nature of the weave. Interlocked weaves have originally been devised to increase dry sifting capacity, and as regards to resistance to thread sliding over one another in the sea, they are inferior to heat-set plain weave. The ratio of the sum of porous area to the total area of gauze, including pores and strands, is called *mesh porosity* (or open surface). The best gauzes for plankton nets are durable, with thin strands, higher mesh porosity, and even meshes that do not deform. Thus, nylon gauze with monofilament heat-set plain weave is the best material.

The characteristics of nylon gauzes suitable for plankton sampling and/or mesh filtration are summarized in Table 2.2.[4] The standards of fabric numbers are based on silk bolting cloth. Grid gauze (GG), Double Extra (XX), Triple Extra (XXX), and so on, refers to the weight of the fabrics. The number often differs for different systems, materials, and weaves, even though the mesh opening may be very similar. Plankton gauze of 330-μm mesh opening is commonly used to sample meso- and macroplankton; it is bolting silk GG54 and nylon NGG52, respectively.

Table 2.2. Characteristics of nylon gauzes suitable for plankton nets and mesh filtration

Fabric number	Mesh opening (μm)	Number of meshes per cm	Mesh porosity (%)
NGG[a] 15	1400	5.4	59
20	1000	7.5	58
26	800	9.3	55
32	600	12.0	51
38	500	14.0	49.5
45	400	17.2	47.5
52	335	20.2	46
58	300	22.7	46.5
NXXX[a] 7	200	31.2	39
9	150	41.7	39
13	100	58.8	34.5
DIN[b] 110–50	50	117.6	34.5
P[b] 30	30	167.0	25
HD[b] 20	20	185.0	14
HD[b] 10	10	190.0	3.5

[a] Monofilament heat-set plain weave.
[b] Monofilament heat-set plain weave, for mesh filtration. DIN, German standard; P, light-weight fabric; HD, heavyweight fabric.

NGG and NXXX are nylon monofilament heat-set plain weave, whereas NXX (nylon Double Extra or normal quality) is nylon monofilament simple locked weave. Although NXX is still widely used for net to collect micro- and mesoplankton today, it should be replaced with NXXX.

A fine net often clogs, leading to deterioration of its collecting capability. It is therefore advisable not to use a plankton net with a mesh size less than 100 μm in the open sea and 200 μm in coastal waters. On the other hand plankton nets sometimes lose organisms by extrusion through the mesh when organisms larger than the mesh opening are compressed against the net and forced through the mesh. For example, copepods of 350 μm in body width can be quantitatively collected by a net with mesh size of 200 μm but not 330 μm (Vannucci, 1968). In practice, for usual towing speeds (0.7–1.0 m·sec^{-1}) it is advisable to use a net with a mesh size of about 75% of the width of the smallest organisms to be sampled.

2.3.3 Cod-End Bucket

The cod-end bucket must be able to prevent damage to organisms while sampling and should facilitate transfer of the catch into a sample bottle. A solid bucket lowers filtering efficiency of the net and often causes "hang-up" of specimens in the net. When a net is towed, clogging usually begins at the cod-end and gradually extends towards the mouth. The bucket should have two to four mesh windows near the top to avoid damage to plankton by excessive compression and back-flowing of water from the bucket during tow and slop-over of the contents on deck after tow. This is especially useful for the collection of living specimens with large nets.

2.3.4 Filtration Efficiency and Open-Area Ratio

When towing over a certain distance, the ratio of the actual volume of water filtered by the net to the theoretical volume passed through the mouth ring of the net without net is called *filtration efficiency*. The equation is as follows:

$$F = \frac{V}{AD} \tag{2.3}$$

where F is the filtration efficiency, V is the volume of water filtered by the net, A is the area of the mouth of the net, and D is the towing distance.

The ratio of the total open area of mesh openings where the water is filtered to the area of the mouth of a net is called *open-area ratio*. The open area ratio (R) is expressed by the following equation:

$$R = \frac{a\,P}{A} \tag{2.4}$$

where A is the mouth area of the net, a is the total area of the net, and P is the mesh porosity.

The filtration efficiency of a net is influenced greatly by the shape and open-area ratio of the net and partly by the structure of the cod-end bucket. Laboratory examination of the flow pattern near nets has demonstrated that filtration affects the water ahead of the net but not for great distances. The severity of angular deformation of the water streamline increases towards the rim of a conical net. If the filtration efficiency of a conical net with a nonporous, mouth-reducing cone is rated to be 1, that of a simple conical net and a conical–cylindrical net (conical net with porous collar) with the same open-area ratio in the conical net part is less than 1 and nearly 1, respectively (Tranter and Smith, 1968).

Theoretically, the filtration efficiency decreases monotonically with increasing towing distance. According to Tranter and Smith (1968), when a conical net of nylon gauze of 330-μm mesh and open-area ratio of 3.2 was towed at $0.7-1.0$ m·sec^{-1}, it maintained a filtration efficiency of greater than 0.85 until it filtered 49 m^3 of coastal water with abundant plankters or 390 m^3 of oceanic water with less plankton. The volume of water filtered efficiently was increased sixfold by doubling the open-area ratio; namely, filtration efficiency did not lower to 0.85 until the net ($R = 6.4$) filtered 300 m^3 in coastal waters and up to 2564 m^3 in the open sea. A net should have an open-area ratio of at least 3.5, and preferably greater than 6.0 if it is used in an area of high plankton density.

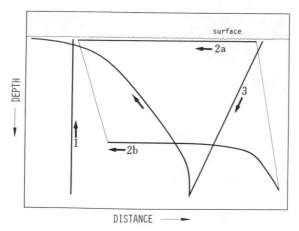

Figure 2.10.   Towing paths of a plankton net. (1) Vertical tow; (2a) Horizontal tow with open net at the surface; (2b) Deep horizontal tow with opening–closing mechanisms; (3) Double oblique tow.

2.3.5 *Sampling Methods*

Vertical, Horizontal, and Oblique Hauls

Figure 2.10 indicates various possible towing paths of a plankton net. In an ordinary vertical haul a net is lowered to a fixed depth from a stationary ship and immediately pulled upwards at a speed of $0.7–1.0$ m·sec^{-1}. An appropriate weight is between $10–15$ kg for a plankton net such as the NORPAC net. Because of its great resistance, the net must be cast at slow speeds so that the wire will not sag or break. This is especially true when a net is lowered from the gunwale of a ship which is rolling heavily. One satisfactory vertical-haul device consists of a pair of nets attached to the wire with a hinge (hence, there is no bridle in front of the nets) so that the mouth rings of the two nets are folded vertically and descend smoothly while lowering the nets (Fig. 2.11).

In a horizontal tow most net types deployed from a moving ship rise rapidly when opened in midwater because of the increase in resistance and therefore samples a certain range of depths. As the wire length increases, the vertical movement of a net increases, which necessitates leaving distance between nets in the upper and lower layers when sampling various layers simultaneously with several nets.

In an oblique tow a heavy weight or depressor is attached some distance below the net. The ship should move slowly (e.g., $2–3$ m·sec^{-1}) while a certain length of wire is let out at a speed of 1 m·sec^{-1}, then rewound at the same rate while the ship decreases its speed (1 m·sec^{-1}) to maintain the net speed as constant as possible through the water. With this method it is not necessary to stop the ship, and sampling is possible even in stormy

Figure 2.11. Simple divice for duplicate samples in a vertical plankton haul (Omori, unpublished). (A) lowering; (B) vertical haul. (1) releasing mechanism; (2) messenger; (3) sliding weight; (4) spreader bar.

weather as the towing operation is relatively simple. However, it is often difficult to control the ship's speed or the speeds of paying out and retrieving of the wire in order to filter the same amount of water from each depth. Also, there is a danger of clogging because of the long towing distance, which causes the filtration efficiency to vary.

Towing Distance and Depth

Towing distance and depth should be determined according to the ecological characteristics of the zooplankton community or assemblage in question. When only the maintenance of filtration efficiency is considered in quantitative sampling a short towing distance is preferred in all sampling methods. The depth reached by a net is estimated by wire angle. In order to measure the exact depth of the tow, a depth recorder[5] or monitor system should be attached to the net.

In quantitative sampling, a vertical haul is often made between the surface and 150–200 m. However, as plankton distributions are uneven, it is better to decide sampling depths by considering the physicochemical and biological structure of the area of investigation rather than sample at an equal depth at each location. In waters where thermocline and/or halocline are present, plankton assemblages differ above and below, and biomass often increases near the discontinuity layer. Therefore, even though the towing depth is constant, samples from nets that pass through the discontinuity layer are often very different from those that do not reach the discontinuity layer.

As stated earlier, the population density of plankton obtained by ordinary sampling only indicates the average value in the water volume sampled. Therefore, the standing stock should be stated with such conditions as "in the layer of 0–150 m" or "over horizontal distance of 600 m." Calculation of the average population density by a vertical haul from 150 m will contain many errors for analysis if the plankton is distributed only in depths shallower than 30 m. Needless to say, one must be aware of the fact that comparison of the average density of a particular organism calculated from a vertical haul between 0 and 30 m depth to that between 0 and 150 m may not be meaningful.

In his research on the change in the concentration and age structure of *Sergia lucens* in Suruga Bay, Japan, Omori (1983) conducted a series of double oblique tows with an Isaacs-Kidd midwater trawl along the fixed lines from the coast to offshore at night. As *Sergia lucens* is usually distributed in the 10–80 m layer during the night, the trawl was towed between the surface and an 80-m depth. Then the average population density was estimated by adjusting the volume sampled at each location using the distributional layers identified by simultaneous observation with a sonar with a frequency of 200 kHz. The equation was

$$N = \frac{n}{V} \frac{L}{D} \tag{2.5}$$

where N is the real population density (m^{-3}), n is the number of individuals per tow, V is the volume of water filtered, L is the towing depth, and D is the depth range of the vertical distribution of the species.

Measurement of the Volume of Water Filtered

A flowmeter[6] is used to measure the quantity of water filtered by a net. The flowmeter has a propeller that rotates with the flow of water and records the number of revolutions (Fig. 2.12). It must be calibrated on a calm day by the following procedure. The flowmeter is attached to the wire with a specially made frame or with the mouth frame of a net from which the net has been removed. The flowmeter is towed vertically from a 100-m depth to the surface at the same velocity used for the actual sampling, and the number of revolutions is recorded. By repeating this operation five times, a reliable average value of the number of revolutions can be obtained. The procedure may be repeated with the net attached to the frame. The filtration efficiency (F) of the net is expressed as follows:

$$F = \frac{N}{N'} \tag{2.6}$$

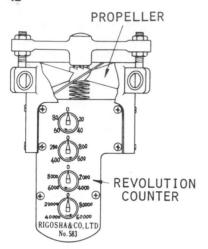

PROPELLER

REVOLUTION
COUNTER

Figure 2.12. Flowmeter.

where N is the average number of revolutions with a net with N' is the average number of revolutions without a net (calibration) when the flowmeter is towed the same distance. If a net with mouth area A is towed over a certain distance D, the volume of water filtered (V) by the net is $V = ADF$. The flowmeter is, in practice, always attached to the net, and it is read after each tow. Therefore, the volume of water filtered is calculated by multiplying the number of revolutions of the flowmeter by the volume of water filtered per one revolution of the flowmeter in the calibration. For a long cruise it is advisable to calibrate the flowmeter at least twice, once at the beginning and once at the end of the investigation.

Thus far, the flowmeter is attached to the center of the mouth frame of the net. But when there is an obstacle such as a bridle in front of the net, the flow of water is significantly disturbed, and the volume of water filtered, measured at the center of the net, becomes less than the amount actually filtered by the net (Mahnken and Jossi, 1967). Therefore, the working group of UNESCO advises placement of the flowmeter at a middle point between the center and the rim of the sampler's mouth (Fraser, 1968a).

Net Retrieval

When a net is raised to the surface, the flowmeter must be secured against wind-driven rotation. The flowmeter and depth recorder are first read. The net is then washed down from the outside so that the sample falls into the bucket. The bucket is then opened and the sample is transferred to a bottle or other container, after which the bucket is closed again for another rinse so that the sample remaining in the net can be obtained. With the NORPAC net this operation is repeated at least twice. When all the sample is obtained, the bucket is removed (or kept open) and the net is washed thoroughly with a hose. The net will then be ready for the next tow.

Nets are always carefully checked after each use and holes and tears must be repaired. Small holes may be repaired temporarily by using a silicone rubber marine sealant.[7] When a net is not used for a while, the net itself should be detached from the frame and bridle, soaked in fresh water, and allowed to dry in the shade. It may then be stored in a cloth bag. Metallic parts, such as shackles and wire clamps, should be oiled and stored separately.

2.3.6 *Various Types of Plankton Samplers*

Nonclosing Nets for Standard Sampling

The principal characteristics of some contemporary nets for usual plankton sampling are shown in Table 2.3. Among them, Kitahara net, Marutoku A net, NORPAC net, and ORI net (Fig. 2.13A) have been widely used in Japan, but the Marutoku A is not recommended because of its low filtering ability.

In quantitative sampling a net with only one kind of mesh must be used. We sometimes see a net with th anterior part made of coarse mesh and the posterior part made of fine mesh. However, because of differences in the filtration efficiency of different meshes, partial clogging and the loss of small organisms through coarse mesh often occur in such a net.

Closing and Opening–Closing Nets

For the study of the vertical distribution and diel vertical migration of zooplankton, it is necessary to collect samples from various layers separately. Samplers have been devised to close either at the mouth or body or at the cod-end bucket. However, cod-end closing devices often have "hang-up" problems, and plankton in the net do not always pass immediately into the cod-end. They remain for a long time in the net and consequently, cause contamination of organisms from other layers.

The closing net is used in a vertical haul in which an open net is lowered to a certain depth and is closed after towing upward through the desired layer. An opening–closing net is used primarily for horizontal or oblique tows in which a net in a closed position is opened at the desired depth and then closed immediately after sampling the desired layer. Although many types exist, only a few samplers have a large sampling capacity, precise opening and closing capabilities, minimal contamination by specimens from other layers, and cause no damage to specimens inside the net. A large sampler capable of sampling simultaneously from more than two or three layers does not exist today.

Messenger-, electric-, sonic-, time-, and pressure-releasing methods are available to open and close nets. Clarke-Bumpus sampler (Clarke and

Table 2.3. Principal characteristics of some contemporary nets[a]

Net	Mouth area (m²)	Form	Mesh opening (mm)	Open area ratio	Source
ORI 100	2.00	Conical–cylindrical	1.00	3.8	Omori, 1965
Indian Ocean Standard	1.00	Conical–cylindrical	0.33	4.3	Currie, 1963
Large tropical Juday	1.00	Reducing cone	0.45	3.1	Bogorov, 1959
CalCOFI Standard	0.79	Conical–cylindrical	0.55	3.2	Smith et al., 1968
Bongo	0.38	Conical	0.50	6.8	McGowan and Brown, 1966
UNESCO WP-2	0.25	Conical–cylindrical	0.20	6.0	Fraser, 1968
NORPAC Standard	0.16	Conical	0.33	3.7	Motoda, 1957
Marutoku A	0.16	Conical	0.33	1.7	Nakai, 1962
Kitahara	0.05	Reducing cone	0.10	4.2	Nakai, 1962

[a] Reducing cone means conical net with a mouth-reducing cone.

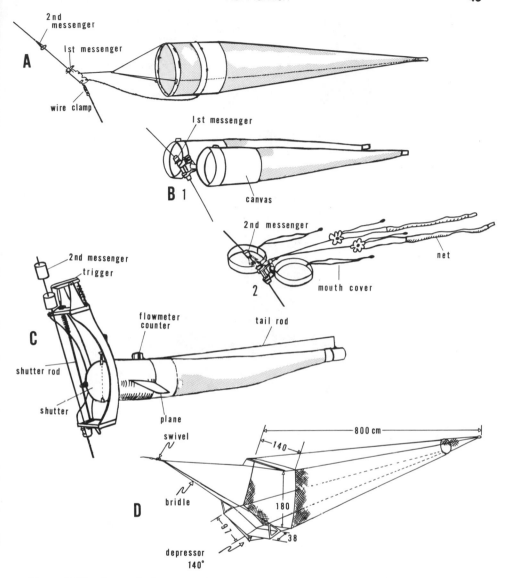

Figure 2.13. Some plankton samplers for horizontal and/or oblique tow. (A) ORI net; (B) Bongo net, (1) net open (2) net closed; (C) Clarke-Bumpus sampler; (D) 1.8-m Isaacs-Kidd midwater trawl (scale in centimeters).

Bumpus, 1950) (Fig. 2.13C), Bongo net (McGowan and Brown, 1966) (Fig. 2.13B), MTD net (Motoda, 1971), and ORI net are closing or opening–closing samplers with messenger-releasing mechanisms. To enable the investigator to maximize his/her decision-making ability as to where and when a net should be opened or closed, the system should be controlled

Figure 2.14. RMT 1 + 8M being launched at sea. The net is lifted by a crane and the weight bar held steady with ropes. (After Roe and Shale, 1979.)

from the surface and have sensors that continuously monitor the operation of the nets and selected environmental variables. Examples of such systems are the Oregon State University IKMT-MPS, which is an Issacs–Kidd Midwater Trawl (Fig. 2.13D) with a five-bar multiple plankton sampler attached to the trawl cod-end (Pearcy and Mesecar, 1971), the MOCNESS multiple opening–closing net (Wiebe et al., 1976), BIONESS net (Sameoto et al., 1977), RMT 1 + 8 with modifications (Clarke, 1969; Baker et al., 1973; Roe and Shale, 1979) (Fig. 2.14), and KOC net (Japan Fishery Agency, 1980). The first three systems are operated from the surface through a conducting cable, whereas the latter systems are opened and closed acoustically by commands from the deck.

Characteristics of the MTD net system (Fig. 2.15), with a mouth area of 0.25 cm^2 and a side length of 1.7 m, are that 5 to 10 nets can be used simultaneously at different layers, all nets can be closed by means of a messenger before retrieval, and the simple method of attaching and detaching nets facilitates repetitive sampling so that 20–30 layers can be sampled in a short time. The ring of the net is fixed to one side of a triangular frame attached to the wire. When the net is lowered, the inclination of towing wire is kept less than 20°, and thus the mouth of the

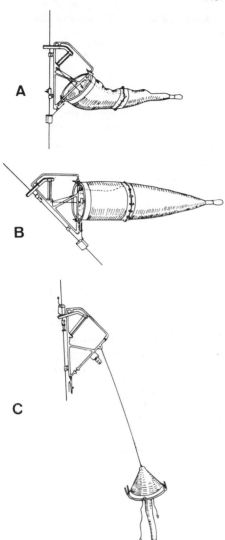

Figure 2.15. Stages in the operation of the MTD net. (A) Lowering the open net; (B) Sampling (horizontal tow); (C) Net closed and retrieval. (Changed after Motoda, 1974.)

net is directed upward so that contamination by plankton from undesired layers is minimized. Nets should be attached to wire as quickly as possible, and after their attachment towing should begin by increasing the speed of the ship until the wire inclination is 45°. To determine the degree of contamination of samples by plankton from other layers, the MTD samples may be compared with samples collected simultaneously with an opening–closing net. The relative effects of contamination may be reduced by lengthening the horizontal towing distance. Used in this manner, this net system is suitable for the quantitative investigation of the vertical distribution of plankton in comparatively thin layers.

Underway and Continuous Plankton Samplers

Underway plankton samplers are designed to be towed from the stern of a ship moving at a speed of 1–7 m·sec^{-1} with the advantage of not having to stop the ship. An underway plankton sampler may also be more effective for the collection of net-avoiding zooplankton. Zooplankton with good swimming ability and well-developed senses can perceive the approach of a plankton net and thus escape it. This net avoidance behavior is especially evident for fish larvae and euphausiids during the day in surface waters (Clutter and Anraku, 1968). Bernard et al. (1973) reported that by increasing the towing speed of a plankton sampler from 1 to 10 m·sec^{-1} the quantity of sample increased gradually and reached a maximum when the speed exceeded 10 m·sec^{-1}. The disadvantage of underway samplers is that the mouth is often less than 5 cm in diameter; therefore, it may not be large enough for fast swimmers and may damage the sample considerably.

Representative underway samplers include the Gulf III (Gehringer, 1952; Bridger, 1958) (Fig. 2.16A), MTD underway plankton sampler V (Motoda, 1959) (Fig. 2.16C), Miller sampler (Miller, 1961), and Jet net (Clarke, 1964; Tanaka et al., 1968) (Fig. 2.16D). Most of them contain a net (some use monel instead of gauze) in a metallic protective tube and are towed with a depressor. The Jet net is said to minimize damage to the sample by decreasing the velocity of water after it enters the sampler.

The distribution of zooplankton over a wide area can be investigated in a short time by continuous underway plankton sampling. With the well-known Hardy continuous plankton recorder (Glover, 1962), the plankton retained on the filtering gauze are wound into a Formalin tank inside the recorder (Fig. 2.16B). Although the sample is compressed between two gauzes and alters the appearance of many organisms, it is possible to distinguish important groups.

The Longhurst–Hardy sampler (Longhurst et al., 1966) is an improved plankton gauze-winding system for research on the detailed structure of spatial distributions. It is attached to the cod-end of a normal plankton net and the gauze is intermittently advanced by battery during vertical or oblique tows. However, as previously noted, "hang-up" is a problem in such a cod-end closing device. In some cases it places a lower limit on the spatial scale that can be accurately sampled (Haury, 1973).

New approaches to studying the spatial distribution pattern are the *in situ* silhouette camera system (Ortner et al., 1981) and *in situ* electronic zooplankton counter mounted on an underway vehicle (e.g., Herman and Dauphinee, 1980; Herman and Mitchell, 1981). With such instruments, zooplankton passing through a net can be counted, sized, and, in sea areas

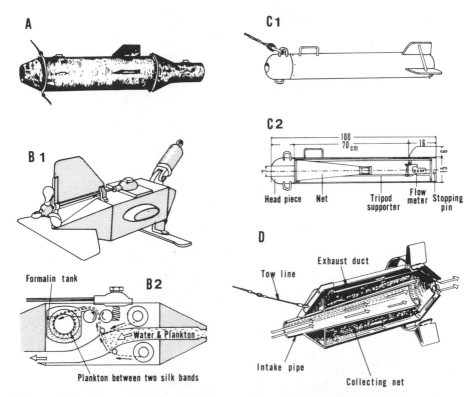

Figure 2.16. Some underway plankton samplers. (A) Gulf III; (B) Hardy continuous plankton recorder, (1) whole profile, (2) a section showing the paths of the two bands of plankton gauze; (C) MTD sampler V (scale in centimeters); (D) Clarke jet net (section).

where few zooplankton species of discrete body sizes predominate, translated to species data. Simultaneous *in situ* environmental data, such as chlorophyll *a* and water temperature, density, and depth, are also obtainable by attaching various sensors (e.g., Variosens fluorometer and CTD unit) to the vehicle.

2.4 Sampling of Micronekton, Neuston, and Epibenthic Plankton

When sampling large plankton and micronekton having considerable swimming ability, net avoidance is a serious problem. A large net without obstacles in front of it that can be towed at a high speed is necessary. The Isaacs-Kidd midwater trawl (IKMT) is equipped with a horizontal bar in the upper part and a depressor in the lower part of the net and is capable of a stable oblique tow at a speed over 2.5 m·sec^{-1} (Isaacs and Kidd, 1953; Aron, 1962). It is an excellent sampler and widely used today. The difficulty of attaching an opening–closing system could be considered a

Figure 2.17.   Isaacs-Kidd midwater trawl with a depth recorder attached to the folding type depressor. See Fig. 2.13D for details.

disadvantage, but this has been overcome (Isaacs and Brown, 1966; Pearcy and Mesecar, 1971). The 1.8- or 3.0-m (vertical dimension of the net mouth) IKMT with a depth recorder attached to a folding depressor (Bercaw, 1966) made of light alloy is very useful.

The Tucker net (Tucker, 1951; Davies and Barham, 1969; Hopkins et al., 1973), RMT net, and their derivatives are rectangular conically shaped nets capable of being towed between 1.5–2.5 m·sec^{-1}. All are designed for horizontal and oblique tows and use a release to open and close the mouth of the net with collapsing of the sides (Fig. 2.18).

In the surface layer, between the water's surface and a few centimeters below, a particular zooplankton community, the pleuston and neuston,

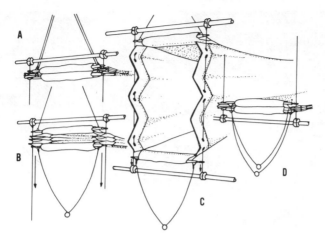

Figure 2.18.   Stages in the operation of the RMT 1 showing the folding of the sides. (A) Closed while paying out; (B) and (C) Opening; (D) Closed while retrieving. (After Baker et al., 1973.)

exist. Because of the difficulties inherent in sampling this very shallow surface layer, it did not attract the notice of planktologists for a long time, but some birds and bats are aware of its existence and have adapted to feeding on it while skimming over the water surface. To sample these plankton, nets have been devised with a wide rectangular or elliptical mouth and suspended from a float at the surface (David, 1965; Hempel and Weikert, 1972; Matsuo et al., 1976; Brown and Cheng, 1981). They have asymmetric bridles which cause the nets to kite away from the ship's bow wave and consequently fish undisturbed water. As an example, a sampler by Matsuo et al. (1976) is shown in Figure 2.19. Ellertsen (1977) designed a net which can sample a number of microhorizons near the surface at once.

Some zooplankton occur in midwater during the night and aggregate close to the seafloor during the day, whereas others migrate into the water column near the bottom at night and reside in the sediment or reef substrata during the day (Russell, 1928; Hesthagen, 1970; Alldredge and King, 1977). They often constitute a particular assemblage in benthic boundary layer. It is not easy to sample these plankton quantitatively, but several apparatuses have been designed. One is to tow close to the flat seafloor by attaching a sled below the net (Omori, 1969b; Rothlisberg and Pearcy, 1977). Beyer's epibenthic closing net (Fig. 2.20) is an example of this type of sampler. Another is the emergence trap. Many traps devised for sampling plankton of the coral reefs, including those of Alldredge and King (1977), Porter et al. (1977), and Hobson and Chess (1979), are either pushed into soft substrates or placed over the bottom with anchors and funnel vertically migrating animals through a pyramid-shaped region,

Figure 2.19. The ORI neuston net. A net, with a rectangular mouth opening, is attached to a frame which holds floats on both sides and is towed from its sides.

Figure 2.20. Beyer's epibenthic closing net. The net was designed primarily for use on muddy bottoms in the Norwegian fiords, but has also been used with some success in the open sea. In this photograph the net with closing band may be seen, above which, outside the net, is a flowmeter. The symmetry of the bridle and the circular framework make the net self-righting, but if it should hit a rock or other obstruction a weak link to the towing bridle breaks and the net closes and capsizes, being brought up by the wire attached to the back end of the frame. Normal closing of the net is by messenger, which also causes the net to close and capsize. (After Holme, 1964.)

concentrating them with plankton netting in a catch chamber. Basal area of the traps ranges from 0.25 to 1.0 m². The sampling effectiveness of different traps has been compared by Youngbluth (1982).

To sample benthic boundary layer zooplankton in the deep sea, Grice and Hülsemann (1970) used single nets that were attached to the submersible ALVIN, opened and closed by ALVIN's manipulator arm, and towed within 30 cm of the bottom, whereas Wishner (1980) attached a MOCNESS-type multisampler below the Deep Tow instrument of the Scripps Institution of Oceanography and towed 10–100 m off the seafloor.

2.5 Sampling of Living Plankton for Laboratory Experiments

Most zooplankters are fragile, and for their use in biological studies of living animals, it is important that the specimens remain undamaged. Figure 2.21 shows two examples of collecting buckets designed for such sampling (see also Reeve, 1981). Their volumes (2–30 liters) are several times larger than that of an ordinary cod-end bucket and have mesh windows in the anterior part.

At the sampling, the bucket is filled full of seawater beforehand so that the net will sink promptly into the water when deployed. A vertical haul or short horizontal tow with a fine-mesh net is preferable in order to avoid damage to the plankton during the haul. The towing speed should be about 0.3 m·sec⁻¹ or less, and when sampling from a large ship, it is better to stop the engine and make use of the drift of the ship caused by the wind. A long tow weakens the plankton in the net. Especially in tropical waters, where the water temperature is high, a considerable shortening of towing time is required. After retrieving the net, the bucket is quickly placed in a container filled with a large amount of seawater, after which the sample is gently transferred to the container. The plankton that remain attached to the net are considered to be damaged and, therefore, should not wash into the container. As plankton are damaged

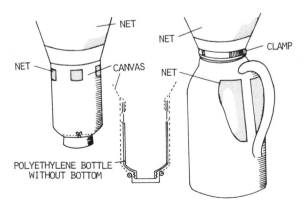

Figure 2.21. Two examples of collecting buckets for living specimens.

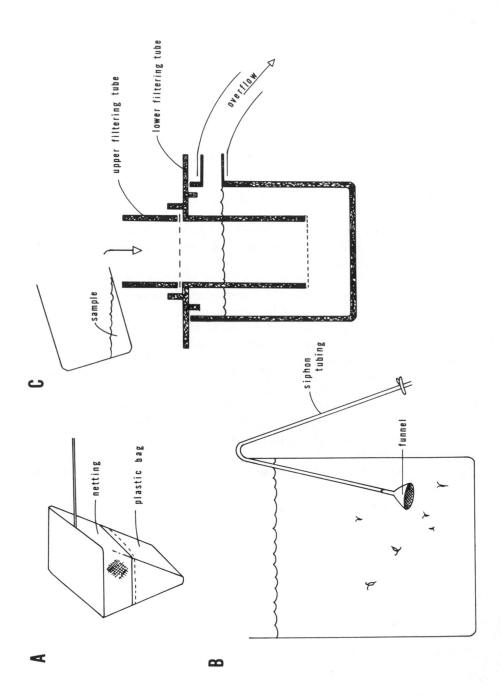

C

upper filtering tube

lower filtering tube

overflow

sample

siphon tubing

funnel

A

netting

plastic bag

B

from direct exposure to the sun, the best sampling time is after sunset when the plankton gather near the surface.

To sample actively swimming macroplankton, a net with a large mouth and good filtration efficiency, such as the ORI-C net (Omori, 1965), should be used. The posterior part of the ORI-C is made of mesh gauze finer than the anterior part in order to reduce damage to the organisms during tow. Childress et al. (1978) designed and used thermally protected cod-end buckets for the recovery of deep-living oceanic plankton. Johnson and Attramadal (1982) devised another type of protective cod-end bucket for use on sledges to collect epibenthic animals. This simply designed bucket can be modified for midwater horizontal tows to avoid the mechanical damage of the specimens and prevent exposure to large fluctuation of temperature, salinity, and light during recovery.

Gelatinous plankton are best captured by divers who introduce them into wide-mouth containers, but some may be obtained with net haul in the manner described above. To transfer specimens to another container, they should be scooped gently with a small beaker or specially designed net (Fig. 2.22A).

When transporting the living zooplankton to the laboratory, their density should be adequately dilute, that is, as low as possible. They are often kept in polyethylene containers with airtight lids. To reduce damage of the organisms due to agitation of the water during transport, the container should be completely filled. Also, every effort should be made to avoid direct sunlight and a change in water temperature. It is therefore advisable to cover the containers with dark cloth. For the purpose of rearing zooplankton, but not for respiration and excretion experiments, the water is best kept 2–4°C lower than the water from which the specimens were collected to decrease mortality. For certain species aeration may be needed if transportation takes a long time.

For sorting specimens, a portion of the sample is gently poured into a transparent container of appropriate size and healthy animals are removed by pipette. The pipettes are made out of glass tubes of various diameters.

Figure 2.22. Simple apparatuses to aid sorting of living zooplankton. (A) Scoop net of which the lower part is replaced with a plastic bag. With this net large zooplankton is transferred with a certain amount of water from one container to the next without exposing them to air. (B) A siphon system to remove undesired small specimens. On one end of the system is fitted a funnel covered with a mesh screen. A gentle siphoning out of the water is essential to avoid larger specimens adhering on the screen. (C) A double filtering system to remove undesired larger and smaller zooplankton simultaneously. A zooplankton sample is gently poured into the upper filtering tube through a coarse mesh screen and passes into the lower filtering tube through a fine mesh screen into a container filled with seawater. The desired intermediate size zooplankton are retained in the lower filtering tube. Upon completion, the upper filtering tube is removed and zooplankton in the lower tube are transferred to the container with a pipette. A whole set is made with Plexiglas or Perspex.

Generally, in warm waters where many small-sized species may be present together, sorting is tedious work. Simple apparatus and methods to aid sorting of live zooplankton are shown in Figure 2.22 (B and C). In some cases the sample may be left for 10–30 min after having been transferred to a container, and the differences in the phototaxis or geotaxis of different species taken advantage of in sorting. Small transparent organisms are easily seen if the background of the container is darkened and light enters through the side. For some species their characteristic swimming behavior as well as body coloration may distinguish them for sorting. Anesthesia may be helpful also (see Chapter 4, Section 4.1.2).

It is known that cladocerans and some neritic copepods produce resting eggs, and their life histories have received attention in studies of the seasonal fluctuation of zooplankton communities. As resting eggs are distributed on or in the bottom sediment, an appropriate sediment sampler, such as the Phleger corer or the Smith–McIntyre grab (see Holme, 1964), is necessary to sample these eggs. Usually, 10 mm of the top layer of the bottom sediment is removed from the sampler with a spoon and filtered through a gauze with a mesh of about 100-μm opening. The residue is then washed off with filtered seawater into a culture dish to sort out the eggs under a dissecting microscope. Resting eggs can also be collected by placing the residue into a dense solution of NaCl or sugar and then centrifuging at 3000 rpm for 5 min; the eggs float to the surface (Onbé, 1973, 1978).

Notes

[1] The area, including complicated coastlines, is measured by means of a planimeter.

[2] Such as DV or DVS type submersible pump, manufactured by Ebara Corp. (Japan).

[3] Now manufactured by several optical firms, such as Carl Zeiss (W. Germany) and Wild Heerbrugg AG. (Switzerland).

[4] Information on plankton gauze is based on the materials NYTAL manufactured by Swiss Silk Bolting Cloth Mfg. Co. Ltd. (Switzerland).

[5] Such as the depth-distance recorder by the Tsurumi Seiki Co. Ltd. (Japan) and time-depth recorder by Benthos Inc. (U.S.A.).

[6] Flowmeter for plankton net is manufactured by Rigosha Co. Ltd. and the Tsurumi Seiki Co. Ltd. (Japan).

[7] Such as the KE45 RTV manufactured by the Shine tsu Chemical Co. Ltd. (Japan) and a similar product by Dow Corning Corp. (U.S.A.).

Fixation and Preservation of Samples

3.1 Fixation

After collecting plankton the sample should be fixed as soon as possible. Apart from some special cases, such as the determination of dry weight or chemical composition, specimens should be fixed within 10 min after collection. We must carefully avoid leaving specimens unfixed. It should also be borne in mind that the condition of specimens will be determined by the initial fixation procedure and that any later modification to the fixative will not generally improve their condition.

3.1.1 *Microzooplankton*

When fixative is added to samples with naked flagellates, the cells are often destroyed or significantly transformed, making their subsequent identification almost impossible. Because of this difficulty, Lohmann (1911) recommended microscopic examination of fresh material immediately after collection. If the material is to be fixed, a common way to preserve specimens is to add 1–5% (V/V) of buffered Formalin (37–40% formaldehyde)[1] to the water sample. To make buffered Formalin, see Section 3.1.2. If the concentration of Formalin is less than 3%, the sample should be refrigerated. Glutaraldehyde, Rodhe's iodine, and osmic acid are also effective fixatives. Glutaraldehyde readily oxidizes to form glutamic acid and as a result lowers the pH to about 5. Therefore, before its use glutaraldehyde is usually neutralized with 1N caustic soda (NaOH) and preserved at 4°C (Trelstad, 1969). The supernatant solution obtained after adding barium carbonate or calcium carbonate to oversaturation can also be used (Greenhalgh and Evans, 1971). To use glutaraldehyde to fix marine microzooplankton, 5–20 ml of fixative, obtained by diluting the original solution to 10% with filtered seawater, is added to 100 ml of the water sample to make a final concentration of 0.5–2.0%. Another method is to dilute the glutaraldehyde to a 5–10% solution by adding sodium cacodylate, $Na(CH_3)_2A_5O_2 \cdot 3H_2O$, and then adding 5–20 ml of this solution to 100 ml of the water sample. The sodium cacodylate buffer, pH 7.4, is made by mixing $0.2M$ HCl and $0.2M$ sodium cacodylate (42.8 $g \cdot liter^{-1}$) with distilled water (2.7:50:147.3 V/V). The sample is kept at around 4°C. Glutaraldehyde inactivates enzymes more rapidly than formaldehyde and is generally considered an excellent fixative, better than formaldehyde, for morphological fixation and enzyme preservation (Hopwood, 1967; Hündgen, 1968).

For the identification of ciliate species it is important to examine the form and/or number of cilia. For this purpose the use of Rodhe's iodine (Holmes, 1962) is recommended. The fixative is prepared by adding 10 g of iodine, 20 g of potassium iodide, and 20 ml of acetic acid to 200 ml of distilled water and mixing well. Add 5 ml of this solution to 100 ml of the sample and store refrigerated, as in the case of glutaraldehyde. Utermöhl (1958) made a fixative by adding 10 g of sodium acetate instead of acetic acid. In this case 1 ml of fixative is added to 100 ml of sample.

Osmic acid is usually added as a fixative to a water sample to make a concentration of about 0.0001% (V/V). For small water samples a few drops of the sample containing the specimens is placed on a glass microscope slide. The slide is turned upside down and held over a bottle containing a 1–2% osmic acid solution (V/V) for 10–20 min to expose it to the vapors. Care should be taken so that osmic acid vapor, which is poisonous, does not leak out of the bottle. Modified Fleming solution is also used as a fixative. It is made by adding 5 g of chromic acid, 2 ml of glacial acetic acid, and 4 ml of a 2% osmic acid solution to 100 ml of filtered seawater and mixing well. Then 5 ml of this solution is added to 100 ml of sample water.

Gold (1969a, 1976) stated that the method used by Karnovsky (1965) is appropriate for the fixation and preservation of marine Tintinnida and other microplanktonic organisms. We have no personal experience in its use and therefore just introduce the preparation of the fixative as stated by Gold. Paraformaldehyde powder (2 g) is dissolved in 25 ml of distilled water heated to 60–70°C. The solution is cleared by adding 1–3 drops of $1N$ NaOH. After cooling this solution, 5 ml of 50% glutaraldyhyde is added. The volume is adjusted to 50 ml with $0.2M$ sodium cacodylate buffer (pH 7.4–7.6) and 25 mg of anhydrous calcium chloride is added. The final pH is 7.2. The solution may be filtered if it is not clear. Since the solution is employed cold, it should be kept refrigerated. For rapid fixation 2 ml of cold fixative is injected into 10 ml of seawater sample or culture medium. The amount of fixing solution added is not critical; cells can be fixed in a full-strength solution using Karnovsky's method if required for some specialized purpose.

3.1.2 *Net Plankton*

Buffered Formalin is the most widely used and highly recommended fixative as well as preservative. One liter of Formalin (preferably analytical reagent grade) is added to 30 g of Borax (sodium tetraborate) and stirred well. Borax will dissolve completely within the first few days but will later gradually precipitate at the bottom of the container. Therefore, the solution should stand for 1–2 months and, before using, any sediment should be removed with filter paper. Since hexamine, rochelle salt, sodium bicarbonate, and potassium oxalate may adversely affect the preservation of some zooplankton species, they should not be used to buffer samples.

The following two solutions are appropriate fixatives: (1) Buffered Formalin (undiluted solution) and, (2) Steedman's fixative (Steedman, 1974, 1976a). The mixed solution (V/V) is as follows:

Buffered Formalin (undiluted solution)	5
Propylene phenoxytol	0.5
Propylene glycol	4.5

To make this solution propylene phenoxytol is dissolved in propylene glycol and then added the Formalin.

In general, after the sample is transferred to a bottle, either solution 1 in the ratio of 5–10 to 90 parts (V/V) or solution 2 in the ratio of 10 to 90 (V/V) is added. If necessary, filtered seawater should be added before adding the fixative so that the level of the sample reaches the shoulder of the sample bottle after fixation. The sample should be gently inverted two or three times so that the fixative becomes evenly dispersed. The pH of both solutions is around 7.6–8.3. If large plankton are few in number or if the volume of specimens is not large, 5% (V/V) Formalin solution will be enough for fixation.

The specimens in the bottle should occupy less than 20% (preferably 10%) of total sample volume. If there are too many specimens, they should be divided into several aliquots; if a shortage of containers occurs during a long cruise, the sample should be fixed in a single bottle with a higher concentration (about 10%) of Formalin solution and transferred to a larger container as soon as possible after returning to port. In bottles containing many specimens, the possibility of physical damage to the specimens is increased. A decrease in pH caused by the oxidation of organic matter in the specimens results in the dissolution of the calcareous shells of Formainifera and Thecosomata (Fig. 3.1).

3.2 Relaxation

When some living specimens are put directly into a fixative, breakage, shrinkage, or other distortions of form may occur. These types of zooplankters are best relaxed before fixation.

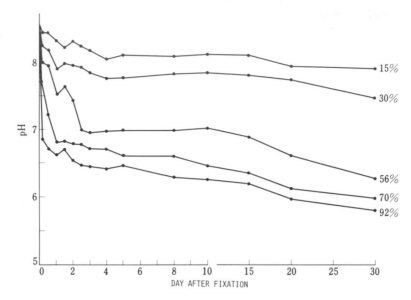

Figure 3.1. Change in the pH of 10% formalin–seawater solution buffered with borax and containing zooplankton, mainly copepods. Percentages are the approximate ratios of the displacement volume of the specimens to the volume of the preservatives ($\times$ 100) (Omori, unpublished).

The following chemicals (including narcotizing agents) are frequently used to render living animals comatose (see also Chapter 4, Section 4.1.2 and Steedman, 1976b, for anesthetic technique): for Ciliata, osmic acid or potassium iodide; for Cnidaria, menthol crystals or magnesium chloride; for most Thecosomata having shells, acetone-chloroform (1% solution), magnesium chloride, propylene phenoxytol, MS-222,[2] or urethane; for Gymnosomata, chloral hydrate or Formalin; for Polychaeta, cold fresh water, magnesium chloride, or Formalin; for Crustacea, menthol crystals, chloroform, urethane, MS-222, or propylene phenoxytol; and for Chaetognatha, menthol or chloral hydrate.

Menthol is used by simply dropping small flakes on the surface of the sample water so that they float. The crystals will gradually dissolve and will anesthetize the plankton. The animals should be touched from time to time until they show no reaction and then put into a fixative.

In the case of magnesium chloride 7–8% (W/V) is diluted in distilled water so that it will be isotonic with the seawater of the sampling area, and this is added to seawater in a ratio of 1:2 or 1:3 (V/V). Animals are placed into this solution; the effect will be evident in about 20 min. Generally, chloral hydrate is used in 0.1% solution.

When 2 or 3 drops of undiluted propylene phenoxytol are dropped into 1 liter of seawater containing plankton specimens, they will gradually dissolve on the bottom of the vessel and anesthetize the organisms in about one day (Owen and Steedman, 1958).

Formalin is used in the following manner (Gohar method; Runham et al., 1965). Animals are placed into a volume of water 50–500 times their own volume and 3 drops of 2.5% Formalin is added for each 100 ml of water every 15 min for the first hour; for the second and third hour 6 and 12 drops, respectively, are added at 15-min intervals. By thus increasing the amount of Formalin, the animals become stationary and are eventually fixed in about 6 hr.

3.3 Prevention of Color Fading of Specimens

The rich variety of colors and color patterns of zooplankton are often adaptations for various purposes, such as concealment or mating recognition. To study such ecological phenomena and to describe and compare species and populations, it may be necessary to measure and describe precisely the color of specimens before fixation. There are several color references for that purpose, among which the *Naturalist's Color Guide* (Smithe, 1974, 1975, 1981) is useful.

There is no way to perfectly preserve the body color of fixed specimens but, by using a phenolic antioxidant such as a 40% emulsifiable concentrate of butylated hydroxytoluene (BHT)[3] or butylated hydroxyamisole (BHA),[4] it is possible to maintain the body color of fish and crustaceans for a reasonable period. The BHT emulsifiable concentrate mixes readily with Formalin and ethanol. When 1–2 ml of BHT is added to 450 ml of 10% Formalin, the body color of a specimen can be maintained for a period of more than one year (Waller and Eschmeyer, 1965). Toyama and Miyoshi (1963) obtained good results by adding 1 ml of a flake of BHA dissolved in 20 ml of ethanol to 1 liter of a 10% Formalin solution.

3.4 Preservation

Specimens can be transferred to preservative about 1 week after fixation. The preservative should be chosen according to the taxonomic group of zooplankton. To preserve Coelenterata, Polychaeta, and Chordata, 70% ethyl alcohol (ethanol) is more suitable than Formalin. For the longer term preservation of Ctenophora, a particular method (described later) is necessary. Generally speaking, buffered Formalin and Steedman's fixative are also used to preserve most kinds of plankton. For the purpose of preservation the concentration of Formalin should be lowered to 2.5–5%. Steedman (1974) advised the use of the following mixed solution (V/V) as a preservative:

Buffered Formalin (undiluted solution)	2.5
Propylene phenoxytol	1.0

Propylene glycol	10.0
Filtered seawater	86.5

When surface seawater from tropical regions is used as a Formalin diluent, a semitransparent deposit of platelike particles may appear in the preservative after several months. This deposit consists of calcite, from the chemical reaction of calcium carbonate, often dissolved at supersaturation level in surface waters, and certain kinds of organic compounds (Omori, 1976a). The following methods prevent the production of such deposits: (1) acidification of the seawater with hydrochloric acid, boiling off the carbon dioxide, and neutralizing the solution with NaOH or (2) removal of carbon dioxide by bubbling N_2 or carbon-dioxide-free air after passing through an alkaline solution through the seawater.

Ethanol is a suitable preservative for many species of zooplankton, but because of its costliness and volatile and inflammable nature, it should be used only under appropriate conditions. A specimen is taken out of the fixative and washed thoroughly in distilled water before being immersed in ethanol; if seawater remains on the specimen, a cottonlike precipitate may occur from calcium and magnesium salts. The specimen is then immersed in a 30% ethanol solution for 10 min, moved to a 50% solution, and after 1 hr transferred to a 70% solution. For the preservation of specimens a 70% solution is adequate; for crustaceans a better result will be obtained if a small amount of glycerin is added to the solution. Ethanol is not advised for Chaetognatha, as it may cause the animals to shrink.

Lastly, the method to fix and preserve Ctenophora will be briefly mentioned. Animals such as *Mnemiopsis* and *Beroe* are fixed apart from other plankton by placing them for 30 min into 99 ml of seawater in which 1 g of trichloracetic acid or *p*-toluenesulfonic acid has been dissolved. Then they are immersed for 5–7 days in 99 ml of seawater in which 1 ml of Steedman's fixative has been diluted. Then they should be preserved in Steedman's preservative. The storage temperature should be between 5 and 20°C (Adams et al., 1976).

3.5 Containers and Storage of Samples

In recent years various plastic containers have come into use, but for various reasons transparent glass bottles are most suitable for the preservation of zooplankton specimens. Glass bottles with wide mouths and polypropylene or rubber linings in a plastic screw-on lid (e.g., mayonnaise bottles) are found in many sizes and are inexpensive and easy to obtain. A glass bottle with a snap-type polypropylene lid is also convenient in preserving specimens, but it is not suitable as a container to be carried to the field. Metal containers and lids should be avoided. Since polyethylene and polypropylene containers are light and durable when empty, they are suitable for use on ship. But because of their opacity it is difficult to observe the sample. Furthermore, polyethylene containers are easily damaged by fixatives and preservatives, causing paraffin to leach out into the sample water. On the other hand polycarbonate containers are transparent and strong but are damaged by polypropylene glycol and lipids of crustaceans.

A Whirl-Pak[5] is a polyethylene bag made in various sizes with a mouth that can be sealed watertight. Because it is light and compact, this bag is useful as a temporary container when it is impossible to transport many glass or polyethylene containers to a sampling location. It is also convenient for shipping specimens. When used for shipping plankton samples, it is safest to put the Whirl-Pak bags in a large, watertight container.

Small specimens should be separated into respective groups or species and stored in adequate-sized glass tubes with preservative and absorbent cotton stuffed tightly in their mouths so that no bubbles remain in the tube. Then the tubes should be sorted according to taxonomic groups or sampling stations and put upside down in large glass containers with preservative and labels. For storage of very tiny specimens such as tintinnids or the appendages of microcrustaceans, Vitro Dynamics' microslides[6] may be used. They are platelike capillary tubes made of hard glass and each has a rectangular cross-sectional opening of a certain size. Specimens and preservative can be sucked up by capillary action or placed by a needle into a microslide. After both ends have been sealed the slides are stored in a large vial. Specimens can be observed under a microscope from both sides of the glass without being taken out of the slide.

It is desirable to store samples, both on board ship and on land, in a dust-free constantly dark and cool place. To transport the samples, it is convenient to put sample bottles in a wooden box with partitions. Also, it is best to seal the lids with vinyl tape to prevent leakage.

During storage evaporation may occur and the pH of the preservative may gradually lower due to the oxidation of organic matter. The best pH of Formalin solutions for the preservation lies between 6.5 and 7.5. To store valuable samples for long periods, care should be taken to check the color and pH of the preservative and the condition of the specimens at least once a month during the first 3 months, and once every 3 months for 1 year following that period. The preservative should be changed if necessary.

3.6 Labels and Log Sheets

Labels are put into the containers when the specimens are fixed. Labeling paper[7] should withstand Formalin, ethanol, glycol, and salt water and should be able to tolerate long immersion in liquids. It should not affect the specimens either. There are papers that contain bleaching agents or starch that eventually cause damage to the specimens. Equal care should be taken with printing ink so that samples are not damaged. Papers used

University of Washington
OCEANOGRAPHY DEPT.

Station .... Date ....

Gear .... Haul No. ....

Time .... Depth ....

Location ....

Collector ....
Form No. UW-Do-P-9

Bottle No: ....

北大水産浮游生物

Vessel: .... Cruise: ....
Area: .... Station No: ....
Lat: .... Long: ....
Date: .... Time: ....
Net: .... Haul: ....
Wire run out: .... Wire angle: ....
Flow-meter No: .... Revolutions: ....
                               Recorded by ....

OCEAN RESEARCH INSTITUTE
University of Tokyo

_____
_____
_____
Locality _____
_____
Date _____
Remarks

北大水産浮游生物

Sample No. .... St. ....

Vessel .... Cruise No. ....

Date & time ....

Lat. .... Long. ....

Net .... Wire out ....

SMITHSONIAN OCEANOGRAPHIC SORTING CENTER
EXPED._____
DATE_____ STA. NO. _____
LOCALITY _____
_____
_____
_____
LAT. _____ LONG. _____
COLL. _____

SMITHSONIAN OCEANOGRAPHIC SORTING CENTER

TAXON:	
PROGRAM:	DATE:
VESSEL:	LAT.
CRUISE:	LONG.
STA. NO.	DEPTH :
GEAR:	SORTER:
REF. NO.	

SI-USNM-180-7-5-66

Figure 3.2.   Some plankton sampling labels.

for labels should be such that they can be written on with pencil and India ink on both sides.

When labels are written in the field, pencils are convenient. However, for labeling preserved specimens, high-quality India ink which will not be affected by Formalin and ethanol is best suited. Ball-point and felt-tip pens should never be used. If typewriter with a carbon ribbon is used, pre-soaking of the ribbon is needed to eliminate leaching of blue ink from the label into the alcohol.

At least the following information should be entered on the label of a field sample:

1. Identity of sampling station (if necessary name of the vessel and cruise number).
2. Sampling date and year.
3. Sampling time.

Plankton Laboratory

Plankton Log Sheet

Cruise No: OS - 26

Method of haul: *Vertical* NORPAC 0-150m

p. 15

Ship: *Oshoro Maru*

Area: *Western North Pacific*

Year: *1973*

Recorded by *Ikeda, T.*

No.	Sta.	Position	Date	Ship's time and (GMT)	Net	Wire out (m)	Wire angle	Flow-meter No.	Reading	Water filtered (m³)	Sample Bottle No.	Total wet weight(gr)	Remarks
1	OS12	09°13.0' Ⓝ S 141°45.0' Ⓔ W	17 Jan. 1973	2100-2109 (1200-1209)	NORPAC 0.33mm	174	25°	R45 213	183	52.9	63	1.66	Removed Diaphus sp. 1 ind (No.12)
2	OS17	14°00.0' Ⓝ S 142°01.0' Ⓔ W	20 Jan 1973	2235-2242 (1335-1342)	NORPAC 0.33mm	163	20°	R45 213	89	25.6	64	0.28	many Sapphirina
		N S E W		⌣									
		N S E W		⌣									
		N S E W		⌣									
		N S E W		⌣									
		N S E W		⌣									
		N S E W		⌣									

Figure 3.3. Example of a plankton log sheet.

4. Sampling depth or wire out.
5. Type of sampler (bottle, pump, net, etc., and, if necessary, mesh size, size of mouth, etc.).
6. Method of sampling (if net sample, vertical haul, horizontal tow, etc.).
7. Volume of water pumped if pump sample; flow meter reading if a quantitative net sample.
8. Number of sample.
9. Fixative used.

For specimens that are to be preserved for a long time, a separate label indicating the sampling location, species name, and, if necessary, the names of the collector and the person who identified the species is put into the preservative. Figure 3.2 shows some labels actually used.

When an opaque sample bottle is used, it is desirable to also paste a label on the outside of the bottle. An easy way is to write the sampling station and date with paint or felt pen directly on the lid or bottle itself. However, this method makes removal of the paint difficult.

Apart from a specimen label a plankton log sheet for the sampling is needed. In addition to those items listed above, such details as position of the sampling station, detailed sampling method, depth, name of the collector, and other information important to the research should be recorded in the log (Fig. 3.3).

Notes

[1] Formalin is a trade name. A 10% Formalin solution is approximately 4% formaldehyde. In order to make an exact solution of 4% formaldehyde, dissolve 4 g of paraformaldehyde in 100 ml of distilled water or filtered seawater and heat the solution for 20–30 min at 70°C.

[2] MS-222 (ethyl-*m*-aminobenzoate methanesulfonate), available from Sandoz Co. and Fisher Scientific Co. (U.S.A.)

[3] Manufactured by the Shell Chemical Co. (U.S.A.) under the trade name Ionol CP-4.

[4] Manufactured by the Universal Oil Products Inc. (U.S.A.) under the trade name Sustane I-F.

[5] Available from NASCO Co. (U.S.A.)

[6] Available from Vitro Dynamics, Inc. (U.S.A.)

[7] In the Smithsonian Institution, a 36 pound "Byron Weston Resistall Liner Ledger paper" is used for labels in the wet collections (Bowman, personal communication).

Observation and Identification

4.1. Observation of Living Animals

Observation of living specimens is the basis for the study of all types of zooplankton. Individual investigators who first watched the plankton through a microscope must have been struck by the mystery of the nature or the beauty of color and forms in a spoonful of water. This kind of naive surprise or joy at looking into the world of these small creatures must have inspired them to study plankton. In studying plankton, one should always try to observe not only the organisms which are the subjects of research but also as many other plankton species in the community as possible, registering in detail their modes of swimming, eating, mating, spawning, defecation, and molting. It goes without saying that this basic knowledge of the behavior of zooplankton made from direct observation enhances the understanding of community structure and production mechanisms.

Our knowledge of zooplankton behavior is still limited and, in particular, little is known about individual response to physico-chemical and biological conditions. It is strongly hoped that more study will be conducted on the area.

4.1.1. Field Observation

The methodology for studying zooplankton by scuba diving and snorkeling has been developed in the last 10 years. *In situ* studies have contributed a great deal of essential information about the lives of planktonic organisms that cannot be learned by collecting the animals with a net. Direct observations of the swimming, feeding, and swarming behavior and color of plankton in the water have indicated how these animals feed, escape predators, and interact with other organisms; reports on the modes of life of gelatinous plankton have been particulary surprising to researchers

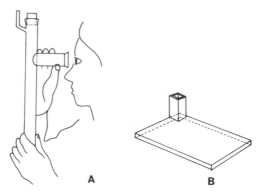

Figure 4.1. (A) Plankton observation tube and viewer; (B) Plankton observation chamber.

(Alldredge, 1972; Hamner, 1974; Madin, 1974; Swanberg, 1974; Hamner et al., 1975). Scuba diving as a tool for plankton research can be criticized for its lack of quantitative precision compared to a plankton net. However, recent technical developments in underwater photography prove the possibility of quantitative analysis. For example, the density of copepods in swarms can be measured with a Nikonos 35-mm camera with electronic flash, 80-mm lens, and extension tubes (Hamner and Carleton, 1979). Regardless, diving observations can greatly complement traditional methods of plankton research. As more is learned about the behavior of individual plankton, blind sampling may be replaced by specific and efficient sampling. The reader is advised to refer to the works of Fager et al. (1966), Hamner (1975, 1977), and Hamner and Carleton (1979).

In order to observe living plankton on board ship, a simple plankton observation tube (Fig. 4.1A) was devised, following the microscopic examination chamber designed by Bieri (1956) and Motoda (1975) (Fig. 4.1B). The observation tube is made of glass, 220 mm in height and 20 mm in outside diameter, with a thin tube 1.5 mm in inner diameter on the shoulder. Organisms are put into the tube with water, and a silicon lid is placed on top of the tube in such a way that there are no bubbles left in the tube. Observation is made through the tube using a fixed-distance magnifying viewer while holding the tube in the hand.

4.1.2. Laboratory Observation

Living microzooplankton can be observed by placing them on a glass slide in one or two drops of water under a coverslip. Pipettes and containers must be prewashed with detergent or chloral sulfate, thoroughly rinsed with distilled water, and dried completely. Glass slides should be soaked in 2% chlorate alcohol (2 ml of 100% hydrochloric acid and 98 ml of ethanol) and wiped thoroughly with cotton gauze before use. Fishing lines of various thickness, a piece of plankton gauze, or Vaseline (petroleum jelly) should be used as spacer between the coverslip and the slide so that animals are not damaged by the coverslip.

Certain species of larval plankton are particularly temperature sensitive and may be adversely affected or die during observation. Dan and Okazaki (1956) succeeded in observing the normal growth of larval sea urchins by creating a humid area on a glass slide in the following simple manner. Filter paper of an appropriate thickness was cut slightly smaller than the coverslip and a hole a few millimeters in diameter was punched in its center. This filter paper was placed on the slide after soaking it in seawater, then a water drop containing the larva was placed in the center of the hole without touching the paper. A coverslip was carefully placed over the filter paper, and by adding seawater from outside the filter paper, a humid area between glass slide and coverslip was maintained. Long-term microscopic observations were made in a constant-temperature room. Today various expensive and sophisticated temperature control chambers for microscopic examination are available.

Under normal conditions it is not easy to make visual observations of microorganisms moving in and out of the field of a microscope. The movements of organisms may be slowed down by artificially raising the viscosity of the water; however, observations under these unnatural conditions should be interpreted with caution. One way to raise the viscosity of seawater is with a solution of methyl cellulose made by dissolving 10 g in 100 ml of seawater, allowing it to stand for 1–2 days and eliminating insoluble materials by centrifugation. The solution should be refrigerated. A few drops of this solution added to a water sample will markedly reduce the swimming speed of an organism. Another approach is to dissolve polyethylene oxide powder in seawater, creating a polymer of high viscosity.

Use of anesthetic agents may be considered. The effect of anesthetics differs not only with various zooplankton groups but also with water temperature and physiological condition; therefore, there is no single treatment appropriate for anesthetizing all species of zooplankton. Experiments should thus be undertaken by changing the chemicals or their concentrations so that the best method can be determined by the researcher. It is important to remember that plankton should not be anesthetized too rapidly. Strong anesthetic action may lead to the animal's death or to physiological impairment, which would make it impossible to achieve the desired results.

Solutions of Janus Green B, chloral hydrate crystals, and urethane (0.1%, 0.1%, and 1% in seawater, respectively) have been successfully used to anesthetize Mollusca and Polychaeta. Nicotine is also an effective and easy anesthetic agent; by placing a few fibers from the filter of a smoked cigarette into the water, the ciliary movements of Ciliata and Mollusca larvae may be halted for a few hours. To halt muscular contraction without halting ciliary movement, a mixture of seawater and chloretone in the ratio of 2500 : 1 (V/W) may be suitable (Bell, 1964).

MS-222 and Tricaine[1] are effective anesthetics for copepods and fish larvae. Solutions of these agents are made by dissolving the chemicals in isotonic potassium chlorate and storing in a dark place (Wallace et al., 1967). Several drops are normally sufficient to anesthetize the organisms in 50–100 ml of seawater.

Generally, for the observation of growth, metamorphosis, or fecundity of zooplankton, several individuals are fixed in a time sequence for later examination. However, if the number of individuals is few, the same animals can be measured over time by the use of an appropriate anesthetic.

Recent developments in video monitoring systems have made observations of zooplankton behavior easier and more accurate. The video techniques are easier than cinematography and are currently being used to investigate feeding, predator–prey interactions, and mating behavior. The behavior of small organisms can be recorded with a video camera attached to a microscope and time-lapse video tape recorder. The advantages of using a time-lapse video recorder are (1) unattended monitoring of activity for up to 4 days and playback in 15 min or less, allowing the observation of rare-occurring events, and (2) frame-by-frame analysis of the movements of plankton and interactions between organisms, enabling measurement of appendage action, search speed, perceptive distance, selection, and so on. The standard video systems record 30 pictures/ second. A high speed video system (up to 200 pictures/second) may be used for the recording of fast-occurring events and their slow motion analysis. A time-date generator automatically prints the time and date on each frame, providing the necessary information during playback for the estimation of rates. For analysis, a high sensitivity TV monitor is needed. Larger organisms can be studied with a video camera equipped with a macrolens instead of a microscope.

The present problem of video monitoring systems lies not in the method but in the systems themselves. There are various types of systems with different sizes of tapes and types of image control (e.g., Beta, VHS, and U-matic) which are generally not interchangeable. Therefore, an investigator must be careful in selecting the monitoring system which meets the purpose of his/her observations.

4.2. Identification of Species

The exact identification of all species is possible only with the cooperation of taxonomic experts. However, a fundamental knowledge of animal systematics, classification, and general zoology of important groups of organisms is necessary for any ecologist. Even in the case of research concerning a particular species, one must be able to distinguish the subject species from other species and know the systematic position of this species in the plankton community or assemblage in the sample.

Much literature and experience is required for the identification of species. It is convenient for prospective researchers to establish a list of references for each animal group to be investigated on the basis of recent papers written by taxonomists. Identification of very common and abundant species in a particular area can be made on the basis of introductory books or illustrated checklists. Care should be taken, however, to ask the help of an expert in confirming the identification of a species that will be the subject of study; an inaccurate identification may spoil an ecological study.

Sample specimens, especially those based on written papers, should be preserved for future use, even after completion of the research. Those identified by experts will be of great help for comparison in future identification and should be carefully preserved as a reference collection.

When a new species is found, a new name should be given in accordance with the International Code of Zoological Nomenclature (Stoll et al., 1964). Type-specimens or type-series, such as holotype or syntype specimens, should be deposited with their labels in a permanent storage of specimens such as the National Science Museum of Japan or the Smithsonian Institution. The registration number for the species should be obtained. It is recommended that one copy of a paper describing a new species be sent to the editor of the *Zoological Record* in the United Kingdom.[2]

In addition to a microscope, drawing and photographic equipment, and containers with the specimens, the following items are necessary for the examination and dissection of zooplankton in the laboratory:

1. Ordinary glass slides and coverslips.
2. Glass depression slides of various sizes and depths.
3. Pipettes.
4. Various forceps of stainless steel (in addition to ordinary forceps, Turtox flexible forceps with blunt tips[3] and rigid finely tipped forceps[4]) for sorting.
5. Slender metal needles for dissection[5] (extra-fine metal needles may be mounted in lightweight wooden handles, metal handles, or glass rods).
6. Chemical reagents (narcotizing agents, preservatives, staining and mounting agents, lactic acid, glycerin, 70% ethanol, etc.).

Work on microcrustaceans may require delicate tools made from finely drawn tungsten wire or pig's eyelashes mounted on a slender wooden handle with beeswax. Selection and use of suitable tools benefits from practice and experience. Metal tools may be honed with the finest-grained sharpening stones under a dissecting microscope.

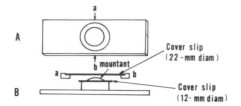

Figure 4.2. A slide for microscopic observation of microcrustaceans. (A) View of upper surface of wooden slide; (B) Cross section of wooden slide and mounting pedestal. (After Humes and Gooding, 1964.)

4.3. Dissection, Staining, and Mounting

Dissection, staining, and mounting procedures, dissection tools, and optical equipment will vary widely depending on the nature of the specimens. In general, dissections are performed under a dissecting stereomicroscope. It is easier to observe small crustaceans like copepods when an individual is immersed in a viscous liquid such as glycerol and propylene glycol on a glass depression slide; the dissection is best performed when the tissues become clear. Lactic acid is a better clearing agent and is less viscous but may decompose internal tissues after prolonged exposure. Specimens should not be directly immersed in highly viscous liquids; first they should be immersed in a 50–70% solution (diluted with distilled water) for a few minutes. When examination requires magnifications higher than $100\times$, flat glass slides of standard thickness should be used. Light staining with methylene blue or lignin pink is suitable for microscopic examination of crustaceans. Coverslips should be supported on a piece of fishing line or finely drawn glass rod of a suitable thickness to prevent compression and distortion of the specimen.

The method of Humes and Gooding (1964) is often used for the observation of microcrustaceans. In this method a hole 15 mm in diameter is drilled in the center of a wooden slide 1.5 mm thick (the same size as that of an ordinary glass slide), and a round coverslip 22 mm in diameter is glued over the edge of the hole on the lower side of the slide (Fig. 4.2). This wooden slide is then inverted and a specimen is immersed in a drop of glycerol or lactic acid on the coverslip. The specimen may be dissected in this position. Afterwards the slide is again inverted so that the specimen is suspended from the coverslip and can be observed through the microscope from above. In this way the specimen will not be pressed when oil immersion is used. For mounting specimens, a small coverslip 12 mm in diameter is placed on a stage, 10 mm in diameter, and a drop of mountant is placed on this coverslip. Then the wooden slide under which the whole specimen or its appendages are suspended is carefully placed on the coverslip under the dissection microscope. The specimens thus obtained have the advantage of being able to be looked at through the microscope from both sides.

Staining may aid in dissecting and studying specimens, especially if microscopic examination is required. Ciliata may be stained on membrane filters after fixation using 1% erythrosin (Jannash, 1958), which is made by dissolving 1 g of erythrosin (powder) in 99 ml of 5% phenol solution. In this operation the staining liquid is dropped on the specimens and drawn through the filter by suction, immediately after which the filtration vacuum is broken, and 5–10 min is allowed for staining. Finally, to eliminate the remaining stain, the specimens on the filter are rinsed with distilled water until the red color of excess erythrosin disappears. Staining with 0.1% (W/V) acid fuchsin is also effective.

Generally successful stains for crustaceans include chlorazol black E, fast green, and acid fuchsin. The stain is usually prepared as a 1% (W/V) solution in distilled water or 70% ethanol. Before staining with ethanol solutions, specimens should be washed in distilled water and partially dehydrated by transfer through successive 1- to 10-min baths (depending on the body size) of 30, 50, and 70% ethanol. These stains are taken up rapidly, though immersion time may vary with the nature as well as thickness of the integument; copepods stain intensively within 30–60 sec after immersion. Specimens should be washed in distilled water after staining. When using ethanol solutions of stain, transfer specimens through a reverse series of dehydrating baths before immersing them in distilled water. The intensity of these stains may be varied by altering the duration of immersion or by soaking in destaining solutions. (see, for example, Pantin, 1964, for details.)

Light staining may be obtained by adding tints to the mounting medium. A few drops of an aqueous solution of methylene blue or lignin pink added to lactic acid (undiluted or 50% aqueous solution) provides a dissecting medium that clears and lightly stains the specimen. Chlorazol black E provides a more intense tint. To prepare this tinting mountant, 1 part chlorazol black E solution (1% by weight in 70% ethanol) is added to 19 parts lactic acid or glycerol. The specimen is then immersed in the tinted medium and soaked until the tint disappears from the mountant and is concentrated in the integument (about 30 min for copepods). If the specimen is overstained, it should be immersed in a solution of 1 or 2 drops of pyridine in 1 ml of 85% ethanol; it will return to its normal condition within a few hours. Chlorazol black E tinting in lactic acid is useful for examining integumental glands and sensilla of crustaceans. It is also suitable for Polychaeta.

Critical examination of integumental characters and topography of microcrustaceans is enhanced by eliminating all internal tissues of the specimens. A bath of 10% KOH by weight in distilled water and heated to about 90°C will remove all soft tissue from a specimen (e.g., copepod) in 1–2 hr. The skeleton should be gently washed with distilled water to remove all traces of KOH and stained for about 10–40 sec with an aqueous solution of 1% chlorazol black E, rinsed in water, and transferred to glycerol for examination and permanent storage.

To observe the corona ciliata or sensory spots of Chaetognatha, specimens should first be stained for a few minutes in a solution of neutral red, methyl green, or alum carmine, washed with distilled water, and then examined through a microscope under reflected light. To make an alum carmine solution, 3–5 g of alum and 2 g of carmine are added to 100 ml of distilled water, boiled for about an hour, cooled, and filtered. Then 1 ml of 100% Formalin should be added to the solution.

Permanent glass slide mounts may be made in natural or synthetic resins or in water-soluble mountants (see Pantin, 1964; Steedman, 1976c). Ciliata can be mounted using the Chatton-Lwoff silver line stain method (Darbyshire, 1973) or the protagol technique (Tuffrau, 1967) in which specimens fixed with Champy and/or Da-Fano solution are permanently mounted on the slide. However, since these involve difficult handling procedures and the authors have no personal experience, we just introduce them here as references. Uhlig's (1972) modification of the Chatton-Lwoff technique also seems useful:

1. Fix the specimens for 2 min in Champy solution[6] and drain off the fixative.
2. Wash twice in Da-Fano solution (marine)[7] and rinse in distilled water.
3. Place specimens on a slide and drain off water.
4. Add 1–2 drops of warm (40°C) 10% gelatin solution to 2 drops of seawater; place this on the animals and mix and drain off rapidly before the gelatin sets (gelatin layer of approximately 200 μm thick).
5. Cool slide until the gelatin has set and place in refrigerator for 5 min in cold 3% silver nitrate solution (keep in the dark).
6. Rinse slide with cold distilled water and place it in a shallow basin with distilled water; water level 2–3 cm.
7. Irradiate the slide for 10–20 min with a UV lamp (longer in sunlight) until the objects become sufficiently dark.
8. Rinse with cold distilled water, run through alcohol series, and embed. For steps 5–7 always keep below 10°C.

Balsam, gum-chloral, glycerine-jelly, and Hoyer's medium have been commonly used in the past as permanent mounts for marine zooplankton, particularly crustaceans. Gum-chloral is made by dissolving 8 g of gum-arabic and 30 g of chloral hydrate in 1 ml of distilled water and adding 1 ml of acetic acid and 2 ml of glycerol, then filtering through filter paper. However, since the durability of the above mountants is not very great, we recommend mounting with nonresinous mounting media, Turtox CMC-10 and CMC-S[8] as follows:

1. Wash slide and cover slip with 95% ethanol.

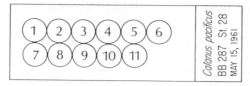

Figure 4.3. The arrangement of appendages of a copepod on a glass slide: (1) antennule; (2) antenna; (3) mandible; (4) maxillula; (5) maxilla; (6) maxilliped; (7) first pair of legs; (8) second pair of legs; (9) third pair of legs; (10) fourth pair of legs; (11) fifth pair of legs.

2. Place a drop of mountant on the slide using a slender glass rod with a beaded end and transfer the dissected structure to the drop. If more than one appendage is to be mounted, they may be arranged in their natural sequence in separate drops as in Figure 4.3 or, if morphologically distinguishable from one another, in a small cluster under one large coverslip. For thicker appendanges support the coverslip as described above to avoid distortion of shape by compression. As a rule, mount appendages serially and in the same orientation.

3. Place the coverslip at one edge of the drop and lower gently to avoid trapping air bubbles in the mountant.

4. Keep the slide horizontal and do not use until dry; drying time may be reduced by heating in an oven at about 35°C. Greater durability and strength will be imparted to the preparation if a quick-drying clear lacquer or fingernail polish is painted around the coverslip when the slide is completely dried. Several layers may be put on at daily intervals to increase resistance to rough handling.

The CMC-10 is a colorless, transparent mountant, and stained specimens sometimes lose their color in it. For such specimens the CMC-S is better to use as it contains acid fuchsin and stains the specimens red after mounting. The CMC-10 and CMC-S both have high viscosity. When their viscosity is too high, a small amount of distilled water or lactic acid should be added for dilution. In the CMC series there are some mountants of lower viscosity and one containing aniline blue; thus, it is advisable to use these media according to the purpose of the observation.

To handle dissected appendanges or to remove air bubbles between the coverslip and glass slide, a slender wooden handle with a pig's eyelash attached to the tip is useful. The coverslip should vary according to the size of the specimen. The authors use round coverslips of 9–12 mm in diameter for mounting small specimens and the appendages of copepods.

4.4. Vital Staining Techniques

Sometimes a sample contains a large number of dead organisms. Knowledge of the number of live and dead zooplankton at the time of sampling is needed for particular investigations. The estimation of mortalities of estuarine zooplankton passed through a condenser of a power plant is an example. The use of neutral red dye, which vividly stains many live plankton, enables a rapid count of live and dead specimens in a large sample at a convenient time after its collection (Dressel et al., 1972; Crippen and Perrier, 1974).

Various staining techniques for live marine zooplankton have been described. According to our experience, the staining technique modified by Crippen and Perrier (1974) (ICDC method) is best to distinguish live and dead specimens by color. Another procedure (CEGB method) developed by Fleming and Coughlan (1978) is also useful, especially when it is necessary to preserve the samples, with their live and dead stain information, for subsequent examination.

The ICDC technique is as follows. A stock solution is made by dissolving 1.0 g of neutral red (powder) in 1 liter of distilled water (i.e., 0.1% W/V). This solution is added to seawater containing animals in a ratio of 2–5 : 100, yielding a final dye concentration of 1–2.5 : 100,000. Staining time is 60 min but varies depending on the type and volume of specimens. After staining, the specimens are fixed with 10% Formalin to which 4 ml of 1N NAc–1N NaAc solution per 100 ml of sample is added. The specimens are then stored overnight at 2–3°C prior to examination. In the CEGB technique the concentration of neutral red in a stock solution is 0.05% (W/V). After staining, the specimens are fixed with alkaline 2% Formalin (adding 200 mg NaOH per 1-liter 2% Formalin) and stored at room temperature. At the time of examination specimens are transferred to a counting chamber or tray and glacial acetic acid is added dropwise until the magenta color is restored. Copepods stained and alive immediately prior to fixation turn a deep magenta after acidification, while dead specimens are light pink to white. If the sample is to be retained after examination, it should be washed to remove excess acid and titrated back to a basic pH with a NaOH solution before adding Formalin.

It is noteworthy that the uptake of stain takes place mainly through the respiratory and/or feeding organs and it varies among species and developmental stages. Calanoid copepods are most easily stained. The stain is also effective for rotifers, polychaetes, and chaetognaths. However, cyclopoid and harpacticoid copepods, barnacle nauplii, and decapod zoeas are often lightly or inconsistently stained. In order to decrease sorting error for live/dead specimen and to determine effect of various environmental stresses on plankton more accurately, it would be better to select the easily stained taxonomic group as an indicator of the zooplankton in the region.

4.5. Silhouette Photography

To preserve zooplankton samples without preservatives in archives, and to facilitate automatic enumeration by means of computer image analysis

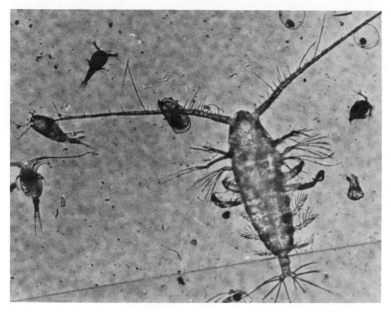

Figure 4.4. Silhouette of live zooplankton collected in the Gulf Stream. (Photograph courtesy of P. B. Ortner.)

techniques, silhouette image preservation techniques are being developed (Ortner et al., 1979).

In these techniques fine-grain positive film is placed directly beneath a dilute sample of live or fixed zooplankton in water and is exposed with a small electronic flash lamp in a dark room. No optics are required for the initial exposure. The photographic system used consists of 200×260 mm^2 Eastman fine-grain positive film 7302, an EG&G electronic flash with xenon bulb,[9] and a plastic container tray. Distance from lamp to film is about 1 m. A water sample containing zooplankton is poured into the tray directly on top of the film to a depth of about 1 cm. One flash with a pulse duration of 3×10^{-6} sec exposes the film. The short exposure effectively stops the motion of the subjects.

Figure 4.4 is an enlarged contact print of a negative produced as described above. Subsequent observation of the film is made with a dissecting microscope or by projected enlargements. Silhouette photography of plankton samples has sufficiently high resolution for quantitative analyses, which could not have been obtained by conventional photographic techniques. An advantage of the present lensless technique is that it can inexpensively produce a permanent record of live organisms; thus, it should be particularly useful to field workers. Recently an *in situ* silhouette camera system with a shutterless 35-mm film drive has been designed which permits assessment of microscale zooplankton distributions (Ortner et al., 1981).

Notes

[1] Tricaine (ethyl-*m*-aminobenzoate methanesulfonic acid), available from Sigma Chemical Co. (U.S.A.).

[2] *Zoological Record:* The Zoological Society of London, Regent's Park, London, NW1 4RY, U.K.

[3] Available from Turtox Co. (U.S.A.).

[4] Jeweler's forceps of stainless steel, such as AA type and No. 5 of Inox.

[5] Minute dissecting needles and needle holders are available from Carolina Biological Supply Co. and Clay-Adams Inc. (U.S.A.).

[6] Champy solution. Mix the following three solutions before use:

3% (W/V) potassium dichromate solution	7 ml
1% (W/V) chromic oxide solution	7 ml
2% (V/V) osmium tetroxide solution	4 ml

According to Uhlig (personal communication), 1% chromic oxide can be replaced by 1% chromic trioxide so that the solution becomes very stable.

[7] Da-Fano solution (marine):

Buffered Formalin (undiluted solution)	10 ml
Cobalt nitrate	1 g
NaCl	1 g
Distilled water	90 ml

Note: Da-Fano solution is only suitable for fixation of certain species and cannot be used generally (Uhlig, personal communication).

[8] Available from Turtox Co. (U.S.A.).

[9] FX6A, manufactured at EG&G Inc. (U.S.A.).

CHAPTER 5

Processing
and Measurements

5.1 Splitting and Sorting

In the analysis of a sample, when the number of individuals in one sample is greater than needed or when a sample is used for several purposes, the total amount of sample is fractionated into several portions. For this purpose a Stempel pipette, Folsom splitter, or Motoda splitter is generally used.

As shown in Figure 5.1, the Stempel pipette is made such that a certain volume (0.1–10.0 ml) of sample is retained in the space between the glass tube and metal piston when the handle is pulled up. This pipette was originally used in studies of phytoplankton. For zooplankton it is used to obtain a subsample of fixed volume to estimate the number of specimens in the entire sample when individuals are smaller than approximately 0.5 mm. Larger individuals must be removed from the sample beforehand. The zooplankton sample in a glass container is diluted to a known volume with filtered seawater and is then mixed gently so that it becomes randomly distributed in the container. The Stempel pipette is then used to remove an aliquot of the sample, which can be transferred to a counting chamber. Any specimens adhering to the outside of the pipette should be washed off so they are not mixed with those inside. The concentration of the diluted sample should be such that 100–300 individuals of the organism in question are contained in the aliquot taken for counting.

The Folsom splitter has a semicircular divider in the center of the drum, such that the sample can be split into two parts when the drum is rotated. The splitter shown in Figure 5.2 is a modification by the authors of the prototype. The diameter of the drum is 22 cm and the width is 9 cm. The drum is placed on a horizontal stand and rotated back and forth several times so that the sample moves between the divided and nondivided sections and the plankton are randomly mixed. The inner side of the nondivided section is facilitated with ridges to mix the sample well. When

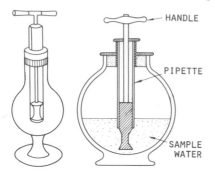

HANDLE

PIPETTE

SAMPLE WATER

Figure 5.1. Two types of Stempel pipettes. (After Kokubo, 1963.)

the sample is actually split, it is necessary to insure that the water level is the same on both sections of the divider. The drum is then further rotated to transfer the sample into separate containers. Any plankton remaining on the walls of the splitter should be washed off into the containers. If the aliquot needs to be divided further, the procedure should be repeated again.

There are two types of Motoda splitters made of Plexiglas (Motoda, 1959). In one splitter a 10-cm square board is attached to the bottom of a 15-cm-high transparent cylinder having an inner diameter of 7 cm (Fig. 5.3A). In the cylinder a 3.5-cm-wide divider is provided to cover one-

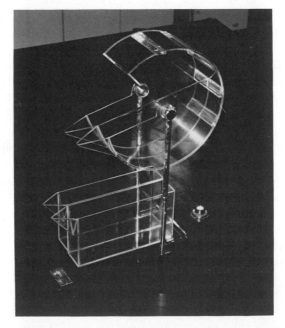

Figure 5.2. A modified Folsom plankton sample splitter.

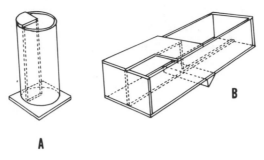

A

B

Figure 5.3. Plankton sample splitter. (A) Cylinder type; (B) Box type. (After Motoda, 1959.)

quarter of the upper area of the cylinder. The other type of Motoda splitter is 40 cm long, 14 cm wide, and 12.5 cm high; half of it is divided by a partition and the upper half is covered (Fig. 5.3B). Both types are used on a horizontal surface, and the sample inside them is divided by tilting the splitter. Half of the sample remains in the splitter and is redivided when necessary. The box type splitter is particularly useful for use aboard ship.

After splitting, the next step in the analysis is to sort and count specimens. These operations may be done simultaneously. If the sample is used to investigate the species composition of the plankton or the ecological roles of various species in an assemblage, complete sorting of the specimens in representative aliquots is necessary. Even if only specimens belonging to particular genera or species are the objects of the research, all organisms of the same class or order to which the particular animals of interest belong would best be sorted.

Samples from extensive research cruises or international cooperative surveys may be sorted into various taxonomic groups, which may later be studied by specialists. The Indian Ocean Biological Center in Cochin, India, and the Regional Marine Biological Center in Singapore were established to provide such a service.

Sorting is monotonous time-consuming work and requires perseverance, but it is a very important basic part of a plankton study. There is a primary and a secondary sorting stage; in the primary stage the sample may be divided into some 30–40 taxonomic groups (Fig. 5.4), whereas in the secondary stage the abundant or important groups of organisms, such as copepods and fish larvae, are further separated into their respective families or genera. Sorting should be carried out in a well-ventilated room. The necessary tools include various types of forceps and slender dissecting needles referred to in Chapter 4, Section 4.3. The preservative is diluted with filtered seawater[1] to 2% Formalin (V/V) to reduce the occurrence of formaldehyde gas. The specimens are separated into a number of sorting dishes according to category with the aid of a dissecting microscope. The

Data Sheet of CSK
Standard Zooplankton Sample

C.S.N. _____941_____

1. Vessel — Umitaka Maru
2. Institution & country — TUF, Japan
3. Cruise no. — No. 37
4. Area of investigation — South China Sea
5. Station no. — Um 6818
6. Position Lat. — N. 06 ° 02.0 '
   Long. — E. 113 ° 00.0 '
7. Date & time (GMT) — 26 October 0110
8. Date & time (LMT) — - -
9. Year — 1968
10. Method of haul — Norpac net, Vertical haul

11. Sample no. — 10
12. Wire length (m) — 170
13. Wire angle (°) — 28 °
14. Depth estimate by wire angle (m) — 150
15. Strained water estimated by flow-meter (m^3) — 25.9
16. Wet weight (gm) per haul — 1.19
17. Wet weight (mg) per m^3 of sea water — 45.95
18. Total number per haul — 8597
19. Total number per m^3 of sea water — 332
20. Remarks — Total sample counted.

Category	Vial No.	Number of animals in vial	Number of animals per haul	Number of animals per m^3 of sea water	Percentage composition (%)
1. Noctiluca	1	0	0	0	0
2. Foraminifera	2	447	447	17.26	5.20
3. Siphonophora	3	429	429	16.56	4.99
4. Other Medusae	4	23	23	0.89	0.27
5. Ctenophora	5	0	0	0	0
6. Nemertinea	6	0	0	0	0
7. Cyphonautes larva (Bryozoa)	7	3	3	0.12	0.03
8. Actinotrocha larva (Phoronidea)	8	0	0	0	0
9. Chaetognatha	9	585	585	22.59	6.80
10. Tomopteridae (Polychaeta)	10	2	2	0.08	0.02
11. Other polychaeta, including larvae	11	134	134	5.17	1.56
12. Cladocera	12	1	1	0.04	0.01
13. Ostracoda	13	232	232	8.96	2.70
14. Copepoda, including larvae	14	4795	4795	185.14	55.78
15. Cirripedia, larvae	15	4	4	0.15	0.05
16. Cumacea	16	0	0	0	0
17. Amphipoda	17	29	29	1.12	0.34
18. Euphausiacea, with Mysidacea	18	50	50	1.93	0.58
19. Luciferidae (Decapoda), including larvae	19	8	8	0.31	0.09
20. Phyllosoma larva (Decapoda)	20	0	0	0	0
21. Other Decapoda, including larvae	21	12	12	0.46	0.14
22. Stomatopoda, larvae	22	0	0	0	0
23. Heteropoda	23	4	4	0.15	0.05
24. Janthinidae	24	0	0	0	0
25. Pteropoda	25	52	52	2.01	0.60
26. Cephalopoda	26	1	1	0.04	0.01
27. Other Mollusca, including larvae	27	57	57	2.20	0.66
28. Echinodermata, larvae	28	89	89	3.44	1.04
29. Appendicularia	29	1397	1397	53.94	16.25
30. Thaliacea	30	199	199	7.68	2.31
31. Pyrosomata	31	0	0	0	0
32. Fish eggs	32	2	2	0.08	0.02
33. Fish larvae	33	17	17	0.66	0.20
34. Halobates (marine insect)	34	0	0	0	0
35. Unidentified forms	35	15	15	0.58	0.17
36. Isopoda	36	1	1	0.04	0.01
37. Amphioxus larvae	37	5	5	0.19	0.06
38. Platyhelminthes	38	3	3	0.12	0.03
39. Tornaria larva	39	1	1	0.04	0.01
40. Residue including small forms unsorted	40	Pyrocystis, Radiolaria, Ceratium.			
Total		8597	8597	331.93	

Figure 5.4. An example of a zooplankton sorting record; CSK international collection. (After Chuang, 1974.)

material should not be maintained in the excessively dilute preservative for more than one working day. If the sorting takes more than one day, 3–5% Formalin and a drop of glycerol should be added to each dish; when not being sorted, the samples should be covered so that the specimens do not dehydrate.

When a fine-mesh net is towed in coastal waters close to the seafloor or above a seagrass or seaweed bed, considerable amounts of detritus and fine sediments may be in the sample. Thus, it is not easy to distinguish microzooplankton, chaetognaths, and appendicularians, all of which are small and/or almost transparent. In such cases phloxine B, rose bengal, or borax carmine may be added to the sample to stain the animals red and facilitate sorting and counting. Borax carmine is prepared by adding 4 g of sodium tetraborate and 3 g of carmine to 100 ml of distilled water, mixing well, and boiling for 30 min, after which distilled water is added to make a volume of 100 ml. A 100-ml solution of 70% ethanol is added and stirred; the solution should stand for a few days and is then filtered. Normally, a quantity of phloxine B or rose bengal is added to make the preservative pale pink; for borax carmine a volume equivalent to half of the displacement volume of the specimens is the normal amount added. The sample should be gently shaken so that the solution mixes well. Borax carmine takes some time to stain the specimens. The color will fade gradually over several weeks and hence is not an obstacle for the later identification of species.

5.2   Counting of Individuals

5.2.1   *Microzooplankton*

The sample is concentrated by the method described in Chapter 2, Section 2.2.2. After the sample volume is measured, a known amount is pipetted onto a glass slide with ruled lines, such as an haemacytometer (0.1 or 0.2 mm deep) for microscopic examination (Fig. 5.5A).

For microzooplankton of relatively large size a Sedgwick-Rafter chamber (McAlice, 1971) is used. The chamber is without rulings, rectangular (50×20 mm^2 or 40×25 mm^2, 1 mm deep) and has a volume of 1.0 ml (Fig. 5.5B). To fill the chamber, set a coverslip diagonally across the chamber, leaving a space open at opposite corners. A 1.0-ml solution of concentrate is delivered to the chamber through one opening with a pipette; the coverslip is slid into position to seal the chamber and the contents are allowed to settle. The chamber should be examined for satisfactory distribution of organisms and individuals tallied in the entire chamber at 100 or 200× magnification or in a given number of randomly selected Whipple fields, of which the area subtended, at the magnification used, is known. Densities are also estimated with the combined plate chambers for sedimentation and counting using an inverted-phase micro-

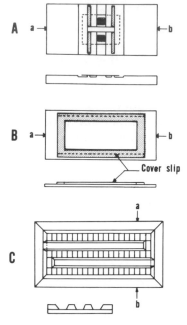

Figure 5.5. Counting chambers. Views of the upper surface and a cross section between a and b. (A) Haemacytometer; (B) Segwick-Rafter chamber; (C) One type of zooplankton counting chamber.

scope fitted with a Whipple grid (Chapter 2, Section 2.2.2.). The specimens should be counted until the number required for the desired level of statistical reliability is attained. In order to reduce the counting error, at least 30–40 fields representing 150–200 individuals should be counted in each subsample (see Venrick, 1978b).

Specimens retained on a membrane filter can be counted on the filter in the following manner. A preliminary estimate of the number of individuals in the water sample is made; an appropriate volume is filtered such that individuals will not overlap on the membrane filter. The filtration may be done by either suction or gravity. If the suction pressure is too strong, animal bodies may deform or squeeze through the filter resulting in an inaccurate quantitative estimate. On the other hand gravity filtration takes longer and often organisms are not distributed randomly on the filter but remain densely concentrated at the outer edge. It is preferable to use a hand-operated suction pump (e.g., Mityvac[2]) and filter at approximately 250 mm Hg. Organisms may be stained at this stage (see Chapter 4, Section 4.3). When a Millipore HA filter is used, it should first be soaked in distilled water. Following filtration, the filter is rinsed briefly with isotonic ammonium formate (6.0–6.5% W/V for salinity of 30–35‰). The filter is then desiccated and refrigerated. When the filter is dry it is placed on a glass slide and a drop of immersion oil is applied to make it transparent, after which a coverslip is placed over the filter. The filter is then examined. A Gelman TCM 450 filter has a low affinity for water and, if used, should first be soaked in 10% ethanol.[3] Before filtration, a small piece of glycerin jelly is placed on the coverslip and melted at 50–60°C. Following filtration, the filter on which the sample was retained is reversed (i.e., sample side down) and placed over the glycerin jelly on the cover glass. After placing one drop of glycerin on the filter, it is kept in the desiccator and refrigerated. The filter becomes transparent when dry; it is inverted again so that the coverslip is on top and is placed on a glass slide. For long preservation apply Laktoseal[4] or nail polish to seal around it. Glycerin jelly is made by diluting 15 g of gelatin, best quality, and 100 ml of pure glycerol with 100 ml of warm distilled water, mixing the solution well. The complete mixture is warmed and stirred; 1 drop of liquified phenol is added as a preservative. It should be tightly sealed for long preservation. The method using glycerin jelly takes more time than the Millipore filter technique, but the specimens are retained in better condition because they do not completely dehydrate on the filter. Therefore, it is especially suitable for the microscopic examination of Protozoa.

5.2.2 *Net Plankton*

Samples are placed in a large culture dish or tray. Megaloplankton and micronekton are first removed with forceps or a small scoop net and counted separately. The remaining sample is transferred to a splitter if necessary. The concentration is determined using a dissecting microscope and a counting chamber whose bottom surface is marked with a 5- or 10-mm grid. There are many kinds of chambers; an example is shown in Figure 5.5C. For counting, the visual field of the microscope is adjusted so that it is about 1.5 times wider than each grid of the counting chamber. It is convenient to use a mechanical stage to move the chamber. Specimens whose identification is difficult should be picked out for closer examination. When many categories are to be counted simultaneously, several counters standing in a row may be used. A label indicating a category should be put on each counter. The researcher should train himself to manipulate the counter by feel in order that the eye never leaves the visual field of the microscope while counting. Recording data on tape is another means of registering the counts.

   The composition of the plankton is most commonly expressed in number of individuals per area or volume of water or in percentage. In the latter method total number of individuals in a sample is roughly estimated first; the sample is then adequately diluted and a part of the diluted sample is put in the counting chamber. In the random sample 100–300 individuals of various categories are counted. The counting is repeated on three to five random samples from which the average percentage composition is calculated. The remaining sample is qualitatively examined, and species that did not appear in the percentage calculation are recorded as minor elements. Besides these methods of expression there is the "C-R method," which is a qualitative analysis with a relative quantitative element. The abundance of a plankton category is expressed according to its frequency of occurrence with a codes such as CC (very common), C (common), R (rare), or RR (very rare). These symbols have no absolute quantitative meaning.

In this way the composition of a plankton community or assemblage in a certain area and the relative occurrence of each species become evident. The dominant species and their ecological positions and variations become clear. Species in taxonomic groups and trophic levels that play important ecological roles are called *key species*. Species that comprise a considerable percentage of the assemblage can be considered as potential key species; this is true for species of more than 30% in subarctic and subantarctic waters and 5–20% in tropical and subtropical waters. If the life histories and trophic relationships of the key species are known, biological processes of the plankton in the study area will become clearer.

There are two types of key species. The first type is abundant throughout the year. The second type appears in great quantity only in a particular period of the year. These two types of species may be treated separately. For example, among the net plankton assemblage in Suruga Bay, Japan, those species that are considered to play an important role in the productivity of the bay are as follows:

Type 1.   From smaller sizes, *Oithona similis, Oncaea venusta, Paracalanus parvus, Clausocalanus arcuicornis, Calanus sinicus* (Copepoda), *Oikopleura longicauda* (Appendicularia), *Sagitta nagae* (Chaetognatha), *Euphausia similis* (Euphausiacea), *Sergia lucens* (Sergestidae).

Type 2.   From smaller sizes, *Noctiluca scintillans* (Dinoflagellata), *Penilia avirostris, Evadne tergestina* (Cladocera), *Oikopleura dioica* (Appendicularia), *Engraulis japonica* larvae (Pisces).

5.3   Measurement of Biomass

Biomass has commonly been expressed by settled volume, displacement volume, wet weight, dry weight, ash-free dry weight (organic weight), or carbon weight. The usual measure after fixing a sample has been its settled volume, displacement volume, or wet weight. The term *biomass* is often inappropriately used synonymously with wet weight. In most cases, however, the biomass is measured to determine the productivity and nutritional condition of the species in question and to assess the role of the species in the food web. In this sense biomass expressed as settled volume, displacement volume, or wet weight is not always adequate because considerable variation occurs in these values due to the manner of treatment and the composition of organisms. Furthermore, the measurement includes ash and other materials of low nutritive value.

In settled-volume measurement the sample is poured into a graduated cylinder or sedimentation tube of 50–100 ml in volume, gently stirred with a glass rod, allowed to settle for 24 hr, and the settled volume read. On the other hand, in displacement-volume measurement, the volume of the total water containing the plankton sample is first measured, after which the water is removed and its volume measured separately. The difference in volume is due to the plankton. The settled and displacement volumes include not only the absolute volume of plankton but also the water between the organisms. In the determination of wet weight the plankton is usually drained with a filter or screen and then the adhering water is blotted off with filter paper.

During this operation some part of the living matter, such as body fluids, may be lost, resulting in an underestimation of the value (Nakai and Honjo, 1961). During fixation and preservation, the volume and weight of a plankton sample decreases with time (Ahlstrom and Thrailkill, 1963; Lovegrove, 1966; Omori, 1978). The loss of volume is small for samples consisting mainly of crustaceans but is considerable for those dominated by gelatinous plankton. There is great variation among species in terms of their content of organic matter per unit volume. Gelatinous plankton, particularly those with shells, cannot be considered in the same way as copepods when measuring displacement volume and wet weight.

Considering these facts, settled and displacement volumes are not to be recommended. The best expression of plankton biomass is dry weight, preferably ash-free dry weight, or carbon weight on live specimens. In any case, when expressing biomass, it is important to indicate the method of measurement.

5.3.1 *Wet Weight*

The weight of plankton is determined after eliminating as much surrounding water as possible. The water in a sample can be eliminated by vacuum filtration. Blotting the sample on filter paper can remove additional water. The blotting paper should be replaced until the paper does not absorb any more water from the plankton mass. Care should be taken not to compress the plankton and damage the specimens to rush dehydration. If the plankton sample consists mainly of crustaceans and its wet weight is less than 5 g, the determination is easy because most adhering water can be removed by suction with a membrane filtration apparatus (e.g., Pyrex filter holder for membrane filters) by attaching it to a hand-operated aspirator or a small vacuum pump. When the vacuum pump is switched on, the pressure-regulating cock should be left open so it does not draw too strongly. The plankton sample without the adhering water is then weighed. On board a ship a small handheld beam balance is most useful for weighing the samples. In general, the value is expressed in $mg \cdot m^{-3}$.

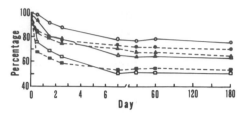

Figure 5.6.   Change of organic weight and chemical composition of zooplankton preserved in 10% formalin–seawater solution buffered with borax. Percentage means the difference in absolute weight when that of the fresh, seawater-rinsed sample is rated as 100%. O●, Organic matter; △▲, Carbon; □■, Nitrogen. Open-*Sagitta nagae*; Solid-*Calanus sinicus*. (After Omori, 1978.)

5.3.2   *Dry Weight*

Live specimens must be used for this measurement as the changes in dry and organic weights as well as chemical composition of Formalin-preserved specimens are considerable (Fig. 5.6). When samples are wanted for both the determination of dry weight and the identification of species, samples should be collected twice at the same site or a single sample should be split quickly and accurately aboard ship so that one-half can be processed for dry-weight measurement and the other half for microscopic examination. When only preserved specimens are available, they should be used about a month after preservation when the weight has become relatively stable. Correction factors must be obtain by comparison of these results with the results for live specimens.

A Pyrex holder for membrane filters or similar filtration apparatus may be used to drain the water. A glass fiber (often referred to as GF) filter or round piece of plankton gauze, with the same or smaller mesh size than that of the sampling net, is weighed and then moistened on the filter holder with distilled water. The sample is added and sucked dry at about 250 mm Hg. When no more water can be eliminated, salt contained in any water remaining between the specimens is eliminated with an isotonic ammonium formate (6.0–6.5% W/V) rinse, which is also removed by suction. For specimens having a hard outer covering, such as relatively large crustaceans, sea salt can be eliminated by a quick rinse with distilled water. The rinse with distilled water should not be prolonged, otherwise the body will lose fluids, resulting in a reduced dry weight (Omori, 1978). When a sample is large, it can be put in a bag of nylon plankton gauze and immersed for a short time in isotonic ammonium formate solution that is then drained with a vacuum pump. After removal of the rinse, the specimens are heat dried to a constant weight in an electric oven at 60°C (Lovegrove, 1962) [Note: Båmstedt (1974) recommends 70°C for drying general zooplankton materials]. Then the sample is kept in a desiccator until weighing. For freeze drying (lyophilization), the best way is to quickly freeze the sample with liquid nitrogen or with dry ice and store it in a desiccator at -20 to $-40°C$ until lyophilized.

5.3.3 *Dry Organic Weight (Ash-free Dry Weight)*

A known weight of the dry sample is ashed to constant weight in a crucible at 450–500°C in an electric furnace. After it is completely ashed, the material is cooled in a desiccator and then weighed. The dry organic weight is obtained by subtracting the ash weight from the dry weight.

When material is very small, using a pipette, they should be placed on a GF filter which has been burned at 450°C and tared with a small amount of seawater. Blank filters should be prepared on which similar amounts of seawater are placed. Both filters are then treated with isotonic ammonium formate by applying a low vacuum, drying them in an oven, then weighing and ashing them. After cooling, the filters should be weighed again. When any weight changes are found in the blank filters during the dry-weight and ash-weight measurements, appropriate corrections[5] are necessary for accurate measurement.

It is noted that the above procedure for estimating dry organic weight by ashing is not valid for ctenophores (and probably coelenterates). Mullin and Evans (1974) working on *Pleurobrachia bachei* and Reeve and Baker (1975) studying *Mnemiopsis mccradyi* found unusually low ratios of carbon to organic matter (3–7% and 9%, respectively, in contrast with the general value of up to 65%). Mullin and Evans (1974) considered that this low ratio of carbon to organic matter was caused by the overestimation of dry organic weight due to "bound" water which is not driven off by usual drying procedures at 60°C.

5.4 Body Length, Volume, and Weight

The measurement of the size of zooplanktonic organism is important for estimating growth and age. By monitoring the change in time of the body length of an individual, its growth rate can be estimated. By using the relationship of body length to age and developmental stage and taking advantage of differences in body length, the age structure of a population can be clarified. There are relations between the size of animals and their volume or weight with which it is possible to make rough estimates of the organic or carbon content of zooplankton living under similar environmental conditions[6] from their size frequency distribution.

The relationship between the length (L) and weight (W) of an animal is generally expressed as

$$W = aL^b \quad \text{or} \quad \log W = \log a + b \log L \tag{5.1}$$

where a and b are constants; the value of b is close to 3. For the relationship between volume and wet weight, the conversion of 1 ml = 1 g can be made since the specific gravity of an animal is close to 1.

The body length can be accurately measured with a micrometer disc placed in the eyepiece of a microscope. Specimens of a few millimeters in size can be measured with a dissecting microscope after being placed on a glass slide on which a tiny rule (about 20 mm in length) has been marked. For the measurement of micronekton, a plastic caliper is also useful. Animals with bent bodies must be stretched; they can be measured using a map ruler along the axis of the animal body magnified by a profile projector or video system. If a body is damaged so that the measurement of its total length is impossible, one can estimate it by measuring the length of only a part of the body (e.g., the carapace or a uropod, which are easily measured and grow in constant relation to body length) and using the appropriate conversion factor.

The volume of small-sized zooplankton can be approximated from measurements of the length, width, and thickness of the animal body. The volume of large individuals, or large numbers of organisms, can be measured by measuring the displacement volume. With large-sized animals, accurate body length to volume relationships can be obtained from individual measurements.

The method of measuring the wet weight has been described, but it is difficult to obtain the wet weight of small-sized individuals by weighing because their values change very rapidly due to evaporation. Another way to estimate wet weight is to convert the displacement volume of the individual to its wet weight.

The dry weight of copepod is usually between 1 and 3000 μg; therefore an electromicrobalance with high sensitivity, such as a Mettler M3 or a Cahn 29, is necessary.

5.5 Chemical Composition

Zooplankton of single or mixed species may be used for chemical composition studies depending on the purpose of the investigation. Formalin-preserved samples are inadequate for chemical composition studies since considerable changes in organic matter (Fudge, 1968) and elemental (C, N) composition (Hopkins, 1968; Omori, 1978; Champalbert and Kerambrun, 1979) are observed in such samples. Fresh, dried fresh, or frozen fresh zooplankton samples are adequate for the analysis of organic matter, while dried fresh samples are exclusively used for elemental analyses.

Frozen samples (generally at $-20°C$) should be stored in airtight containers, otherwise the samples are dehydrated during storage and the wet weight underestimated. Fresh samples can be dried following the procedure in Section 5.3.4, but rinsing the samples with isotonic ammonium formate should by avoided. Lyophilization is the best method for drying frozen samples.

The amount of zooplankton sample required for a single determination varies depending on the kind of organic matter or element. In the case of small species a number of individuals may need to be combined, whereas in the case of large species a ground or sliced portion of the animal may be analyzed and the results summed up.

5.5.1 *Total Protein, Carbohydrate, Chitin, and Total Lipid*

There are various analytical methods for the determination of protein, carbohydrate, chitin, and lipid in animal material. Raymont et al. (1964) proposed standard methods to measure these organic components in fresh or frozen marine zooplankton samples. On the other hand, with dried and homogenized zooplankton samples, Båmstedt (1974) found that the methods of Raymont et al. (1964) for chitin and protein are not satisfactory, while the method for carbohydrate is applicable. Båmstedt used slightly different extraction procedures for the lipid determination but made no comparison with the method of Raymont et al. (1964). We will now describe the analytical procedures of Raymont et al. (1964) and the supplemental methods of Båmstedt (1974) for dried and homogenized zooplankton samples. The methods of Raymont et al. require fresh or frozen zooplankton samples weighing 20–250 mg for the single determination of each class of organic matter, whereas the dry weight of specimens needed for Båmstedt's method is 5–80 mg.

1a. Total protein (Raymont et al., 1964). A weighed sample is homogenized with 1 ml of distilled water and 4 ml biuret reagent in a homogenizer until all the purple-stained particles dissolve. The homogenate is filtered through a GF filter to remove chitinous debris (normal filter paper cannot be used because it contains cellulose and interferes with the reagent). Extinction of the clear filtrate at 540 nm is measured with a spectrophotometer. The concentration of protein is read from a standard curve derived from the analysis of known amounts of bovine serum albumin dissolved in 1 ml of distilled water and 4 ml of biuret reagent. For both samples and standards, the measured extinction is corrected by that of a reagent blank prepared concurrently (1 ml of distilled water and 4 ml of biuret reagent). The biuret reagent is made up as follows: 1.5 g of $CuSO_4 \cdot 5H_2O$ and 6.0 g of NaK tartrate $4H_2O$ is dissolved in about 500 ml of distilled water in a 1-liter flask, after which 300 ml of a 10% NaOH solution is added and constantly swirled; then enough distilled water is added to make a 1-liter solution. The reagent should be stored in a polyethylene bottle.

1b. Total protein (Båmstedt, 1974). A weighed sample is placed in a centrifuge tube with 1 ml of $1N$ NaOH and 4 ml biuret reagent. The contents are mixed and the centrifuge tube is placed in a boiling water bath for 10 min. After cooling for 10 min, the contents are centrifuged (3000 rpm, 10 min). The supernatant is removed with a pipette, placed into a test tube, and shaken with 1.5 ml of diethylether to eliminate any cloudiness due to lipids. The test tube should stand for 2 hr. The extinction of the lower aqueous phase is measured at 550 nm with a spectrophotometer. General procedures for the correction with a reagent blank and the reading of protein concentration from a standard curve are the same as mentioned above.

2.   Carbohydrate (Raymont et al. 1964).   A weighed sample is placed in a glass test tube with 1 ml of distilled water, 1 ml of 5% phenol solution, and 5 ml of concentrated H_2SO_4. The last solution is very viscous and should be delivered from a 5-ml automatic dispenser. After 20 min the extinction is measured at 490 nm with a spectrophotometer. The concentration of carbohydrate is read from a standard curve derived from analyses of known amounts of glucose dissolved in 1 ml of distilled water, 1 ml of 5% phenol solution, and 5 ml of concentrated H_2SO_4. For both samples and standards the measured extinction is corrected with that of a reagent blank (1 ml of distilled water, 1 ml of 5% phenol solution, and 5 ml of concentrated H_2SO_4).

3a.   Chitin (Raymont et al. 1964).   A weighed sample is boiled in a glass test tube with a 50% NaOH solution for 2 hr and left overnight. In this way the protein in the sample is digested, leaving only chitinous exoskeletons. The exoskeletons are washed with distilled water, and the absence of protein is confirmed with biuret reagent as mentioned above. The exoskeleton residue is washed with dilute HCl and then with distilled water, dried in a tared crucible (60°C, 24 hr), and weighed. The weight of this dried material is assumed to be chitin and ash. The weight of chitin is taken as the loss in weight due to ashing at 500°C.

3b.   Chitin (Båmstedt, 1974).   A weighed sample is placed in a centrifuge tube and treated with $1N$ HCl and $4N$ NaOH successively in the proportion of 1 ml each per 10 mg sample—first with $1N$ HCl in the centrifuge tube in a boiling water bath for 3 min, then with $4N$ NaOH in the boiling water bath for 20 min. The residue in the centrifuge tube is washed with distilled water, 96% ethanol, and finally diethylether using successive centrifugation (3000 rpm, 10 min) and decantation. The final residue is rinsed with distilled water into a tared cruicible, dried (60°C, 24 hr), and weighed. The chitin weight is obtained from the difference in the dry weight of the residue before and after ashing at 500°C. [The main disadvantage of the method of Raymont et al. (1964) for dried and homogenized samples is the difficulty of centrifuging samples in viscous 50% NaOH solution.]

4a.   Total lipid (Raymont et al. 1964).   A weighed sample is homogenized in a 2:1 chloroform:methanol (V/V) mixture to extract the lipid. After filtration, the filtrate is washed with $0.5N$ KCl solution to remove nonlipid contaminants and break the bonds of acidic lipids. The upper phase (KCl wash solution) is removed and the lower phase (lipids in solvent) is transferred to a tared container for evaporation of the solvent in a stream of nitrogen. The container with the lipids is weighed, subtracting the weight of the container to obtain the weight of lipids.

4b. Total lipid (Båmstedt, 1974). Lipids from a weighed sample are extracted into a tared glass dish with chloroform of 20 times the weight of the sample. The residual lipid is extracted with 2:1 chloroform–methanol (V/V) mixture. This extract of residual lipids is washed with a 0.9% NaCl solution of 30 times the weight of the extract, centrifuging (3000 rpm, 10 min) this biphasic solution and pipetting the lower phase to combine with the chloroform extract. The extracted sample is dried in a desiccator for 2 days and weighed.

The composition of some marine zooplankton analyzed by the above methods is shown in Table 5.1. It is apparent that the composition changes both within and between species, depending possibly on various internal and external conditions of the animals. Among the organic constituents, both protein and lipid generally comprise major organic fractions of zooplankton, but the amount of lipid is highly variable. It is noted that most previous studies of the organic composition of single zooplankton species were made on samples collected from boreal and temperate waters, or deep waters. This reflects the fact that the sorting of single zooplankton species is relatively easy for samples from cold-water environments. Data for samples from warm-water environments, particularly for shallow or neritic species in the tropical and subtropical regions, are highly desired.

5.5.2 Carbon, Nitrogen, Hydrogen, and Phosphorus

Among these four elements, carbon (C), nitrogen (N), and hydrogen (H) are now determined simultaneously by automated gaschromatographic elemental analyzers. A weighed zooplankton sample is combusted in a constant amount of pure oxygen and the evolved CO_2, H_2O, and N oxides (later reduced to N_2) gases are detected and registered electronically as C, H, and N. Detailed accounts on the principles, operation, and so forth, can be seen in Belcher (1977). The main advantages of a CHN analyzer over organic matter analyses are its simple operation and the smaller sample size (usually 1–3 mg of dry weight). One type of analyzer, the Yanagimoto CHN Corder Model MT-3,[7] is shown in Figure 5.7. This analyzer takes 15 min to complete one analysis (2–3 mg of dry weight), but since two series of mechanisms can alternate at an interval of 7.5 min, two samples may be measured every 15 min.

As the combustion temperature is high (about 900°C), the C, H, and N determined with the CHN analyzer are the total amounts in both organic and inorganic fractions of zooplankton. With regard to C, Curl (1962) found measurable quantities of inorganic C (as carbonate) only in *Limacina* sp. and an isopod *Idotea metallica* (2–3% of dry weight). Amino acids are known to contain most of cellular N (65–90%) in marine zooplankton

Table 5.1. Composition of organic matter and ash measured simultaneously in marine zooplankton[a]

Species		Protein	Lipid	Chitin	Carbohydrate	Ash	Sampling season and location	Source
Copepods								
Euchaeta norvegica	male	40.7–55.4	29.2–42.9	4.4–6.4	0.8–1.9	4.4–7.8	One year at Korsfjorden, Norway	Båmstedt and Matthews (1975)
	female	35.6–46.5	26.4–47.3	3.7–6.4	1.0–2.2	4.4–11.2		
	C$_V$	38.5–48.3	25.9–43.4	3.5–9.3	0.9–2.4	4.4–11.8		
	C$_{IV}$	43.0–52.9	20.3–21.8	—	1.8–2.7	—		
	eggs	36.4–48.8	49.9–57.5	0.9–1.0	1.7–3.8	3.2–3.3		
Chiridius armatus		(46.1)	(24.2)	(7.0)	(2.0)	(7.4)	One year at Korsfjorden, Norway	Båmstedt (1978)
Euphausiids								
Euphausia superba		50.1–55.1(52.2)	11.3–27.1(16.0)	3.8–5.2(4.3)	3.1–5.5(4.4)	13.0–18.3(16.2)	Dec.–Jan. off Antarctica	Raymont et al. (1971b)
		37.0–48.5	16.1–34.4	3.3–5.7	2.1–4.1	11.1–26.6	Dec.–Feb. off Antarctica	Ferguson and Raymont (1974)
Meganyctephanes norvegica		52.3–73.5(61.1)	9.6–26.6(17.5)	3.0–4.4(3.4)	1.2–2.1(1.7)	10.3–12.9(11.5)	One year at Korsfjorden, Norway	Båmstedt (1976)
		49.5–62.0(56.5)	10.2–29.3(18.4)	3.5–5.5(5.4)	1.8–2.3(2.0)	11.0–15.3(13.6)	Nov.–May off southern Norway	Raymont et al. (1969)

	51.1–61.5(56.6)	12.1–29.0(17.2)	3.7–5.7(4.2)	1.5–2.9(2.0)	11.7–21.0(16.1)	Jan.–Nov. in Scottish waters	Raymont et al. (1971a)
Mysids							
Boreomysis arctica	32.6–43.3(38.0)	25.7–42.2(33.7)	2.3–4.0(3.2)	1.2–2.9(1.9)	12.1–17.2(14.7)	One year at Korsfjorden, Norway	Båmstedt (1978)
Neomysis integer	(70.9)	(13.1)	(7.1)	(2.4)	(7.9)	May–Mar. in Southampton waters	Raymont et al. (1964)
Decapods							
Acanthephyra purpurea	(60)	(12)	(6)	(2)	(19)	Sept. at Gulf of Aden	Raymont et al. (1967)
Gennadas clavicarpus	(62)	(15)	(5)	(3)	(16)		
Sergestes sp.	(58)	(29)	(3)	(2)	(17)		
Chaetognaths							
Eukrohnia hamata	(39.1)	(32.1)	(0.1)	(1.5)	(18.3)	One year at Korsfjorden, Norway	Båmstedt (1978)

[a] All figures are percent of dry weight. Ranges of values are not on individual determinations but on means of batch samples from designated sampling dates or grand means over the sampling periods (the latter figures are in parentheses).

95

Figure 5.7. Yanagimoto CHN Corder, Model MT-3. (1) Main unit; (2) Auto-sampler; (3) Data processing system.

samples (Corner and Cowey, 1964). Thus, it would appear that the major sources of total C and N measured in zooplankton are from organic compounds. Although no relevant information is available for H, parallel changes in H content with N or C content (Ikeda, 1974) suggest that it is derived chiefly from organic compounds.

The phosphorus content of zooplankton is determined as inorganic phosphate after the digestion of a zooplankton sample in acid. The following digestion procedure is cited from Syzper et al. (1976). A weighed sample is placed in a digestion tube with 1 ml of 50% (V/V) H_2SO_4 and incubated in a water bath at 90–100°C for 1 hr. The digestion tube is cooled briefly in tap water. The solution is diluted with distilled water, neutralized with 5 ml of 20% (W/V) KOH, and made up to a known volume. The concentration of inorganic phosphate in the solution is determined following the procedure in Chapter 8, Section 8.4.

Some relations between the elemental (C, H, N, P) and proximate (protein, carbohydrate, lipid) components are noteworthy. As no precise information is available on the elemental composition of organic matter in marine zooplankton, general figures for animal organic matter are quoted from Rogers (1927) and shown in Table 5.2. Among the three proximate components, protein and lipid are characterized by higher N and P content, respectively, while carbohydrate contains negligible amounts of N and P. Figures for C and H are similar for these three types of organic matter. Since carbohydrate is less abundant than lipid and protein in marine zooplankton (see Table 5.1), it can be concluded that protein and lipid are the major contributors of C, H, N, and P. Thus, the ratio

Table 5.2. Average elemental composition (% of dry weight) of organic matter and ratios of elements within each proximate component in aminal[a]

Elements	Protein	Lipid	Carbohydrate
C	51.3	69.05	44.44
N	17.8	0.61	—
H	6.9	10.00	6.18
P	0.7	2.13	—
Ratios of elements			
C/N	2.9	113.2	—
H/N	0.4	16.4	—
N/P	25.4	0.3	—

[a] After Rogers (1927). The ratios better distinguish the different types of organic matter.

of protein to lipid may be estimated from such elemental ratios as C/N, H/N, and N/P. For example, while the C/N ratio for protein is about 3 and that for lipid is infinitely great, since the amount of nitrogen is extremely small compared to carbon, the C/N ratio of animals increases with the lipid/protein ratio. Because of possible variations of the types and amounts of protein and lipid between species, such estimates must be made with caution when comparing diverse zooplankton species. Protein has been calculated from N by multiplying by 6.25, assuming the nitrogen content of protein to be 16%, not the 17.8% shown in Table 5.2. However, Raymont et al. (1968) analyzed the amino acid composition of *Neomysis integer* and gave a figure of 13.3% as the N content of this animal.

Table 5.3 presents the results of C, H, N, and P analyses on several zooplankton species from the Antarctic Ocean measured by the methods mentioned above. Interspecific variation in the abundances of these elements are obvious. Intraspecific change, depending on the size of specimens, is also seen in *Euphausia superba*. In general, the order of abundance of these elements is C>N>H>P. Omori (1969a) and Ikeda (1974) made extensive determinations of the C, H, and N content of various zooplankton species living in geographically different locations and found the carbon content as high as 65% of the dry weight in some boreal copepods. These high values for carbon reflect a large accumulation of lipid in the body. In tropical and subtropical zooplankton the carbon content is always lower than 45%. In contrast to carbon the nitrogen content is rather stable and usually around 10% of the dry weight or less. Beers (1966) measured the C, N, and P content of major zooplankton groups over a year in the Sargasso Sea. The average yearly carbon content ranged from 7.2% of the dry weight for siphonophores to 41.6% for copepods. Nitrogen was 9–11% of the dry weight. The average yearly

Table 5.3.   Carbon, nitrogen, hydrogen, and phosphorus contents of some Antarctic zooplankton collected during the austral summer[a]

Species	Average dry wt. ind^{-1} (mg)	% of dry weight			
		C	N	H	P
Ctenophores					
Beroe sp.	401.6	9.0	2.3	2.3	0.09
Pteropods					
Limacina antarctica	6.45	33.3	8.1	4.8	0.53
Clione antarctica	25.3	40.0	7.7	6.5	0.26
Cleodora sulcata	54.8	36.5	8.5	5.4	0.51
Polychaetes					
Tomopteris carpenteri	9.73	34.8	8.7	5.8	0.45
	137.9	34.9	8.3	5.8	0.35
Copepods					
Calanus propinquus	1.04	43.6	12.5	6.7	0.69
Metridia gerlachei	0.265	45.3	11.4	6.9	0.63
Amphipods					
Parathemisto gaudichaudii	3.60	35.9	8.6	5.6	0.82
	13.1	41.5	7.4	6.4	0.64
Viblia antarctica	12.1	40.3	8.2	6.0	0.86
Hyperia gaudichaudii	105.3	46.8	7.1	7.5	0.92
Euphausiids					
Euphausia triacantha	22.8	41.2	11.6	6.7	0.81
Euphausia superba	27.0	41.1	11.0	6.4	1.23
	76.8	44.7	10.2	7.0	0.89
	143.8	46.6	9.9	7.3	0.75
	237.7	47.4	10.2	7.3	0.81
	353.8	47.5	10.3	7.3	0.78
Salps					
Ihlea racovitzai	10.9	10.1	2.8	2.1	0.16
Salpa thompsoni	114.6	4.7	1.2	1.7	0.09

[a] Two size groups of *Tomopteris carpenteri* and *Parathemisto gaudichaudii* and five size groups of *Euphausia superba* were presented separately (modified after Ikeda and Mitchell, 1982).

phosphorus content ranged from 0.14% of the dry weight for siphonophores to 1.48% for euphausiids–mysids.

5.6   RNA and ATP

5.6.1   *RNA*

Theoretically, the rate of protein synthesis (= growth) in organisms should be closely related to the RNA (ribonucleic acid) content in the cells or organisms because RNA is considered to be a necessary precursor to protein synthesis. Sutcliffe (1970) examined the interspecific relationship between RNA concentration and growth rate for widely different organisms, from microorganisms to mammals, and found a good positive correlation between these two parameters. The same positive correlation has been observed intraspecifically in several marine microcrustaceans, including the zooplankton species *Euchaeta elongata*, *E. norvegica*, and *Meganyctiphanes norvegica*, but at the same time intraspecific variation was found to be considerable (Dagg and Littlepage, 1972; Båmstedt and Skjoldal, 1980). Thus, the usefulness of RNA measurements in estimating growth rates of various marine zooplankton has not been established as yet, but it may warrant future investigation as a simple tool for zooplankton production estimates in the field.

The RNA content of marine zooplankton is known to change depending on various variables such as age, size, species, season, and so on, and a general range of a few μg to several tens of $\mu g \cdot mg$ dry wt^{-1} is recorded for various mesoplankton from Norwegian waters (Båmstedt and Skjoldal, 1980).

The RNA extraction procedure described below is the Schmidt–Thannhauser method, modified by Munro and Fleck (1966a, 1966b) and used by Dagg and Littlepage (1972) for zooplankton samples.

1. **Sample preparation.** Immediately after sorting with a small screen and forceps, the live animals are quickly rinsed with distilled water dropped into liquid nitrogen, and then lyophilized. The dry weight of samples may be estimated at this stage. Analysis of RNA should be made as soon as possible. Changes in RNA content during freeze storage ($-20°C$) has been reported in some samples (Munro and Fleck, 1966a, 1966b).

2. **Reagents.** Perchloric acid (PCA) solutions of $1.2N$, $0.6N$, and $0.2N$ and KOH solution of $0.3N$.

3. **Procedure.** Because RNA is highly susceptible to heat degradation, its extraction must be performed at the temperature of $0–4°C$ by keeping samples on ice (except step d in the following).

   a. Samples of 2–10 mg of dry weight (or 10–50 mg of wet weight) are homogenized in 4.0 ml of ice cold distilled water with a tissue grinder and the homogenate is placed in a test tube. The inside of the grinder is rinsed with 1.0 ml of ice cold distilled water and this rinse is added to the homogenate.

   b. In a centrifuge tube containing 4.0 ml of the homogenate, 2.0 ml of ice cold $0.6N$ PCA solution is added. The sample is centrifuged for 10 min at $0–4°C$ (6000 rpm). The supernatant is decanted.

   c. The precipitate is washed twice with 2.0 ml of ice cold PCA solution by centrifugation (6000 rpm, 10 min) and decantation.

   d. A solution of 4.0 ml $0.3N$ KOH is added to the precipitate and incubated at $37°C$ for 60 min.

e.   A solution of 2.5 ml ice cold 1.2N PCA is added. The solution is centrifuged (6000 rpm, 10 min) at 0–4°C and the supernatant is saved. The precipitate is washed twice with 2.0 ml of ice cold 0.2N PCA solution. All three supernatants are combined and the extinction is measured with a spectrophotometer at 260 nm. The concentration of RNA is measured from a standard curve which is established with yeast RNA. In this analysis reagent blanks are prepared with 4.0 ml 0.3N KOH and processed as above.

4.   Note.   When one wants to express the RNA content on a protein basis rather than per dry weight, 1 ml of the homogenate from step a can be taken for protein determination by the biuret method in Section 5.5.1 or the Lowry method (Lowry et al., 1951). For zooplankton samples of less than 2–10 mg of dry weight, the quantities used can be scaled down to one-tenth (i.e., step b starts with 0.4 ml of the homogenate).

5.6.2   *ATP*

As metabolic energy depends partially on the available source of chemical energy stored in ATP (adenosine triphosphate) in cells, the concentration of this energy-rich intermediate has been considered as an index of metabolic activity of organisms. ATP was originally measured to estimate the living microbial biomass (Holm-Hansen and Booth, 1966; Holm-Hansen, 1969) and its metabolic activity (Hobbie et al., 1972). For zooplankton Balch (1972) found that the ratio of ATP to body carbon of *Calanus finmarchicus* is constant irrespective of prolonged starvation of this animal. On the other hand Traganza and Graham (1977) observed a large fluctuation of the ATP/C ratio in mixed zooplankton. From an extensive seasonal survey of ATP in zooplankton from Korsfjorden of western Norway, Skjoldal and Båmstedt (1977) concluded that the zooplankton's ATP is correlated with their seasonal reproductive activity but not with their metabolic activity. This poor relationship between ATP content and metabolic activity has also been shown by Ansell (1977) in his study of various marine mollusks. ATP contents recorded for 18 zooplankton species from Korsfjorden are summarized in Table 5.4. It is seen that the ATP content is highly species specific and varies with the season. Skjoldal and Båmstedt (1977) failed to find any correlations between ATP content and body dry weight within species.

Thus it appears that the ATP content of zooplankton is related to neither metabolic activity nor biomass. Instead, ATP may be taken as a sensitive index of the physiological status of zooplankton, such as reproductive activity and physiological stress, within the same species (see Anraku and Kozasa, 1978). Also, results can be used to estimate the "adenylate energy charge," a sensitive indicator of physiological states of organisms, by measuring AMP (adenosine monophosphate) and ADP

Table 5.4. ATP content of zooplankton species from Korsfjorden, western Norway, sampled in February and April–May 1974[a]

Species	Range of individual dry weight (mg)	ATP (μg·mg dry wt^{-1}) 27 February			29 April–2 May		
		Mean	n	SD	Mean	n	SD
Cnidaria							
Aglantha digitale	10–19	3.06	5	1.85	—	—	—
Polychaeta							
Tomopteris helgolandica	2.5–12.2	1.69	1	—	2.85	6	0.99
Crustacea							
Euchaeta norvegica, females	2.3–3.2	5.62	5	0.65	7.79	5	0.63
males	1.1–1.8	7.13	5	0.42	6.93	5	0.57
C_V	1.0–1.4	6.75	5	1.87	6.56	5	1.76
C_{IV}	0.22–0.39	—	—	—	8.68	5	1.82
Calanus finmarchicus	0.53–0.62	—	—	—	7.81	5	0.41
Calanus hyperboreus	0.8–1.9	6.32	3	4.29	4.84	3	0.72
Chiridius armatus	0.51–0.83	5.78	7	0.68	3.86	7	0.49
Metridia longa, females	0.24–0.35	5.20	7	1.27	3.86	5	0.40
Boreomysis arctica	3.7–16.9	3.46	2	0.19	6.48	27	1.99
Hemimysis abyssicola	1.6–3.9	—	—	—	6.90	5	0.67
Meganyctiphanes norvegica	5.8–8.85	2.84	4	0.13	8.47	5	0.93
Pasiphaea multidentata	15–1193	1.03	2	1.07	5.66	14	2.73
Pasiphaea tarda, eggs	4.0–4.3	0.09	3	0.06	—	—	—
Pontophilus norvegicus, larvae	1.8–2.3	—	—	—	0.98	5	0.38
Munida sp., larvae	0.33–0.65	—	—	—	0.78	3	0.27
Chaetognatha							
Eukrohnia hamata	1.7–8.4	1.42	10	0.62	2.48	9	0.66
Eukrohnia bathypelagica	1.9	1.53	1	—	—	—	—
Pisces							
Argentina sp., eggs	2.5–3.1	0.22	5	0.02	—	—	—

[a] After Skjoldal and Båmstedt (1977).

(adenosine diphosphate) after enzymatic conversion to ATP (see Chapter 8, Section 8.7).

In the following procedure ATP is extracted in a boiling Tris/HCl buffer[8] and subsequently assayed with firefly luciferin-luciferase and an ATP photometer (Holm-Hansen and Booth, 1966).

1. Sample Preparation. As the ATP concentration in cells can change very rapidly through reactions with transphosphorylase and ATP-ase, the use of fresh undamaged zooplankton and the quick killing of them in boiling Tris/HCl buffer are essential. Storage of fresh-frozen samples at $-26°C$ caused a significant loss in ATP even over a few days (Skjoldal and Båmstedt, 1977). Since seawater salts interfere with the luciferase reaction, as much seawater must be removed from samples as possible.

    2.   **Reagents.**   Tris/HCl buffer: 7.5 g of tris(hydroxymethyl) amino-methane is dissolved in 3000 ml of distilled water. The pH is adjusted to between 7.7 and 7.8 by the dropwise addition of 20% (V/V) HCl. About 150 ml of this solution is dispensed into 250-ml Erlenmeyer flasks fitted with loose caps and autoclaved for 15 min. This buffer may be kept for several months. The contents of each flask is used for one batch of samples and standards.

    Luciferin-luciferase enzyme preparation: Crude, refined, and purified enzyme preparations are commercially available, but the crude enzyme can be used satisfactorily for most samples. The enzyme preparation obtained from Sigma Chemical Co. (Cat. #FLE-50) contains a lyophilized water extract from 50 mg of firefly lanterns and is rehydrated with 5.0 ml Tris/HCl buffer. After standing at room temperature for 2–3 hr, the suspension is gently centrifuged (about $300g$)[9] for 1 min. The supernatant is placed in a dry test tube and left standing at room temperature for another 30–60 min; within the next 3 hr it is used.

    3.   **Procedure.**

a.   The zooplankton sample is dropped into the boiling Tris/HCl buffer and the ATP is extracted at 100°C for 3–4 min. The volume of Tris/HCl buffer depends on the sample size but is usually several ml to several tens of ml. To facilitate extraction, intact specimens may be broken up by forceps. The extract is cooled to room temperature and the volume exactly measured. The extract may be stored for many months in the dark at -20°C. The extract is thawed in running tap water just before ATP determination.

b.   A 0.20-ml solution of enzyme preparation is pipetted into a scintillation vial and the background light emission is counted with an ATP photometer for 60 sec. Then 0.50 ml of the extract is added to this enzyme extract by means of a pipette, the contents are mixed, and the light emission is counted for 60 sec with the ATP photometer. The difference between these counts (CPM), representing the light emission induced by ATP in the sample, is compared with the CPM of known amounts of ATP standard prepared from crystalline ATP (di-sodium salt) dissolved in Tris/HCl buffer.

    4.   **Calculation.**

$$\text{ng ATP·ml sample}^{-1} = \frac{\text{CPM sample}}{0.5 \text{ ml}} \times \frac{\text{ng ATP·ml standard}^{-1}}{\text{CPM standard}/0.5 \text{ ml}} \qquad (5.2)$$

$$\text{ng ATP·mg dry wt}^{-1} = \text{ng ATP·ml sample}^{-1} \times \frac{V}{W} \qquad (5.3)$$

where V is the volume of extract (ml), and W is the dry weight of the sample (mg). The dry weight can be estimated from subsamples.

5.7 Calorimetry

While the calorific content of zooplankton can be estimated from organic matter analyses (protein contains 4.2 cal·mg^{-1}; lipid, 9.5 cal·mg^{-1}; carbohydrate, 4.2 cal·mg^{-1}—from Prosser, 1973b), it can be determined directly with an oxygen bomb calorimeter. For bomb calorimetry a zooplankton sample is dried, homogenized, pressed into a pellet, and completely combusted in pure oxygen. The heat liberated from the combustion of organic matter in the sample is read manually with a precise thermometer or automatically recorded with a potentiometer. In the latter case the calorimeter and potentiometer must be calibrated with benzoic acid of known calorific content. The result can be expressed as cal·mg dry weight^{-1} or cal·mg organic weight^{-1}. The temperature in the bomb calorimeter reaches over 500°C during the combustion, so that various salts such as calcium carbonate are considered to be vaporized. Thus, the weight of the residue is less than the ash weight from normal procedures (Section 5.3.5). It is therefore necessary to determine the organic weight independently when the unit of cal·mg organic weight^{-1} is used.

There are several types of oxygen bomb calorimeters available commercially. Cummins and Wuycheck (1971), Paine (1964, 1971), and Richman (1971) discuss details of analytical procedures and technical problems of bomb calorimeters. For plankton samples Richman (1971) recommended a Phillipson microbomb calorimeter (Phillipson, 1964), which requires relatively small amounts of samples (2–100 mg of dry weight). When the amount of sample is insufficient, additional material of known calorific content, such as benzoic acid or Nujol oil, can be added; the calories of this filler material is subtracted from the final result.

Calorific determination with the bomb calorimeter is usually time consuming, and the results of replicate samples are often scattered. To reduce this scatter, errors inherent with the preparation of samples, especially pelletizing and the weighing of filler materials, when used, must be minimized. The end product of combusting proteinaceous materials is largely N_2 in the oxygen bomb calorimeter, while it is NH_3 for most aquatic invertebrates, including zooplankton. This means that the value obtained from the bomb calorimeter overestimates the calories biologically available. Kersting (1972) estimated this extra calorific value to be 5.9 cal·mg N^{-1} so that it can be corrected for knowing the nitrogen content of the sample. In the case of *Daphnia magna*, a freshwater Cladocera, whose nitrogen content is 8% of the dry weight, the corrected calorific content is 10% less than that of uncorrected (Kersting, 1972).

Notes

[1] The seawater may be slightly diluted with fresh water if significant evaporation is anticipated.

[2] Manufactured by Neward Enterprises Inc. (U.S.A.).

[3] This process is not required when a Gelman GA8 filter is used.

[4] Available at Edward Gurr Inc. (U.K.).

[5] Consider that W is the tare weight of a filter and W' is the weight of the filter with feces. The weight of the feces is

$$(W' - W) + (Wb' - Wb)$$

where Wb' and Wb are the weights of a blank filter measured at the same time as the measurements of W' and W, respectively.

[6] Relationships between the body length, weight, and volume vary considerably depending upon environmental conditions. Therefore, one should be careful not to use a single conversion factor for organisms from different conditions.

[7] Manufactured by Yanagimoto Mfg. Co. (Japan).

[8] For ATP extraction methods other than that using boiling Tris/HCl buffer, see Lundin and Thore (1975).

[9] Relation between g and rpm is

$$g = \frac{\pi^2 l(\mathrm{rpm})^2}{900 \times 980}$$

where l is the radius of a circle.

CHAPTER 6 _____

Rearing and Culture
_____

Laboratory maintenance and culture techniques have contributed to the study of zooplankton ecology by providing valuable information on physiological processes and the growth of zooplankton under controlled conditions. The earliest attempts at laboratory rearing of marine zooplankton date to the 1910s (Allen and Nelson, 1910; Crawshay, 1915). At present many zooplankton species have been brought into the laboratory. However, the most successful rearing has involved neritic species. Only few oceanic and deep-water species have been reared.

In this chapter we describe a general procedure for laboratory maintenance, rearing, and culture of net zooplankton species and the effects of experimental conditions on them. The techniques of culturing food organisms, such as bacteria and phytoplankton, for rearing zooplankton are also explained. Finally, we discuss the measurement of an animal's resistance to external physicochemical factors. Readers may find more detailed accounts of the cultivation of various taxa of zooplankton in Kinne (1977) and Paffenhöfer and Harris (1979). The books of Needham et al. (1959) and Sato and Ito (1961), which mention general techniques for laboratory culture of invertebrates, may be helpful in establishing the appropriate techniques to culture the zooplankton of interest.

6.1 Rearing Apparatus

Laboratory procedures and equipment may differ, depending on the purpose of the study. If the aim is to observe mating behavior, feeding habits, molting frequency, or intraspecific variation in morphology with growth, "maintenance" or "rearing" is sufficient. The test animal is kept alive for the duration of the experiment under the best possible conditions. Care extends for a given developmental stage (maintenance) or a number of stages up to one generation (rearing). On the other hand, if the aim is to study genetics, reproduction, or life histories, the "culture" of animals over several generations is necessary.

6.1.1 *Materials and Cleaning*

Materials used in the construction of rearing and culturing apparatus must be nontoxic to animals and highly resistant to seawater. Transparent materials facilitate the observation of animals during their cultivation. In general, borosilicate glass from a known source (trade names: Pyrex, Kimax, Jena, etc.) is most suitable, although consideration should also be given to plasticware such as Teflon, polyethylene, polypropylene, and polymethylmethacrylates (Plexiglas, Perspex) without additives and colors. Many metals and alloys dissolve in seawater and produce highly toxic ions (such as copper, zinc, and chromium) and thus should not be used for even parts of culturing devices. Plumbing for water circulation can be made of Plexiglas or polyvinylchloride. Tubing made of silicone rubber is best, while natural rubber is totally unsuitable and should not be used.

All apparatus, including pipettes for the transfer of specimens from one container to another, should be cleaned thoroughly before every use. Glassware is washed first with a nonionic detergent rinsed sequentially with an acid solution (5–10% HCl), hot running water, and pure water (distilled water or deionized water), then dried completely. Sulfuric dichromate has often been used in the past as a cleaning solution; however, chromate ions adsorb onto glass surfaces where they are very difficult to remove and are now known to be toxic. Plasticware can be cleaned with detergents and rinsed with pure water. Acid treatment after cleaning plasticware is desirable if it does not damage the material. Deionized water usually contains organic materials leached out from the ion-exchange resin, although these may have a beneficial effect by complexing trace elements (Bernhard, 1977). Containers and other equipment which once contained a toxic substance, such as Formalin, cannot be used in principle. Rearing containers should be covered to prevent excessive evaporation. Each cover should be marked on top so that settled dust and harmful substances are not inadvertently introduced into the rearing containers.

6.1.2 *Size, Shape, and Function*

When the observation of individuals and their development is necessary, one or at most a few animals are normally kept in individual containers with water. A tackle box or a few finger bowls may be used as the containers. Tackle boxes made of transparent Plexiglas with 15–20 partitions are available commercially and especially suited for the observation of zooplankton development. Usually, one individual or a clump of eggs is placed in each enclosure with about 100 ml of clean seawater. The animals are carefully transferred into another tackle box with fresh seawater and food every 1 or 2 days by the use of a wide-mouth pipette. As many zooplankton are fragile, care must be taken so that the animals do not become damaged when they are moved. Dead bodies and exuvia are collected during transfers and preserved for later examination. Rearing in a tackle box or finger bowl should be done in a temperature-controlled room because of the small quantity of water in the vessels.

For the long-term rearing of marine plankton in static water, frequent replacement with fresh seawater is essential to prevent pollution of water by bacteria and the metabolic wastes of animals. When a number of individuals are reared in a single vessel containing static water, their phototactic behavior may often cause them to aggregate in the region nearest to the light source unless they are kept in total darkness or diffuse light. Such concentration may cause increased cannibalism and may result in local oxygen depletion. Also, this behavior separates the animals from their food and, in extreme cases, causes starvation in the presence of an abundance of unavailable food. In static water nonmotile foods such as diatoms will settle to the bottom of the vessel relatively rapidly and become unavailable to the animals.

Therefore, it is often desirable to create water movement within the rearing vessels; a number of methods have been used to achieve this. The classic technique is that of the "plunger-jar," with which the water in a vessel is gently agitated by a glass or plastic plate which is slowly raised and lowered by means of a pulley fixed to a motor (Fig. 6.1A). There are various systems similar to the plunger-jar, with vanes and paddles moving the water. One shown in Figure 6.1B was used by Frost (1972) for feeding experiments with *Calanus*. The stirrer, driven by a 1-rpm motor, rotates and oscillates up and down in a 4-liter beaker. However, the plunger-jar technique has two disadvantages: (1) the moving plate in the vessel is liable to damage delicate specimens and (2) the water circulation which it provides tends to be intermittent.

Movement of water has also been often provided by bubbling air through the water column, simultaneously achieving some aeration. However, this technique is physically harmful to zooplankton and may decrease rates of growth and survival. The circulation of water in the vessel is always uneven. The movement is extremely strong within the stream of bubbles but relatively weak in the remaining part.

In order to avoid the problems of providing water circulation, Omori (1979) used a simple device in his rearing experiment of larvae of *Sergestes similis*, of which an improved type is shown in Fig. 6.1C. Larvae were reared in a 2.8-liter Fernbach flask. Water was gently circulated in the flask by a simple aeration device with which the larvae could be suspended in the water without coming in direct contact with air bubbles. The "Planktonkreisel" developed by Greve (1968) is another useful system. Figure 6.2 shows a modified type. Water gently circulates in the vessel through a sand and gravel filter at the bottom, is lifted above the level of

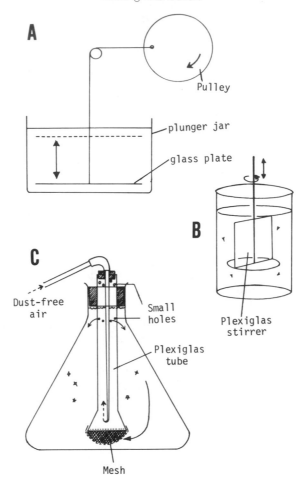

Figure 6.1. Three devices for rearing zooplankton.

the general water surface by a stream of air bubbles which is separated from the animals, and, ultimately, returns via a jet which opens close to the bottom and side of the vessel. Four species of ctenophores have been reared in this system (Greve, 1970).

Figure 6.3 shows the culture apparatus devised by Paffenhöfer (1970). A container, 3–8 liters in volume, is attached diagonally to a universal joint which is driven by an electric motor. The container rotates at a speed of 0.5–2.0 rpm causing damaged algae and fecal pellets as well as dead animals and molts to accumulate at the center of the bottom. It has been suggested that *Calanus* ceases feeding at food concentrations below 50–70 μg C·liter^{-1}. However, Paffenhöfer successfully reared *C. helgolandicus* from egg to adult in phytoplankton concentrations as low as 15–20 μg

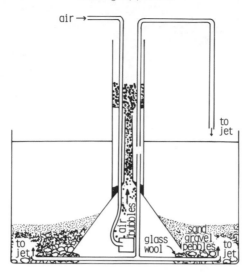

Figure 6.2. Planktonkreisel. The upper jet points horizontally and its tip is just above the surface of the water; all jets point clockwise (or counterclockwise). The two (or more) lower jets incline slightly upwards (at about 15°) and project a little above the surface of the sand. (After Rice and Williamson, 1970.)

$C \cdot liter^{-1}$ using this apparatus. He also succeeded in rearing multiple generations of *Oikopleura*, an animal which due to its fragile body had been considered very difficult to keep in the laboratory.

In a marine laboratory where fresh seawater of the desired temperature and salinity is available in abundance, an "open-circulation" of running water can be used. In such a system seawater is circulated in the aquarium with an inlet at the bottom and an outlet at the top on the opposite side. It is generally useful to set up several aquaria connected with pipes or tubes to achieve seawater circulation, either in parallel or in series. The aquarium series should terminate with a level regulator (Fig. 6.4) to maintain a uniform water level throughout.

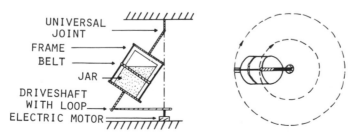

Figure 6.3. Paffenhöfer's culture apparatus. Left, side view; right, viewed from above. (After Paffenhöfer, 1970.)

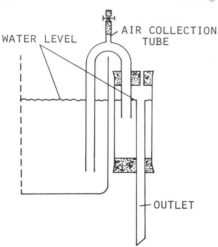

Figure 6.4. Overflow with level regulator. (After Flüchter, 1972.)

A simple apparatus to keep zooplankton with open-circulation is a Plexiglas cylinder 10–15 cm in diameter and 15–30 cm long of which the lower end is covered with a nylon mesh. It is suspended in an aquarium of running water with a float and contains a number of animals (Fig. 6.5A). Another example of an aquarium with open circulation is shown in Figure 6.5B. The usefulness of this simple arrangement has been proven by Dr. M. Reeve of the University of Miami who successfully used it to rear various zooplankton species, including *Sagitta hispida* and *Mnemiopsis mccradyi*, over several generations. One aquarium contains 20–40 liters of seawater and is made of Plexiglas. Seawater enters the aquarium continually and drains gently without damaging the animals. The key to

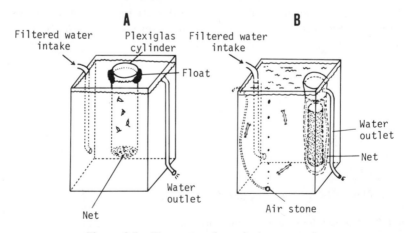

Figure 6.5. Two types of zooplankton aquaria.

the successful use of this aquarium is to regulate the volume of air from the air stone placed at the bottom to both aerate and slowly stir the water in the aquarium. As both *S. hispida* and *M. mccradyi* are carnivorous species, *Acartia tonsa* collected from the sea was provided as food. One problem in such flow-through culture vessels is keeping the animals away from the region of the water outlet. In the aquarium shown in Figure 6.5B, the animals are separated from the outgoing water stream with a bag made of fine mesh. A rigid support is inserted inside the bag to maintain its shape.

If the water is taken directly from the sea, aeration and elaborate hygiene measures are unnecessary and, consequently, open-circulation always yields the best results in keeping animals healthy. We adopted this open circulation system for the maintenance of several types of oceanic plankton aboard ship and obtained a much higher survival rate than with the still-water system. Even if there is a limit to the use of seawater and a closed system is necessary, a flow of water in the culture vessels can be achieved by using a pump to lift water into a large tank from where it will eventually flow back into the culture vessel.

Recently, several types of expensive and sophisticated apparatus have been developed for mass cultivation of zooplankton. These employ the "closed-circulation" concept in which a fixed amount of water is reused after successive treatments, such as filtration, aeration, sterilization with ultraviolet light, and cooling (see Zillioux, 1969; Anraku, 1973). However, as an apparatus becomes large and complicated, problems often occur. A fully satisfactory system has not yet been developed.

6.2 Rearing Conditions

Major factors that greatly influence the rearing of zooplankton include water quality, temperature, salinity, pH, dissolved oxygen, light, food, and density of animals as well as the size of container. Some conditions, such as physicochemical factors of the environment and kinds of food, can be found from field surveys of the species of interest, but other important information, such as the appropriate container size and amount of food to be provided, are not obtainable from field surveys. Thus, trial and error is unavoidable in arriving at acceptable procedures, but considerable effort can be saved by referring to previous results on related species.

6.2.1 Water Quality

Seawater from the location where the animals occur in the field is best and often used. It is filtered through filter paper or a membrane filter with a pore size greater than 1 μm. In recirculating systems seawater may be passed through glass wool and/or cartridge filters, of which various types and pore sizes are commercially available. Natural seawater contains various classes of organic matter, some of which may inhibit the growth of reared organisms, especially unicellular algae used to feed zooplankton. For this reason natural seawater is best filtered through activated charcoal.[1]

Natural seawater from various locations differs in its suitability for rearing a variety of invertebrates, even when chemical differences cannot be distinguished by normal analytical techniques. Generally, for rearing with static water or closed-circulation systems, oceanic water is more suitable than shore water as it contains less oil or other organic matter that may be harmful to the animals. The water is collected in large quantity at one time and kept in a dark cool place for about a month after filtration through a fine paper or glass fiber filter. This "aged seawater" is again filtered and aerated prior to use for rearing. The containers to store water are normally made of glass, polyethylene, or polyvinylchloride, but insufficient experimental work has yet been done to establish whether or not these materials are harmless. For rearing sensitive organisms, filtered seawater may be sterilized by heating for 30–60 min at 80°C. This procedure is adequate for killing all organisms other than some bacteria.

When a large quantity of seawater cannot be obtained easily, artificial seawater can be used. However, it is generally advisable to add at least one-third to one-fourth natural seawater; certain organic constitutents of natural seawater, absent in artificial media, are of importance for rearing many animals. The composition of natural seawater was discussed in detail by Lyman and Fleming (1940). There are various formulas for artificial seawater in the literature (see Tyler, 1953, and Provasoli et al., 1957) and a number of commercial preparations are now available. Artificial seawaters vary in detailed composition; therefore, users should pay attention to the components, especially chelators and trace metals used in phytoplankton culture media. We suggest one, ASP-M, consisting of McLachlan's (1964) relatively simple modification of the well-known ASP medium (Provasoli et al., 1957), shown in Table 6.1. This water is suitable not only for rearing animals but also as the basic seawater for the phytoplankton culture medium f/2, which will be described later.

To reduce the colonization of water by bacteria and the growth of bacteria on the bodies of reared zooplankton, antibiotics such as penicillin or streptomycin are sometimes used. However, the effect of different doses has not been well studied for many animals. Their careless use may even harm test organisms. Neunes and Pongolini (1965) noted that when *Temora stylifera* was reared at 18°C, the addition of 6.5 mg·liter^{-1} of penicillin and streptomycin in alternating shifts of one week was more effective than the addition of greater quantities of these antibiotics at one time. As most doses are efficacious for only one day or so, we suggest frequent changes of water with sufficient care rather than the use of antibiotics.

In early studies of zooplankton culture experimental additions were made of EDTA and TRIS [tris(hydroxymethyl) aminomethane] as chela-

Table 6.1. **Artificial seawater medium ASP-M,
pH adjusted to 7.5**[a]

Compound	Amount	Concentration (mM)
NaCl	23.38 g	400
KCl	0.75 g	10
$MgSO_4 \cdot 7H_2O$	4.93 g	20
$MgCl_2 \cdot 6H_2O$	4.07 g	20
$CaCl_2$	1.11 g	10
$NaHCO_3$	0.17 g	2.0
H_3BO_3	0.25 g	0.4
Distilled water	1000 ml	—

[a] After McLachlan (1964).

tors to remove toxic metals and to stabilize (buffer) the pH of the seawater, and several kinds of vitamins. However, the effectiveness of these chemicals is not clear, and in fact, they have not been used in more recent zooplankton studies (see Paffenhöfer and Harris, 1979).

6.2.2 *Temperature and Salinity*

Numerous experiments have been conducted to clarify the effects of temperature and salinity on the mortality, growth, and feeding of zooplankton. Each zooplankton species has absolute ranges of both temperature and salinity outside of which development does not proceed. Generally, oceanic species tend to be less tolerant than neritic species of salinity changes, especially during their larval stages. Growth and feeding rates increase as the temperature increases to a certain limit within the range of thermal tolerance. The rates are highest at the optimum salinity. To maximize the growth rate, therefore, rearing close to the upper limit of the tolerated range of the temperature may be considered. In reality, however, the plankton is near its physical limit at this elevated temperature, the water can easily be colonized by bacteria, and consequently successful rearing may be difficult. As temperature decreases, the duration of development increases. Low temperature may inhibit the molting of crustaceans without actually being lethal. With regard to the effects of salinity, Katona (1970) showed that at 14°C, the generation length of the neritic copepod *Eurytemora herdmani* in culture varied with salinity. One generation was completed in 21 days at a salinity of 33‰ and in 17 days at 15‰. The body length of female *E. herdmani* showed a linear increase with decreasing salinity and attained a maximum at 15‰, but this trend was less pronounced in males.

6.2.3 pH and Oxygen

The pH of seawater is considered to be one of the main factors that limits the growth rate of plankton; therefore, with the exception of certain species of Protozoa that prefer a particular pH for growth, the value should always be maintained at 7.0–8.5. Care must be taken, particularly when zooplankton are cultured with high concentrations of phytoplankton, to maintain this pH. Matsudaira (1957) reported that carbonate assimilation by phytoplankton caused a rapid rise in pH of the rearing water (close to pH 11) resulting in high mortality of *Sinocalanus tenellus*.

Oxygen content should be kept near the saturation level for good health of cultured organisms. Aeration may not be needed when animals are reared in a large amount of water or at low density, but it is absolutely necessary when a large number of individuals is kept in static water or in a closed-circulation system for a long period. Zillioux (1969) reported mass mortality of *Acartia clausi* when oxygen concentration dropped to 3.2 ppm ($= 2.2$ ml $O_2 \cdot$liter^{-1}) in a high-density culture with over 380 copepods·liter^{-1}.

6.2.4 Light

As with other factors mentioned above, light intensity, quality, and day–night photoperiods most suitable for the culture of zooplankton would be those closest to the animal's natural conditions. Extreme deviations from this normal range may cause problems. For example, Matsudaira (1957) found that direct sunlight killed all specimens of *Sinocalanus tenellus* and complete darkness prohibited reproduction of this animal. However, light conditions used in previous studies of zooplankton in culture vary greatly. Paffenhöfer (1970) reared *Calanus* with light periods of 12 or 14 hr, whereas Mullin and Brooks (1967) kept the same species in a semidark condition. Hirota (1972) reared *Pleurobrachia* in darkness and Zillioux (1969) kept *Acartia* at a light intensity of 650–1300 lux. So far, very little experimental work has been conducted to determine the effects of light intensity, wavelength, and duration on the behavior and growth of zooplankton. Many epi- and mesopelagic species perform diel vertical migrations, and there are suggestions that their feeding rhythm and other behavior respond to the changing illumination of their natural habitat. Furthermore, diel cycles may influence endocrine activity, which in turn affects the mating, ripening, and release of eggs or broods, hatching, molting, and death of zooplankton. The effect of light conditions and diel cycle must be studied more extensively.

6.2.5 Food

Food in use today for the culture of marine zooplankton can be divided into five groups:

1. Cultured phytoplankton (mainly diatoms and flagellates) or mixtures of bacteria, ciliates, and phytoplankton.

2. Cultured zooplankton [rotifers, larvae of mollusks (e.g., oyster), cladocerans, barnacle nauplii, young copepods, etc.].

3. *Artemia salina* (brine shrimp) nauplii.

4. Yeast and nonliving foods (flour, starch, rice bran, soy bean lees, algal powder, dried *Chlorella*, and food synthesized for penaeid shrimp culture).

5. Natural zooplankton.

Yeast and nonliving foods have recently been tested extensively by Japanese fisheries biologists as staple food sources for the mass culture of zooplankton. Natural zooplankton is often used for rearing larval fish but can also be used as food for other carnivorous zooplankton. The general procedure includes net sampling with 50–100 μm mesh, filtration with 500–1000 μm mesh to remove predators, letting stand at lower temperature and aerating, and use as food. Sediment in the container, including feces and dead plankton, should be removed each morning by siphoning. Some of the apparatus shown in Figure 2.23 may be used for the handling of food plankton. Food plankton which gathers under light can be collected with a dip net or trap at night. Naturally, the condition of living food collected in this way is always better than that sampled by towing a net. The primary advantage of wild zooplankton as a food is its natural composition, but the ease of collection will determine whether wild plankton is suitable for a particular study. Its disadvantage is the lack of control of its composition and availability. In addition to the above food materials, some bacteria and pieces of fresh seaweeds, such as *Ulva*, have been used to rear epibenthic copepods of the genera *Tigriopus* and *Tisbe*.

Food suited to support the growth of zooplankton in the laboratory is limited and varies from one zooplankton species to another. Even for the same species the suitability of various foods changes during the course of development. According to Nassogne (1970), phytoplankton smaller than 6–7 μm or wider than 16 μm lowered the rate of increase of a population of *Euterpina actifrons* in culture. This was probably due to the fact that adults of *Euterpina* did not capture food particles smaller than 6–7 μm and the nauplii could not ingest particles larger than 16 μm. *Acartia clausi* and *A. tonsa* can grow when fed *Isochrysis galbana* (4–6 μm), but they cannot reach sexual maturity unless the larger food *Rhodomonas* sp. (9–14 μm) is provided (Corkett, 1970). *Sergestes similis*, like other Penaeidea and Euphausiacea which retain the nauplius phase, does not feed as a nauplius. Feeding begins on unicellular algae during the protozoea phase, and after entering protozoea II, they take *Brachionus* and *Artemia* nauplii. Zoea and postlarvae are largely carnivorous, feeding on *Artemia* nauplii and copepods in the laboratory (Omori, 1979). Because feeding selectivity changes during

Table 6.2. Volume and carbon contents of important food species for rearing zooplankton[a]

Species	Volume ($\mu m^3 \cdot ind^{-1}$)	Carbon (pg C$\cdot ind^{-1}$)
Emiliania huxleyi	14	3.7
	20	4.1
Pavlova lutheri	50	6.1
Isochrysis galbana	80	5.1
Thalassiosira pseudonana (or T. guillardii)	71	37
	77	27
	120	15
	168	19
	220	40
Phaeodactylum tricornutum	120	11
Dunaliella tertiolecta	300	43
	316	52
Chaetoceros ceratosporum	220	27
	390	48
Chaetoceros septentrionalis	310	100
Skeletonema costatum	110	120
	170	95
	312	33
Tetraselmis maculata	310	80
Dunaliella salina	400	60
Amphidinium carteri	740	118
Prorocentrum sp.	780	125
Syracosphaera elongata	1,380	229
	1,610	267
Syracosphaera carterae	1,250	196
	1,760	168
Thalassiosira weissflogii	1,390	139
	1,500	187
	1,720	122
	1,840	128
	2,150	118
	2,750	253
	3,500	193
	4,360	292
Thalassiosira rotula	3,480	288
Thalassiosira nordenskioldii	8,100	850
Scrippsiella trochoidea	8,250	1,270
Ditylum brightwellii	26,000	2,300
	29,000	1,700
	32,000	813
	35,000	911
	38,000	4,200
	54,600	1,370
	89,400	1,220
	120,000	1,485

Table 6.2. *(Continued)*

Species	Volume ($\mu m^3 \cdot ind^{-1}$)	Carbon (pg $C \cdot ind^{-1}$)
Rhizosolenia setigera	13,200	820
	37,000	1,240
	174,000	2,140
	150,000	3,040
Coscinodiscus centralis	970,000	73,000
	1,100,000	89,000
	3,800,000	84,000
	3,900,000	75,000
Coscinodiscus concinnus	6,200,000	111,000
	5,290,000	117,000
Coscinodiscus wailesii	16,000,000	150,000
Brachionus plicatilis	2,260,000	53,300
	3,100,000	71,000
Artemia nauplius	29,800,000	520,000
	41,000,000	701,000

[a] Compiled by the authors using data from Parsons et al. (1961); Mullin et al. (1966); Mullin and Brooks (1970b); Taguchi (1976); and the authors' measurements.

development and growth, successful rearing has been accomplished by providing a mixture of several types of food differing in size and shape. Nassogne (1970) found that the fecundity of *Euterpina* reached its maximum when a mixture of foods was used. The quality of a food is also very important, as the nutritive value of different types of food varies considerably. However, with the notable exception of Provasoli et al. (1959), who studied the nutritional requirements of *Artemia salina* and *Tigriopus japonicus*, little is known about the specific nutritional requirements of various marine zooplankton. As a reference for choosing appropriate planktonic foods for rearing experiments, Table 6.2 shows the volume and carbon contents of plankton that are often fed to zooplankton. It must be noted that even in the same species both the volume and carbon contents vary greatly, depending on the culture conditions.

As the concentration of food increases, feeding and growth rates become higher. In addition, both fecundity and the size of the mature individuals increase. However, as there is always an upper limit to the feeding rate (see Chapter 7, Section 7.3.4), excessive food may be unwise for it may cause clogging of the gut and inefficient digestion. We observed this while rearing *Penilia*. Water quality quickly deteriorates, especially when an excess of nonliving food is given, due to oxygen depletion and fouling of the water by decomposing food or dead specimens. On the other hand some zooplanktonic species appear to cease feeding when the density of food falls below certain levels. For a particular organism threshold density seems to differ with the physical condition of an animal, the nature and size of the food, and the size of the container. Reduced-food condition decreases the rate of development and prolongs larval life.

6.2.6 Other Problems

In the natural environment zooplankton do not normally encounter solid surfaces. Their delicate bodies and setae are easily damaged by contact with such surfaces. Thus, large aquaria are always better than small ones for minimizing the susceptibility of animals to damage. Paffenhöfer (1970) showed that the feeding of *Calanus* is reduced in small containers. The size of the container even affects the mating behavior of copepods. According to Mullin and Brooks (1967), copulation of *Rhincalanus nasutus* was observed when it was reared in a 19-liter tank but not in a 4-liter tank.

Another problem almost specific to the rearing of crustacean plankton is entrapment in the surface film of water and/or stranding above the waterline on the walls of the container, especially those which are small and hence have a high surface-to-volume ratio. This results in especially high mortality of marine cladocerans, copepod nauplii, some hyperiid amphipods, and decapod larvae raised in static water. To avoid this problem, use special culture vessels with a large volume of water and very small surface area or with a fine mesh immersed below the water surface to restrict the animals. An alternative solution is to use a surface-tension-reducing agent in the seawater. One of the authors (M.O.) could reduce or nearly prevent the stranding of *Evadne nordmanni* and *Podon* spp. by adding 250–500 mg·liter^{-1} of bovine serum albumin (plus 10 mg·liter^{-1} of streptomycin to prevent bacterial growth) to the seawater; however, it was difficult to keep the water in good condition during long experiments. Sandifer et al. (1975) found that the addition of ≥25 ppm (W/V) of polyethylene oxide virtually eliminated stranding deaths of larval grass shrimp *Palaemonetes* spp. in culture containers. In addition, the polyethylene oxide seemed to reduce other causes of mortality (natural death), but the development time increased slightly with increasing concentration of the agent.

6.2.7 Balanced Aquarium

When unfiltered seawater in a glass vessel is left in a sunny place by the window of the laboratory and gently agitated and aerated, bacteria, ciliates, and benthic microalgae (diatoms and cyanophyceae) begin to propagate and adhere to the inner wall of the vessel within a few days. If a few individuals of common zooplankton, such as rotifers and neritic copepods, are introduced in such an environment, they grow and breed. In the water physicochemical and biological cycles are maintained adequately and the water quality does not deteriorate. This condition can be maintained for a long time with only little care. A culture system that develops such a stable, self-regulating, and balanced condition is called a *balanced aquarium*. Even in a fairly large aquarium containing 0.5–1 ton of water, such a stable ecosystem may be created. With continuous gentle aeration and the daily addition of about 5 mg·liter^{-1} of marine yeast, cloudlike or cottonlike microbial flock occurs in the lower part of the aquarium, and various small creatures in the water become apparent within a week. Copepod species such as *Longipedia, Tigriopus*, and *Tisbe*, as well as *Oithona*, can be reared over generations. The water condition is stable for some months with little water change and the occasional addition of yeast, and the animals can be used for experimentation.

The balanced aquarium, though not under complete control, suggests many things about the conditioning of the water used to rear marine zooplankton, particularly for mass cultivation. Some microorganisms, though they do not serve as food for most zooplankton, may play a role in maintaining water quality because they accelerate the cycling of materials in the aquarium. It is extremely difficult to maintain a monospecific culture for a long rearing period. Even for herbivorous species which can be reared on only a single autotrophic food type, it is better to introduce one or more organisms into the culture, including detritus feeders and bacteria consumers (some ciliates, rotifers, and benthic copepods), to establish biological associations in the water. These organisms may not only consume the excess bacteria but also serve as an additional food source. It also happens that two species propagate much better when reared together in a mixed culture than when reared separately because the conditions produced by one species are particularly favorable to the other species, and vice versa. Muller and Lee (1969) failed to establish bacteria-free cultures of four Foraminifera species fed one or two species of algae and observed that some bacteria were required for the prolonged survival of the foraminiferans.

6.3 Techniques for the Culture of Food Organisms

The food organisms of marine zooplankton consist of bacteria, diatoms, flagellates, ciliates, and a variety of larvae of mollusks and crustaceans. Live organisms such as rotifers and *Artemia* are also suitable foods, although marine zooplankton rarely encounter these organisms in natural waters. Mass cultivations of zooplankton are being attempted in many places for maricultural purposes. Various problems with the rearing of food plankton are documented by Anraku (1979). Also, the possibility of utilizing residues from alcohol fermentation of rice and domestic refuse as the food of mariculture species is being investigated in Japan with some success. These residues are processed into microbial flocks containing mixtures of bacteria, unicellular algae, and protozoa and used as food for shrimp larvae and fish larvae as well as some food plankton. Cultivation methods for some live foods for meso- and macroplankton will be discussed in order of increasing size of the organism.

6.3.1 *Bacteria*

Although free bacteria are common elements in all natural waters, it is generally considered that they constitute only a minor portion of the food resources available to marine zooplankton; however, they enhance the nutritive value of the detritus they colonize.

The cultivation of bacteria as food for zooplankton consists of the following:

1. Sampling bacteria from the natural environment.
2. Incubating these bacteria, generally by the spread-plate method (Buck and Cleverdon, 1960).
3. Isolating bacteria colonies.
4. Incubating the isolates in enrichment broth.
5. Harvesting the cultured bacteria.

Then, strains may be screened for potential food value.

Various culture media have been devised and successfully tested in microbiological studies. There is also a variety of methods of culturing bacteria. The methods are beyond the scope of this book and are described in many textbooks on microbiology. Here we suggest a technique that can be applied to feeding and rearing experiments with zooplankton. Needless to say, sterilization (autoclave at 120°C, 15 psi for 15 min) of the seawater, culture medium, containers, pipettes, and other equipment, as well as the most painstaking cleanliness in handling the cultures under laboratory conditions, are necessary.

A culture of mixed bacteria is obtained by adding untreated natural seawater to the following medium (Yasuda and Taga, 1980):

Casamino acids (Vitamine-free), DIFCO	0.25 g
Soil extract[2]	1.0 g
Bacto-agar	15.0 g
Artificial (or aged) seawater	100 ml
pH adjusted to 7.6	

Bacteria is incubated for 3 days in the dark at 20°C in the above medium, and some colonies are isolated. Then, these isolated colonies are incubated for 5–7 days at 20°C in 125-ml Erlenmeyer flasks stoppered with cotton plugs, or in the apparatus shown in Figure 6.6, using Medium 2216E broth (Oppenheimer and ZoBell, 1952) or a similar broth (Yasuda and Taga, 1980). The composition of these enriched broths are:

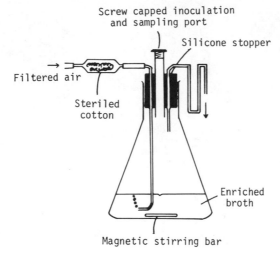

Figure 6.6. Culture apparatus for food bacteria. (After Taga and Yasuda, 1979.)

Medium 2216 E

Bacto-peptone	5.0 g
Bacto-yeast extract	1.0 g
Ferric phosphate, FePO$_4$	0.1 g
Seawater	1000 ml
Final pH 7.6	

Yasuda and Taga medium

Ammonium sulfate	0.5 g
Glucose	1.0 g
Potassium phosphate, K$_2$HPO$_4$	0.005 g
Artificial seawater	1000 ml
Final pH 7.7	

The cells are harvested by centrifugation at 4,000–12,000 rpm for 15–30 min when the bacteria culture is in the exponential phase of growth. The packed cells are resuspended in sterile seawater and provided at a given concentration as food in the rearing container of the zooplankton.

6.3.2 *Phytoplankton*

Today one or more planktonic algal species belonging to each of the following genera can be cultured to feed marine zooplankton and other animals.

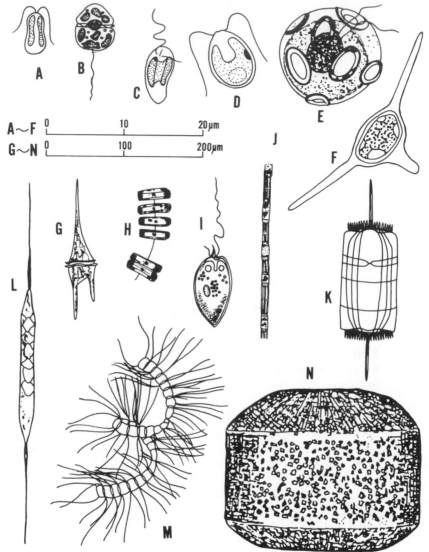

Figure 6.7. Some representative phytoplankton used for rearing zooplankton. (A) *Isochrysis*; (B) *Gymnodinium*; (C) *Pavlova*; (D) *Dunaliella*; (E) *Emiliania*; (F) *Phaeodactylum*; (G) *Ceratium*; (H) *Thalassiosira*; (I) *Prorocentrum*; (J) *Skeletonema*; (K) *Ditylum*; (L) *Rhizosolenia*; (M) *Chaetoceros*; (N) *Coscinodiscus*. (A, C, and D, after Throndsen, 1980.)

Class Bacillariophyceae: *Coscinodiscus, Chaetoceros, Ditylum, Lithodesmium, Phaeodactylum, Rhizosolenia, Skeletonema, Thalassiosira*

Class Haptophyceae: *Crisosphaera, Dicrateria, Emiliania, Isochrysis, Pavlova (Monochrysis)*

Class Prasinophyceae: *Pyramimonas, Tetraselmis (Platymonas)*

Class Chlorophyceae: *Brachiomonas, Chlamydomonas, Chlorella, Dunaliella*

Class Dinophyceae: *Amphidinium, Ceratium, Gymnodinium, Peridinium, Prorocentrum*

Besides size, the shape and structure of an algal cell are important factors in determining its potential value as food for a certain zooplankton species. To assist the reader in selecting algal species to feed marine invertebrates, we show some representative phytoplankton in Figure 6.7.

Methods for culturing phytoplankton are explained in detail in various textbooks (e.g., Guillard, 1975; Iwasaki, 1975). The relatively simple and reliable method of cultivation we describe mainly follows Guillard (1975). To learn more about the culture of algae, we refer the reader to Fogg (1965).

There are a number of enriched natural and artificial seawater media of similar composition. Differences mainly concern the addition or modification of vitamins, metals, chelating agents, or ammonium ion, and many media are surely as good as the ones described here. In this section two media are specified because of the authors' experience with their use: Erdschreiber solution and enriched seawater media $f/2$.

The classic nutrient medium, Erdschreiber solution (Föyn, 1934) is still widely used by those who culture the algae as food. The formula for the solution is as follows:

$NaNO_3$	100 mg
$Na_2HPO_4 \cdot 12H_2O$	20 mg
Soil extract	50 ml

The above components are added to "aged seawater" to make 1 liter of final culture medium. The "aged seawater" is filtered and preserved in the manner described in Section 6.2.1. Dissolved organic matter should be eliminated with activated charcoal.

The $f/2$, based on enrichment f of Guillard and Ryther (1962), has been successfully used for years as a semisynthetic nutrient solution in phytoplankton studies. Enrichment $f/2$ is made as follows; the amount of each component per liter of final medium is indicated in parentheses.

Stock solution I (major nutrients)

$NaNO_3$	7.5 g	(75 mg)
$NaH_2PO_4 \cdot H_2O$	0.5 g	(5 mg)
$Na_2SiO_3 \cdot 9H_2O$	1.5–3.0 g	(15–30 mg)
Distilled water	100 ml	

Silicate may be omitted for organisms other than diatoms, and, if necessary, $Na_2SiO_3 \cdot 9H_2O$ is dissolved by heating it.

Stock solution II (trace metals)

$CuSO_4 \cdot 5H_2O$	0.98 g	(0.01 mg)
$ZnSO_4 \cdot 7H_2O$	2.20 g	(0.022 mg)
$CoCl_2 \cdot 5H_2O$	1.00 g	(0.01 mg)
$MnCl_2 \cdot 4H_2O$	18.0 g	(0.18 mg)
$Na_2MoO_4 \cdot 2H_2O$	0.63 g	(0.006 mg)

Each of these trace metals is dissolved in separate 100-ml volumes of distilled water and

$FeCl \cdot 6H_2O$	3.15 g	(3.15 mg)
Na_2EDTA	4.36 g	(4.36 mg)

are dissolved in 900 ml of distilled water. Then 1 ml of each of the five individual trace metal solutions (5 ml in total) is added as well as enough distilled water to bring the solution to 1 liter.

Stock solution III (vitamins)

Thiamin·HCl	20 mg	(0.1 mg)
Biotin (primary stock solution)	1.0 ml	(0.5 µg)
B_{12} (primary stock solution)	0.1 ml	(0.5 µg)
Distilled water	100 ml	

This stock solution should be kept in screw-capped test tubes, autoclaved, and then stored in a freezer. In order to make the primary stock solution of Biotin, 10 mg of Biotin in crystalline form is dituted in 96 ml of distilled water. Vitamin B_{12} primary stock solution is made by dissolving 1 mg of B_{12} crystals in 1 ml of distilled water. Both primary solutions should be slightly acidic if they are to be autoclaved and should be kept sterile and frozen. The *working solution* (final medium) *f*/2 is prepared as follows:

Stock solution I	1 ml
Stock solution II	1 ml
Stock solution III	0.5 ml
Seawater ("aged")	997.5 ml

The culture media are then autoclaved at 120°C for 15 min at the pressure of 15 psi. Autoclaving often causes a precipitate in media. Although this precipitate usually does not affect the growth of algae, it may be avoided by autoclaving the phosphorus, iron, silicate, and vitamins separately from the remaining nutrient–seawater mixture. Precipitates also tend to form when a large volume of culture medium is autoclaved. Such precipitation is unpredictable when near-shore seawater is used because of the variation in seawater quality. The use of offshore water of constant quality or artificial seawater of known composition is recommended to minimize this problem. Alternatively, seawater may be prefiltered through Whatman GF/C filters and pasteurized at 85°C for 90 min and, after cooling to room temperature, nutrients are added by filtration through a membrane filter (see Perry, 1976).

Phytoplankton species selected from culture collections or isolated from dominant microalgae assemblages, where the animals of interest grow well in nature, are used as inocula to make the stock culture in several 125-ml Erlenmeyer flasks or culture tubes. Stock cultures should not be aerated. When the stock culture grows sufficiently, 10–50 ml (less in the case of diatoms and more in the case of flagellates) are transferred with a sterile Pasteur pipette to a 2-liter Erlenmeyer flask or 2.8-liter Fernbach flask for further culture. If a larger quantity of phytoplankton is needed, a 10–20-liter container should be used. The medium and apparatus must be autoclaved before inoculation in the same manner as those for the stock culture. Flasks and other containers should be stoppered with cotton plugs. Large containers may have a screw cap which should be loosened slightly for gas exchange. A cover of alminum foil is usually placed over a stopper for additional dust protection.

The cultures are placed on shelves in a temperature-controlled chamber or in an air-conditioned room at about 20°C since virtually all important food algae will grow somewhere between 15 and 25°C. The auxiliary temperature safety controls should be set so that the chamber temperature does not go beyond safe limits. In general, the maximum growth of phytoplankton is achieved in salinities between 20 and 30‰. Cultures should be illuminated by daylight fluorescent tubes from above and below perforated shelves, so that the light intensity at the center of the culture vessels is always 2000–3500 lux. When the illumination cycle is set for a photoperiod of 14 hr of light, the intensity should be about 4500 lux. Natural light through the laboratory windows will suffice for cultures, but it is often difficult to control temperature. Cultures in small flasks and tubes grow well if shaken by hand twice a day. If culture is greater than 2 liters, aeration is needed through a glass tube containing $CaCl_2$ and stuffed with cotton to supply CO_2 for plant growth, stabilize the pH, and keep algae in suspension. An example of a large culture apparatus is shown in Figure 6.8.

The growth phases of unicellular algae in a limited volume culture consist of a resting phase, an exponential phase, a phase of decreasing specific growth rate, a stationary phase, and a declining phase (Fig. 6.9). The stock cultures are transferred before half the life span of the culture has passed. For example, diatoms in flasks should be changed once a week and flagellates in tubes every 2 weeks. In mass cultures in big tanks, the period between inocula will vary for different species, although a week to a month is considered adequate. Harvesting algal cultures for food should

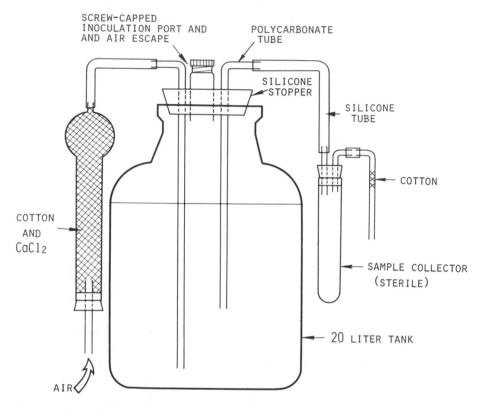

SCREW-CAPPED
INOCULATION PORT AND
AND AIR ESCAPE

POLYCARBONATE
TUBE

SILICONE
STOPPER

SILICONE
TUBE

COTTON

COTTON
AND
CaCl₂

SAMPLE COLLECTOR
(STERILE)

20 LITER TANK

AIR

Figure 6.8. Diagram of apparatus for culturing phytoplankton. (After Guillard, 1975, and personal communication.)

occur during the late phase of exponential growth until the early stationary phase when the cells are nutritious and the density is great so that a reasonable volume will suffice to feed the animals. By starting several cultures on different dates, a constant harvest of algal cells can be maintained for routine use. Cultures should occasionally be observed microscopically to check their purity and health.

6.3.3 Ciliates

Protozoan ciliates represent a class of organisms which feed on marine bacteria, phytoplankton, and/or other microorganisms. They in turn may be preyed upon by larger animals such as copepods. Because of their numerical abundance in the sea (see Chapter 1) and high nutrient turnover rate, these protozoans are considered to play an important role in marine food chains. However, thus far they have not often been used experimentally to feed meso- and macroplankton. This is due mainly to the difficulty

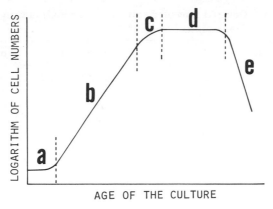

Figure 6.9. Characteristic growth diagram of a one-cell alga in a culture of limited volume. (a) Resting phase; (b) Exponential (logarithmic) phase; (c) Phase of decreasing specific growth rate; (d) Stationary phase; (e) Declining phase.

in cultivating free-living species and their lack of taxonomic identity, except for species living in coastal sediments. We hope that future developments for rearing marine ciliates will assist in elucidating their role in the marine ecosystem. In this section we refer to two successful cultures of marine planktonic ciliates.

Hamilton and Preslan (1969) cultured *Uronema* sp. from an open ocean water sample taken aseptically. *Uronema* was isolated from sterile batch cultures in *f*/2 medium incubated at 20°C in dim light for about a month, until the numbers of ciliates increased. *Uronema* was fed with a bacteria strain isolated from the original sample water and incubated in shaken flasks of 2216E broth. The cultured marine bacterium *Serratia marinorubra* was found equally effective as food. The stock cultures were transferred weekly by adding 0.1 ml (about 10^4 cells) of the previous week's culture to 10–20 ml of fresh bacterial suspension (about 10^9 bacteria) and incubated at 20°C in the dark. Under optimum temperature (20–25°C), salinity (17–43‰), and pH (7.8) conditions, the ciliates transferred from stock cultures to 150 ml of bacterial suspension (about 10^8 bacteria·ml^{-1}) in a 250-ml Erlenmeyer flask increased in population density from 1×10^3 to 1.5×10^4 cells·ml^{-1} after a 40-hr incubation.

Gold (1968, 1971) cultured *Tintinnopsis beroidea*. A modified method for the stock culture of this ciliate is described below. Individuals were isolated from natural seawater with a Pasteur pipette, transferred to a 3-cm diameter Petri dish containing a small portion of sterile seawater, and fed a small quantity of yeast and flagellates once every 2–3 days or when food was completely grazed down, as shown by inspection. The container volume was increased successively, and the quantity of medium remained one-fifth of this container volume. A mixture of penicillin and strepto-mycin, 50 μg·ml^{-1} of each, was added to control the bacterial assemblage.

Table 6.3.   **Defined seawater substitute for maintaining**
               Tintinnopsis beroidea[a]

Compound	Amount	Compound	Amount
NaCl	18 g	$Na_2SiO_3 \cdot 9H_2O$	0.2 g
$MgSO_4 \cdot 7H_2O$	5 g	Metals mix[b]	1 ml
KCl	0.6 g	TRIS	0.5 g
Ca (as Cl)	0.1 g	Distilled water	1000 ml
$NaNO_3$	0.5 g		
K_2HPO_4	30 mg	pH 7.5–7.8	

[a] After Gold (1968).
[b] 1-ml metal mixture contains Na_2EDTA (10 mg), $FeCl_3 \cdot 6H_2O$ (0.5 mg), H_3BO_3 (10 mg), $MnCl_2 \cdot 4H_2O$ (1.5 mg), $ZnCl_2$ (0.1 mg), and $CoCl_2 \cdot 6H_2O$ (0.05 mg).

The culture was maintained in a dark place at 10–12°C. When *Tintinnopsis* reproduced sufficiently, after a few days, it was transferred to 500-ml Erlenmeyer flasks containing 100 ml of seawater culture medium (Table 6.3) and food. The culture was maintained at 10°C with alternate light (18-hr) and dark (6-hr) periods. The low temperature appeared to be essential for successful rearing of the tintinnids. The culture medium was autoclaved at 120°C for 15 min and 50 $\mu g \cdot ml^{-1}$ each of penicillin and streptomycin were added before inoculation. The flasks were covered with Parafilm or corked during the cultivation. Food consisting of *Rhodomonas, Isochrysis, Platymonas*, and other flagellates were provided twice per week. Subcultures of *Tintinnopsis* were prepared at 3–4 week intervals when growth appeared to be optimum. The period required for ciliate populations to double was 2.5 to 6 days. More than 1000 cells·ml^{-1} were routinely obtained. Gold (1971) attempted the mass culture of the species in 2.5-liter culture flasks at 10°C with the same culture medium and food. In order to separate the ciliates from their algal foods and concentrate them, he filtered the culture medium through a Nuclepore filter of 8 μm pore size. Complete recovery of the ciliates was obtained by this procedure. In a similar manner *Favella campanula* was cultured at 20°C (Gold, 1969b).

Besides these cultures, Uhlig (1965) conducted rearing experiments with sessile, littoral, heterotrich ciliates of the genera *Diafolliculina* and *Eufolliculina* under nonsterile conditions, and with the harpacticoid copepod *Tisbe holothuriae* in stagnant Seitz-filtered[3] seawater that had been heated to 90°C for 1 hr. The ciliates were fed *Dunaliella* sp. and *Cryptomonas* sp. The growth of the ciliate and harpacticoid populations were balanced and quite stable. *Uronema marinum* has been cultivated axenically in chemically defined media by Hanna and Lilly (1970, 1974) and Lee et al. (1971) for nutritional and growth studies.

6.3.4 *Rotifers*

Most species of Rotatoria are distributed in the freshwater environment and there are only a few marine species. Those organisms can be collected by a small plankton net, concentrated by their characteristic phototactic behavior, and sorted out by pipetting. Among the marine rotifers, *Brachionus plicatilis* has been the most widely used for feeding crustacea and fish larvae in mariculture. It can be reared by the following process:

1. Stock culture: suitable containers (500-ml or 1-liter flasks) are filled with filtered seawater of 25–33‰ salinity and inoculated with a few individuals of *Brachionus*.
2. Aeration should be provided and if possible the water should be gently circulated. The temperature should be maintained at 10–20°C.
3. Algae such as *Chlorella, Dunaliella,* or *Tetraselmis* should be provided as food. The concentration of food should be adjusted to the needs of the organisms in the culture, but in general 10^5–10^6 cells·ml^{-1}·day^{-1} should suffice.
4. When the density of *Brachionus* in the stock culture reaches around 20–30 ind·ml^{-1} (normally the number of individuals doubles within 12–24 hr), they should be collected and transferred to mass culture containers that are up to 1 ton in size. Fresh (frozen) or powdered marine yeast (2–5 μm in cell size) is better than *Chlorella* and other algal food when the density of *Brachionus* exceeds 100 ind·ml^{-1}. The food concentration should be 7×10^4 cells (fresh yeast)·ind^{-1}·day^{-1} during the early period of culture but later reduced to 3×10^4 cells·ind^{-1}·day^{-1}. The yeast should be suspended in the cultures.
5. *Brachionus* can be reared up to 1000 ind·ml^{-1}. When the density exceeds 200 ind·ml^{-1} in static water with marine yeast, a part of the culture can be used as food. Because the population size recovers under favorable conditions, the population can be thinned several times at intervals of a few days. Part of the seawater in the culture should be replaced with fresh seawater at the time of thinning.

According to Furukawa and Hidaka (1973), it is difficult to raise the density of *Brachionus* over 200 ind·ml^{-1} with cultured *Chlorella*, but the fecundity is greater when *Chlorella* rather than yeast is used. A high concentration of *Chlorella* often causes the pH of cultures to rise and may suppress the reproduction of *Brachionus*. Hirayama and Ogawa (1972) stated that the physiological activity of *Brachionus plicatilis* becomes maximal when fed *Chlorella* at a temperature higher than 22°C, a salinity of 7.8‰, and a pH of about 8.0. However, Furukawa and Hidaka (1973) suggested that better results would be obtained in mass cultures with marine yeast as food and at temperatures of 11–17°C and pH of 7.1–7.5. Maximum yield should be about 1200 ind·ml^{-1} (see also Theilacker and McMaster, 1971, and Hirata, 1974, for mass culture techniques).

6.3.5 *Artemia (Brine Shrimp) Nauplii*

The use of nauplii of *Artemia salina* as food for zooplankton and crustacea larvae is widely known. Commercially available resting eggs can be kept in the dry state for long periods and, when exposed to seawater, will hatch within a few days to produce highly motile nauplii. Products from China, Thailand, and the United States (California and Utah) are imported to Japan. Despite the fact that *Artemia* nauplii are never encountered by marine carnivorous zooplankton under natural conditions, they have proven to be a useful food. However, *Artemia* nauplii are not always suitable for the entire development of a given species. Some crustaceans and fish larvae either fail to grow or die when fed *Artemia* nauplii exclusively. Also, there is evidence that the hatching rate and nutritional value of *Artemia* nauplii depends on their origin.

Incubation instructions are usually provided with the containers of eggs and there are several types of commercially available funnels for hatching eggs. When the eggs are immersed in seawater, they hatch into nauplii in about 1.5 days at 25–30°C. The water is stirred with continuous and strong aeration so that all the eggs circulate well; otherwise, many eggs and empty egg capsules attach to the inner surface of the container. The hatching rate may exceed 70% when the eggs are fresh and of good quality. The nauplii have a characteristic of phototaxis and, therefore, can be collected easily by a pipette after their attraction by light to one side of the container. *Chlorella*, *Chlamydomonas*, soy bean powder, and yeast should be used as food. *Artemia* will reach the adult stage in 1.5–2 months and will then bear eggs.

6.4 Measurement of Physiological Tolerance to Physical and Chemical Conditions

The tolerance of zooplankton to external physical and chemical factors, such as temperature, salinity, hydrostatic pressure, and oxygen, are species-specific hereditary traits. A few simple methods are discussed below for measuring individual tolerances to a variety of conditions. To obtain a complete picture of the tolerance of a species, we must examine both females and males and the stages of development. In some forms the specific determination requires the study of several populations from different biotopes or groups of individuals which have undergone prelim-inary adaptation to differing external conditions.

For orientational purposes only one factor should be initially altered in the trials, with other external factors kept at constant optimal values. If the test species is exposed to variable factor combinations in nature, however, it will be necessary to test further the tolerance to concurrent changes in several external factors. It is true, for example, that for some euryhaline species tolerance to temperature may vary in relation to salinity (see Costlow et al., 1960).

Generally, tolerance can be expressed in terms of the value of the limiting factor at which 50% of the specimens die within a chosen test period or in terms of the average survival times within the lethal factor range.

When comparative measurements of the thermal and/or salinity tolerances of several species are desired, animals which have been kept at or were previously adapted to average biotope conditions are transferred into seawater of a uniform lethal temperature and/or salinity, or combination of both factors, and the sequence of their deaths is observed. The survival times of a series of individuals in separate containers at lethal temperatures or salinities may also be determined. Dead specimens must be removed from the test containers immediately. In order to determine the temperature and/or salinity acclimatization (nongenetic) capacity of the particular species, it is necessary to test groups of animals previously maintained at various temperature and/or salinity conditions of the habitat. In this manner it is possible to determine specific differences. The upper limits of thermal tolerance for two neritic species of copepods from Chesapeake Bay, U.S.A., were measured by Heinle (1969). It was found that the upper limits were near the normal temperature of the habitat during the summer and were affected slightly by the acclimation temperature of the copepods. Effects of acclimation on the lethal upper temperatures have also been studied on *Acartia tonsa* by Reeve and Cosper (1972). It is difficult to measure the lower temperature tolerance of species inhabiting cold waters because they often survive until the freezing point ($-1.9°C$) has been attained. However, it is possible to supercool pure seawater carefully to $-5°C$ without producing ice.

Insight into the thermal tolerance of an animal may be obtained by observing the behavior and ultimate mortality of specimens in water in which the temperature is slowly and continuously raised or lowered from the average biotope temperature by 1°C every 15 or 30 min.

For precise comparative measurements of the effect of salinity, it is best to work with gradations of 2–3‰. Seawater is diluted with distilled water to lower the salinity, and higher salinities are obtained by adding the same salt mixture used in preparing artificial seawater to normal seawater (see Section 6.2.1). Smaller quantities of seawater of high salinity can be prepared by evaporation of the seawater.

The development of vital staining techniques (Chapter 4, Section 4.4) enables the examination of large numbers of marine organisms for determination of mortalities. Thermal and/or salinity tolerances of zooplankton can also be measured by means of oxygen consumption and excretion rates (Chapter 8), ATP concentration (Chapter 5, Section 5.6), and ciliary beat. It must be noted that a fairly large change in salinity may produce a reversible shock effect with temporary ciliary arrest or a reduction in activity and subsequent slow recovery.

The pressure tolerance of marine plankton can determine their depth distribution because water pressure increases by about 1 atm for every 10 m in depth. We differentiate between stenobathic and eurybathic species on the basis of their depth distributions. Species that perform remarkable diel vertical migrations are able to withstand changes in water pressure of some tens of atmospheres in a short period. Pressure tolerance is measured by using a pressure cylinder made of high-strength aluminum or stainless steel. Pressure is applied by means of a hydraulic hand pump, in the manner of a strong trunk jack, through which seawater is pumped from a stock container into the pressure cylinder. Well-aerated seawater with a normal gas content and uniform pressure is used. Air bubbles must be avoided. The duration of pressure experiments is limited by the oxygen supply in the volume of seawater brought under pressure. As a rule the oxygen content should not fall below 80% of the air saturation value by the end of the experiment (see Chapter 8, Section 8.2). It is very important that the increase or decrease of pressure takes place very slowly at the beginning and end of the experiment. The survival capability of animals is judged on the basis of pressure tolerance after an adequate recovery period. Experiments with eurythermal species of various temperatures indicate that their pressure tolerance is greater under warm conditions.

According to Teal and Carey (1967), the respiration rate of epipelagic euphausiids is independent of hydrostatic pressure but decreases greatly with decreasing temperature. Conversely, for mesopelagic crustaceans, such as *Thysanopoda monacantha*, *Sergia splendens*, and *Acanthephyra purpurea*, increased pressure increases respiration at lower temperatures more than at higher temperatures. In these species the decrease in respiration rate caused by a lower temperature at depth is off-set by an increase due to the higher pressure. These contrasting results suggest that it is to the advantage of meso- and bathypelagic species to be able to maintain their activities throughout their range of vertical migration, by day as well as by night (Teal, 1971).

Minimum oxygen requirements and lower oxygen concentration limits can be measured in respiration rate and survival experiments with flowing seawater of varying oxygen content. Low-oxygen seawater can be made by bubbling nitrogen through seawater for a varying period. The test water must be sealed off from the overlying air until the experiment begins. If the tolerance to low oxygen tension is to be examined in long-term trials, it is advisable to bring the test water into equilibrium with suitable air–nitrogen mixtures in large pressure gas cylinders. The gas mixture flows from the cylinder in a slow stream of fine bubbles through the seawater of the test aquaria. The water surface of these aquaria must be carefully protected from contact with outside air by a covering of glass or plastic.

Recently, the effects on estuarine zooplankton of industrial chemicals, such as chlorine and crude oil, and their passage through power plant cooling systems have been studied extensively. The methods of *in situ* and laboratory measurements of tolerance and mortality are basically the same as those described for testing sensitivities to temperature and salinity. Vital staining techniques are commonly used to determine mortalities in the field (e.g., Carpenter et al., 1974; Heinle, 1976).

Finally, it must be noted that the data obtained by a series of experiments should be statistically treated for meaningful interpretation. If there are any differences between the mean values and standard deviations of two similar test series, the degree to which the differences observed are significant must be established. There are various ways to statistically test for significant differences between conditions; the chi-square, Student's *t*, Wilcoxon rank, and Mann-Whitney *U* tests are examples. The reader should consult textbooks on statistics for biologists, such as Siegel (1956), Tate and Clelland (1957), Snedecor and Cochran (1967), and Sokal and Rohlf (1969), for appropriate methods.

Notes

[1] For charcoal treatment to eliminate organic matter, 10 g of decolorizing activated charcoal for every liter of seawater is used. The charcoal is pretreated by shaking it for 10 min with 500 ml of a 5% (W/V) solution of sodium chloride (analytical reagent quality) in distilled water. The suspension is filtered through a Whatman No. 1 or equivalent filter, and the charcoal is transferred to fresh sodium chloride solution. The above procedure is repeated twice. Then the washed charcoal is added to natural seawater, shaking it for 30 min. The seawater is filtered through Whatman No. 2 paper and then aseptically through a membrane filter of 0.3 μm pore size. The filtered seawater is stored aseptically.

[2] Soil extract: approximately equal weights of garden soil (without recent applications of fertilizer and insecticide) and distilled water are autoclaved at 120°C for 1 hr at 15 psi and allowed to stand for 1–2 days. The supernatant is removed by decantation and filtered with a membrane filter (pore size of 0.8–1.0 μm). The extract is reautoclaved, if necessary, and stored sterile in the cold.

[3] A bacterial filter made of asbestos and used to sterilize solutions without the use of heat.

CHAPTER 7 _____

Feeding

7.1 Food Habits of Zooplankton

Knowledge of the food habits of a zooplankton species is a step toward evaluating the functional role of that species in pelagic food webs. The terms *herbivore, carnivore,* or *omnivore* are commonly used to refer to zooplankton which feed primarily on plants (phytoplankton), animals (zooplankton), or a mixture of these food types, respectively. The terms *monophagy* or *polyphagy* refer, respectively, to zooplankton feeding on a single prey type or on many prey types.

A common method for investigating zooplankton food habits is the examination of gut contents. However, the results of such examinations are often misleading. First, fragile and easily digestable foods are difficult to identify in gut contents and are thereby often overlooked. Second, in the course of sampling zooplankton with a net, large zooplankton may "eat" or "swallow" small zooplankton and other material concentrated in the cod-end of the net. This artifact, called *net feeding*, occurs frequently when smaller mesh nets are used to collect large zooplankton species. For example, Angel (1970) reported that the gut of an ostracod, *Conchoecia spinirostris*, was packed with copious quantities of freshly eaten, black pigmented tissue originating from myctophid fishes caught in the same net haul. Finally, identifiable material such as diatom frustules may originate from either materials ingested by the predator of interest or from food ingested by its prey. If these problems inherent to gut content examination are carefully avoided, qualitative and quantitative examinations of gut contents will provide important information of the nature of the food and feeding behavior of zooplankton in the sea.

To reduce the possibility of *net feeding*, one can kill or narcotize animals immediately after they enter the net. Williamson (1962) glued a slip of paper with a layer of mercuric iodide to the inside of his sampling gear at the time of deployment. Mercuric iodide not only kills the catch but also prevents its decay over short time intervals.

On the other hand, if the animal of interest is large (e.g., some macroplankton and micronekton) and commonly occurs in the guts of

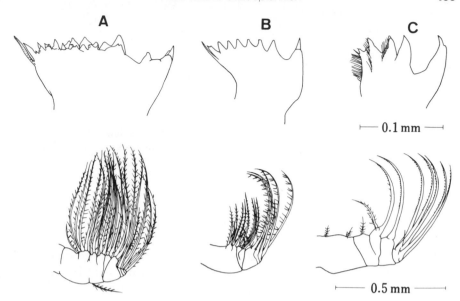

Figure 7.1.   The cutting edges of the mandible (upper) and maxilla (below) of copepods.
(A) *Calanus finmarchicus* (herbivore); (B) *Centropages typicus* (omnivore); (C) *Tortanus discaudatus*
(carnivore). (Modified after Anraku and Omori, 1963.)

large predators, it may be possible to obtain precise information on the
food habits of the species from examination of the guts of freshly eaten
specimens taken from the guts of the predators. Examples include the
investigation of the food habits of the sergestid shrimp *Sergestes similis*
obtained from albacore stomachs (Judkins and Fleminger, 1972) and the
gonostomatid fish *Vineiguerria nimbaria* from Bryde's whale stomachs
(Kawamura and Hamaoka, 1981).

As an experimental approach to evaluate prey–predator relationships
in a natural zooplankton community, Smith et al. (1979) used highly
radioactive preparations (^{14}C or ^{3}H) of phytoplankton and zooplankton.
These labeled prey were introduced to natural zooplankton assemblages
in a container in order to identify the specific predators of these prey.

The food habits of some groups of zooplankton can be determined
from the morphological structure of their feeding appendages. For
example, in the Copepoda the maxilla of species designated as herbivores
(e.g., *Calanus* and *Eucalanus* spp.) contains many setae, each bearing fine
setules, to facilitate the removal of small phytoplankton cells from suspension (Fig. 7.1). The paired mandibular blades of these species have molar
teeth suited for masticating fine particles. In contrast, species designated
as carnivores (e.g., *Candacia* and *Tortanus* spp.) have maxilla with a few,
strong setae suited for the capture and holding of prey animals. The
mandibular blades of carnivorous species have robust, piercing teeth.

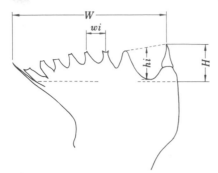

Figure 7.2. Schematic illustration of the "edge index" for the cutting edge of a mandible. (After Itoh, 1970.)

Species considered to be omnivorous (e.g., *Centropages* spp.) have feeding appendages intermediate between herbivorous and carnivorous species (see Anraku and Omori, 1963).

To quantitatively relate the morphology of feeding appendages and the food habits of calanoid copepods, Itoh (1970) expressed the roughness of teeth on the mandibular blade with the edge index (E_i) using the equation

$$E_i = \Sigma \left(\frac{w_i}{W} \frac{h_i}{H} \times 10^4 \right) \frac{1}{N} \tag{7.1}$$

where W and w_i are the widths of the entire blade and of each tooth, respectively; H and h_i are the longest distances between the tip and base of the blade and the depth between adjacent teeth, respectively; and N is the number of teeth (Fig. 7.2). From the calculation of E_i for various copepod species, Itoh (1970) suggested that herbivorous species are characterized by E_i less than 500, omnivorous species between 500 and 900, and carnivorous species more than 900.

The history of food habit studies on marine zooplankton can be traced to the early 1920s (Lebour, 1922, 1923). Since that time the accumulated information from gut contents, the structure of feeding appendages, and laboratory feeding experiments indicate that most zooplankton species, excluding typical carnivores, are euryphagous or omnivorous (see Raymont, 1963; Mullin, 1966). Unlike terrestrial animals exclusive herbivory is rarely seen in marine zooplankton. Food habits of common groups of marine zooplankton are summarized as follows:

1. **Coelenterata.** Mostly carnivorous, but some (e.g., Rhizostomeae) feed on phytoplankton as well as microzooplankton. In general, they eat any animal coming in contact with their tentacles. Some species have the ability to select the prey of most nutritional value to them.

2. **Ctenophora.** Carnivorous. They feed on a variety of prey from small copepods to fish larvae and euphausiids. Cydippida capture prey with their tentacles, while most Lobata use their tentacles and oral lobes.

Cestida feed by trapping small prey on tentacles that lie over the sides of the body (Harbison et al., 1978). *Beroe* preys almost exclusively on other planktonic ctenophores (Reeve and Walter, 1978).

3. **Heteropoda.** Carnivorous.

4. **Pteropoda.** Thecosomata are considered to be ciliary mucus feeders. Gymnosomata are carnivorous.

5. **Polychaeta.** All holoplanktonic polychaetes, with the exception of Poeobiidae, are carnivores. *Tomopteris* has been observed to bite prey animals and suck liquids from their bodies. Alciopids feed on copepods and thaliacians, probably hunting by sight with their large, complex eyes (Fauchald and Jumars, 1979).

6. **Cladocera.** Although little is known, they seem to be omnivorous suspension feeders. *Penilia* tend to prefer phytoplankton, while *Evadne* and *Podon* prefer microzooplankton as food.

7. **Ostracoda.** Mostly omnivorous. *Conchoecia* feeds on dead cope-pods and fecal pellets. Deep-sea species such as *Gigantocypris* are considered largely carnivorous.

8. **Copepoda.** Species belonging to Calanidae and Eucalanidae are mostly omnivorous and herbivorous, whereas other species are either omnivorous or carnivorous. Experimental studies on food selection reveal a broad spectrum of feeding (see Mullin, 1966; Petipa, 1979); even for those considered herbivorous, predation on *Artemia* nauplii and other small prey has been observed. Many calanoid copepods have been commonly described as *filter-feeders*, but they are more correctly termed *suspension-feeders* or *particle-feeders* (see Paffenhöfer et al., 1982).

9. **Mysidacea.** Mostly omnivorous. The major food source for neritic species is bottom sediment rather than particles suspended in the water column.

10. **Amphipoda.** Typically carnivorous. Some species, like *Phronima*, use the outer gelatinous crust of their thaliacean prey as living space or as spawning substrate after eating the body contents.

11. **Euphausiacea.** Species belonging to the genera *Nematoscelis*, *Nematobrachion*, and *Stylocheiron* are carnivorous, while other species are omnivorous suspension feeders. The antarctic krill, *Euphausia superba*, seems to be mostly herbivorous, though it can eat small animals.

12. **Chaetognatha.** Typically carnivorous.

13. **Appendicularia.** Herbivorous and detritivorous. They feed in an unique way—by secreting an external gelatinous house containing an elaborate feeding filter (Fig. 1.7C). Undulatory movements of the tail pump water through the house, and phytoplankton and other nanoplankton in the water are trapped on the feeding filter and sucked into the mouth of the animal. When the incurrent filter becomes clogged with particulate matter, the house is discarded. A single animal may build and discard houses every few hours. The feeding filter of *Oikopleura dioica* consists of arrays of crossed filaments about 0.04 μm thick and contains

pores about $0.24 \times 0.07 \ \mu m^2$ in size with a mesh porosity about 55% and probably enables the species to feed efficiently on particles smaller than bacteria (Flood, 1978).

14. Thaliacea. Omnivorous; typically particle-feeding. They filter food in the pharynx with a continuously renewed mucus net. The mucus feeding net fills about half the body cavity and is continuously moved posteriorly along the gill bar, rolled into a cord, and carried into the esophagus by cilia (Fig. 1.7D). Feeding is apparently nonselective and particles ranging from about 1 mm to less than 1 μm are collected. Digestion appears to be incomplete (Madin, 1974).

7.2 Units of Feeding Rate

The following units are commonly used to express the feeding rate of zooplankton:

1. Filtering (filtration) Rate and Clearance Rate. The volume of water swept clear per animal per hour or day. Filtering rate refers to the total volume of water filtered by the animal over a particular time, whereas clearance rate refers to the volume of water from which food particles were completely removed over that time. Filtering rate is equal to clearance rate only when the filtering apparatus traps all food particles in the water which passes through it. From a methodological viewpoint, most filtering rates of zooplankton measured previously do not represent the "real" filtering rate but an "apparent" filtering rate, better termed *clearance rate*.

2. Grazing Rate. The number of algal cells eaten per animal per hour or day. The use of this unit is restricted to phytoplankton feeders and, in these animals, has often been used as a synonym for filtering or clearance rate.

3. Predation Rate. The number of food animals eaten per animal per hour (or day). This unit should be used for carnivorous zooplankton but is sometimes used for grazers as they are predators of algae.

4. Ingestion Rate. The weight or calorific content of food ingested per animal per hour or day. This unit can be determined from either the grazing or the predation rate by knowing the weight or calorific content per food organism. It can also be determined from either the filtering or clearance rate if the average concentration of food particles in the water during the incubation period is known. Ingestion rates are usually expressed in terms of dry weight, organic matter, carbon, or nitrogen.

Obviously, the filtering or clearance rates are a measure of the animal's capacity to search for food rather than its actual feeding rate. The clearance rate for carnivorous zooplankton is defined as the result of the feeding effort of a predatory animal and the unsuccessful escapes of prey animals (Landry, 1978). The filtering or clearance rates for different cells and/or prey animals are, therefore, difficult to compare, even for a single animal that feeds omnivorously, because of differences in the efficiency with which different particles are retained by, and/or their escape from, filtering apparatus. Grazing and predation rates also lack a common base for comparison when different kinds of food organisms are used as these will most likely have different weights and/or calorific contents. Thus, the ingestion rate is the most useful measure for comparing the feeding rates of zooplankton with different feeding habits or modes.

7.3 Experimental Measurement of Feeding Rate

In experimental feeding studies zooplankton [usually the same developmental stage(s) of a single species] are incubated in seawater containing food. The feeding rate can thus be calculated from the observed decrease in the amount of food in containers with grazers relative to that in control containers without grazers during the incubation. In early experimental work on the feeding of marine zooplankton, Fuller and Clarke (1936) and Fuller (1937) measured filtering rates of *Calanus finmarchicus* placed in a suspension of the diatom *Nitzschia closterium* and carmine powder. Since then, with the refinements of experimental techniques, many studies of zooplankton feeding have been conducted, especially with suspension-feeding copepods (see Marshall, 1973). The balance, tracer, and marking methods for measuring the feeding rate of zooplankton are now described.

7.3.1 *Balance Method*

1. Incubation. Wide-mouth glass bottles or beakers with a capacity of 300–1000 ml or greater can be used for incubation containers in feeding experiments. The bottles are filled with seawater containing a known concentration of food particles (natural particle assemblages, laboratory-cultured algal cells, *Artemia* nauplii, copepod larvae, etc.). A known volume of seawater must be used in each bottle, but in contrast to respiration experiments (Chapter 8, Section 8.1.1), the elimination of air space in the bottles is not critical.

A few or several tens of animals are transferred with a small amount of seawater into the bottles. In order to avoid contamination with foreign particles, it is necessary to rinse the animals with filtered seawater by transferring them from one beaker to another before placing them in the bottles. Control bottles without animals are needed, especially when a change in the concentration of food particles not due to grazing or predation during incubation is anticipated due, for example, to the growth of algal cells. The period of the experiment may range from a few hours to a day or more. Although the number of individual food particles in controls is unlikely to change during an experiment when animals are used as a prey, control bottles should be prepared to assess the efficiency

Figure 7.3. Mixing devices for feeding experiments with zooplankton. (A) The "ferris" wheel. (1) Bottles; (2) Rotating disc; (3) Motor. (B) The "roller." (1) Container; (2) Rolling bar; (3) Motor. Containers on both mixing devices are notated at a rate of 1.0–1.5 rpm.

in recovering and the error in counting food organisms. The incubation room should be dark or the bottles wrapped with aluminum foil or black cloth or tape to minimize the change in algal cell concentration due to growth and to avoid local aggregation of both predators and prey due to their phototactic behaviors. During the experiment, a homogeneous suspension of food particles is necessary. Figure 7.3 shows two devices to facilitate the homogeneous suspension of food particles in feeding experiments with zooplankton. One is the "Ferris" wheel, which turns the containers end-over-end, and the other is the "roller," which rotates the jars on their sides. The choice of a mixing device will vary from one zooplankton species to the other, but needless to say, the test device will be the one which does not disturb the normal behavior of the animals yet maintains a random distribution of food. Since temperature is an important variable that affects the feeding rate of animals, it should be controlled during incubation.

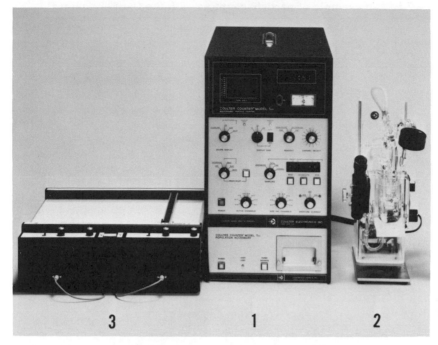

Figure 7.4. Coulter counter Model TAII. (1) Main unit; (2) Sample stand; (3) X–Y recorder.

After the experiment water samples are taken from the experimental containers. The condition of the animals is examined to determine whether each is as active after as before the incubation. The animals are placed on a piece of netting or filter paper to remove water adhering to their bodies and stored in a desiccator or an oven prior to measuring their dry weights. Further determinations of carbon or nitrogen may be made on these dried animals (see Chapter 5, Section 5.5.2).

2. Counting of Food Particles. The number of particles, such as algal cells per unit volume of sample seawater, may be counted microscopically using a Sedgewick–Rafter chamber or haemocytometer. The Utermöhl method with an inverted microscope (Chapter 5, Section 5.2.1) is also commonly used for counting algal cells in natural seawater. For larger particles, like *Artemia* nauplii or larvae of some animals, counting chambers described for net zooplankton (Chapter 5, Section 5.2.2) can be used.

Recently, electronic particle counters have been commonly used in feeding experiments of marine zooplankton. There are various types of instruments commercially available at present. For mechanism and operation procedures the reader is referred to the manual provided by the manufacturers. Figure 7.4 shows a model TAII Coulter counter, the model most widely used in plankton studies. This instrument is capable of

Table 7.1. An example of particle counting with the Coulter counter[a]

Channel number	ESD (μm)	Volume (μm^3)	FSW only No.	FSW only % vol.	FSW and cells No.	FSW and cells % vol.
1	—	—	—	—	—	—
2	2.52	8.38	290	0.4	15,223	0.6
3	3.17	16.8	66	0.2	625	0.1
4	4.00	33.5	52	0.3	460	0.1
5	5.04	67.0	33	0.4	390	0.1
6	6.34	134	13	0.3	320	0.2
7	8.00	268	9	0.4	285	0.4
8	10.1	536	8	0.7	544	1.4
9	12.7	1,072	5	0.9	7,614	40.4
10	16.0	2,146	4	1.5	4,567	48.5
11	20.2	4,290	5	3.7	313	6.6
12	25.4	8,580	5	7.5	12	0.5
13	32.0	17,170	6	17.9	3	0.3
14	40.3	34,300	5	29.8	5	0.8
15	50.8	68,700	3	35.9	0	0
16	—	—	—	—	—	—
				$\Sigma = 99.9$		$\Sigma = 100.0$

[a] The cells of the diatom *Fragilariopsis vanheurkii* were suspended in seawater filtered through HA Millipore filters (FSW). Counting was made with a 140-μm aperture tube on 2 ml of sample (manometer mode). Active channels were selected as 2 to 15. The number of particles in each channel was converted to volume by multiplying by the respective particle volume, and then the percent volume of particles in each channel was calculated (channels with percent volume greater than 10% are usually taken into the calculation of cell number and average cell volume of particles). The increase in particle numbers in channels other than 10 and 11 is due to debris in the culture. In this example

$$\text{cell number} = \frac{7614 + 4567}{2} = 6091 \text{ cells·ml}^{-1}$$

$$\text{average cell volume} = \frac{(7614 \times 1072) + (4567 \times 2146)}{7614 + 4567}$$

$$= 1474.7 \ \mu m^3 \cdot \text{cell}^{-1}$$

and ESD is

$$\sqrt[3]{1474.7 \times (3/4\pi)} \times 2 = 14.1 \ \mu m$$

simultaneously counting 16 size classes of particles with diameters between 2–40% of the aperature of an orifice tube. The size of the particles is expressed as equivalent spherical diameter (ESD) regardless of shape. The range of particle size (ESD) measurable with this counter is 0.3–800 μm with the appropriate orifice tubes (up to 2000 μm). Counting and sizing of particles with the particle counter are extremely efficient compared to visual methods. It can be used on board ship. The particle counter gives information of the *in situ* size spectrum of suspended particules in natural water, which is useful for feeding experiments. Table 7.1 shows the results of a count with the Coulter counter and calculations of the average volume of diatom cells and ESD.

However, the results from electronic particle counters are often misleading in feeding experiments. First, the counter cannot distinguish food particles from other particles, such as fecal pellets and detritus, when the sizes (ESD) of both particles are similar. Second, the average particle volume is biased when the shape of a particle is far different from a sphere. Harbison and McAlister (1980) examined this bias using the cells of five diatom species and comparing the results from the Coulter counter with those from visual measurements (Table 7.2). Agreement between the results of the two methods is fairly good for some diatoms, but great discrepancies exist for other diatoms, especially for *Rhizosolenia setigera*, with its pencillike shape. Deason (1980) put further limits on the application of electronic particle counters in feeding experiments of *Acartia hudsonica* fed the chain-forming diatom *Skeletonema costatum*. Because the chains are easily broken into smaller chains by the animals, the number of particles (broken and unbroken chains) provides no accurate basis for the calculation of filtering rate and, therefore, ingestion rate. In this situation ingestion rate can be obtained from a visual count of algal cells, not chains, when algal growth is negligible during the incubation [$k = 0$, see Eq. (7.2)]. These results stress the need for careful application of electronic particle counters to zooplankton feeding experiments.

As a different technique, the concentration of food particles can be estimated by weighing retained particles (Corner, 1961). However, the technique is usually less accurate than counting because of the difficulty of eliminating other particles such as fecal pellets. Chlorophyll *a* measurement may also be used for this purpose when phytoplankton is used as food (Adams and Steele, 1966). Hargis (1977) compared the filtering rate of *Acartia clausi* on natural phytoplankton by using the Coulter counter and chlorophyll *a* measurements and found no essential difference between these two sets of data.

3. Calculation of Filtering, Grazing, and Ingestion Rates. Let C_0 be the concentration of algal cells at the start of an incubation and C_t and C_{tf} be the concentrations in the control and experimental bottles, respectively, after t hours. The change in cell concentration in the control bottles is due to growth:

$$C_t = C_0 e^{kt} \quad (k, \text{ growth coefficient}) \tag{7.2}$$

Assuming that the filtering rate of animals during the incubation period is constant, the change in cell concentration in experimental bottles can be expressed as

$$C_{tf} = C_0 e^{(k-f)t} \quad (f, \text{ feeding coefficient}) \tag{7.3}$$

Table 7.2. Dimensions of centric diatoms measured visually and with Coulter counter[a]

Diatoms	Visual				Coulter counter	
	Average diameter (μm)	Average per valvar length (μm)	Average vol. cell^{-1} (μm^3)	Average ESD (μm)	Average vol. cell^{-1} (μm^3)	Average ESD (μm)
Thalassiosira pseudonana	3.36 ± 0.78	4.32 ± 0.81	38.3	4.18	55.4	4.73
Thalassiosira sp.	4.94 ± 0.87	7.35 ± 1.4	141	6.45	156	6.68
Thalassiosira fluviatilis	10.5 ± 1.5	15.0 ± 2.3	1,290	13.5	959	12.8
	17.4 ± 4.5	18.2 ± 2.8	4,300	20.2	1,690	14.8
Rhizosolenia setigera	21.8 ± 1.8	368 ± 92	137,000	64.2	63,500	49.5
Coscinodiscus sp.	62.4 ± 6.6	51.6 ± 7.5	158,000	67.1	119,000	61.0
	109 ± 15	83.3 ± 29	781,000	114	454,000	95.4

[a] Average dimensions (±SD) with visual method based on measurements of at least 50 separate cells. The two sets of *Thalassiosira fluviatilis* and *Coscinodiscus* sp. measurements were made with different cultures, several months apart. (After Harbison and McAlister, 1980).

Filtering rate F (volume of water filtered·animal^{-1}·hr^{-1}) is defined as

$$F = \frac{Vf}{N} \tag{7.4}$$

where V is the volume of seawater in the experimental bottle and N is the number of animals in the bottle. Equation (7.3) can be rewritten as

$$\frac{\ln C_0 - \ln C_{tf}}{t} = -(k - f)$$

Substituting this into Equation (7.4), we obtain

$$F = \frac{V}{N}\left(\frac{\ln C_0 - \ln C_{tf}}{t} + k\right) \tag{7.5}$$

When $k = 0$ (i.e., $C_0 = C_t$),

$$F = \frac{V}{N}\left(\frac{\ln C_0 - \ln C_{tf}}{t}\right) \tag{7.6}$$

or

$$F = \frac{V(\log_{10} C_0 - \log_{10} C_{tf})}{0.4343Nt}$$

The average concentration of algal cells $(\overline{C})$ in the experimental bottles during the incubation period is obtained from Equation (7.3):

$$\overline{C} = \frac{\int_0^t C_0 e^{(k-f)t}\, dt}{t} \tag{7.7}$$

$$= \frac{C_0(1 - e^{(k-f)t})}{-(k - f)t}$$

$$= \frac{C_{tf} - C_0}{(k - f)t} \quad \left(\text{or} \quad \frac{C_{tf} - C_0}{\ln C_{tf} - \ln C_0}\right)$$

Therefore, the grazing rate (G, number of cells eaten·animal^{-1}·hr^{-1}) is calculated from

$$G = \frac{Vf}{N}\frac{C_{tf} - C_0}{(k - f)t} \tag{7.8}$$

When $k = 0$,

$$G = \frac{V(C_0 - C_{tf})}{Nt} \tag{7.9}$$

Ingestion rate (I, weight of cells eaten·animal^{-1}·hr^{-1}) is obtained from G as

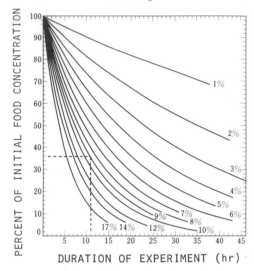

Figure 7.5. Hypothetical curves for the change from initial food concentration with incubation period (hr) at designated, constant filtering rates of zooplankton (expressed as a percent of container volume·hr^{-1}). (After Paffenhöfer, 1971.)

$$I = Gw \qquad\qquad (7.10)$$

where w is the weight of a single algal cell measured as wet, dry, carbon, or nitrogen weight or the caloric content of a cell.

To facilitate calculation of filtering rate, Paffenhöfer (1971) expressed Equation (7.6) graphically (Fig. 7.5). In this graph C_{tf} is expressed as the percentage of C_0 on the Y-axis, and the X-axis denotes the duration of the experiment (hr). When C_{tf} decreased to 36% of C_0 in 11 hr, for example, the graph indicates that animals in the experimental bottles filtered 9% of the total volume of seawater in the bottles per hour. Then the filtering rate per animal can be easily calculated by knowing the number of animals placed in the bottles.

Compared with herbivorous zooplankton, few feeding experiments with carnivorous and omnivorous species have been conducted, due probably to the fact that preparation of natural prey animals is not easy. To calculate the predation rate, essentially the same equations [Eq. (7.8) and (7.9)] as for the grazing rate can be applied. In predation experiments the number of prey attacked generally includes all missing and dead prey. In the above equations k would be meaningless and usually zero when all prey were recovered by the end of incubation (any differences between C_0 and C_t indicate recovery efficiency of prey).

7.3.2 Tracer Method

Marshall and Orr (1955a, 1955c) first used radioisotopes (^{32}P and ^{14}C) as a tracer of food algae in their study of the feeding of *Calanus finmarchicus*.

In their experiments the radioisotope (0.25–2.0 mCi of $H_3{}^{32}PO_4$ per liter, or 50 μCi of $NaH^{14}CO_3$ per 150 ml) was added to the culture media of the microalgae (diatoms and flagellates). Upon harvesting, the radioactive algal cells were washed by repeated centrifugation (2500 rpm) and resuspension in filtered seawater. The labeled algae were fed to single *C. finmarchicus* in 70-ml bottles for 15–24 hr. During this incubation period bottles were wrapped with a black cloth and attached to a rotating wheel (see Fig. 7.3). At the end of the incubation the animal was removed, washed with nonradioactive water, and homogenized for radioactivity measurements with a GM counter. Fecal pellets and eggs, if present, were also collected and processed in a similar way.

Theoretically, the filtering rate of the animal can be calculated from Equation (7.6) by substituting the specific activities $(cpm \cdot ml^{-1})$ in aliquots of water sampled at the start (C_0') and at the end (C_{tf}') of incubation for C_0 and C_{tf}, respectively. However, Marshall and Orr (1955a) found that the indirect estimate of C_{tf}' from the following equation is much more accurate than the direct estimate:

$$C_{tf}' = C_0' - \frac{\sum C_e}{V} \tag{7.11}$$

where, $\sum C_e$ is the total specific activity (cpm) in the body of the animal and all particulate matter produced by the animal during the incubation period and V is the volume of seawater in the experimental bottle (ml). Grazing rate G and ingestion rate I can be calculated from the following equations:

$$G = \frac{\sum C_e}{C_{0,i}'t} \tag{7.12}$$

and

$$I = \frac{\sum C_e}{C_{0,w}'t} \tag{7.13}$$

where $C_{0,i}'$ and $C_{0,w}'$ are the specific activities based on a single algal cell and per unit weight of algal cells (in terms of dry matter, carbon, nitrogen, or calories), respectively.

Sorokin's (1966, 1968) ^{14}C method requires additional reincubation of the animal with nonlabeled food to permit defecation of the undigested, labeled food from the gut of the animal, thus allowing separation of food ingested into the fractions assimilated and not-assimilated (undigested). Lasker (1960) dissected out the stomach and intestine of *Euphausia pacifica* at the end of incubation with ^{14}C-labeled algae for the same reason.

The tracer methods outlined above have an apparent advantage over the balance method in their high sensitivity. In fact, measurements of feeding rate by a single zooplankton in a large volume of water is possible using the method of Marshall and Orr (1955c). However, there are several sources of potential error in these methods: (1) changes in the specific activities of algal cells, animal bodies, and fecal pellets during the experiment and (2) recycling of radioisotopes within the algae–animal–water system in the experimental bottles during the incubation period.

A change in the specific activity of algae is not a difficult problem and can be accounted for by preparing appropriate control bottles without animals. If the specific activity was found to change, Equation (7.5) may be used instead of Equation (7.6) to calculate the filtering rate. Lasker (1960) observed that the $^{14}C:C$ ratio of the algae (*Dunaliella primolecta*) decreased 3–4% during his 24-hr feeding experiment with *Euphausia pacifica*. A loss in radioactivity from an animal body, excluding defecation, may be caused by respiration (for ^{14}C) or excretion (for ^{32}P). However, dissolution and incomplete recovery of fecal pellets are major sources of isotope loss and thus of underestimation of feeding rates. These losses may be minimized by shortening the incubation period to minimize the animal's production of radioactive fecal pellets (Rigler, 1971). For example, Zillioux (1970) was able to measure the rate of feeding of *Acartia tonsa* on microalgae labeled with ^{14}C in 1-hr grazing experiments because he confirmed that the residence time[1] of food in the gut of *A. tonsa* is longer than 1 hr (see also Peters and Rigler, 1973). In this short incubation technique rates of filtering, grazing, and ingestion are calculated from Equations (7.6) and (7.11), (7.12) and (7.13), respectively, but ΣC_e refers to activity in the animal's body only.

The problem of recycling of radioisotopes is the most serious, although it has been overlooked in many feeding experiments (Conover and Francis, 1973). In experimental bottles radioisotopes such as ^{14}C and ^{32}P recycle between algae and water and between animal and water, in addition to the transfer from algae to animal during feeding. To obtain meaningful estimates of feeding rates of animals in this complicated system, a detailed kinetic analysis of each transfer process of the radioisotope is essential. Conover (1978 p. 245) rephrased this problem: "The problem can be pictorially described as trying to measure the increase in volume of a liquid which is being added at unknown depth to a graduated cylinder in which there is some unknown degree of mixing and which has some unknown number of outflows of unknown diameter located at unknown distance above the bottom, by comparing the concentration of a tracer in the liquid being added with that in the vessel at a later time. The measurement of tracer loss at the usual overflows, such as excretion or respiration, is insufficient to assess total loss of liquid because almost certainly the concentration (specific activity) leaving will be different from that being added. The problem is further complicated by the probable presence of smaller intermediate pools, with unknown behaviour, between the source of liquid (or food) being added and the bulk of liquid which we wish to determine (growth, assimilation, etc.)."

Unfortunately, there presently are no simple and reliable radioisotopic methods for the study of zooplankton feeding that completely eliminate this recycling problem. However, efforts to minimize this problem were made in the experiments of Griffiths and Caperon (1979) and Daro (1978).

Griffiths and Caperon (1979) used highly radioactive algae grown in culture media in which all dissolved inorganic carbon (DIC) in seawater had been replaced with ^{14}C-labeled DIC (see Smith and Wiebe, 1977). Highly radioactive algae thus obtained were fed to mixed zooplankton for 0.5–2 hr, and the zooplankton were sorted into species for the measurement of specific activity. They observed neither metabolic loss of the radioisotope from zooplankton nor exchange of the isotope between algae and water. Under such conditions ingestion rates of zooplankton on algae are determined from the activity in zooplankton body divided by the specific activity in terms of algal carbon or nitrogen.

Daro (1978) added $NaH^{14}CO_3$ (40 $\mu Ci \cdot liter^{-1}$) directly into 1-liter glass bottles filled with natural seawater containing phytoplankton and zooplankton. After incubation of the bottles for 1–2 hr in the light (10,000 lux), zooplankton and phytoplankton were separated from the seawater using 50- or 100-μm nettings and 0.45-μm filters, respectively. Phytoplankton retained on the nettings with zooplankton were removed carefully under a dissecting microscope and added to the fractions collected on the filters. The feeding rate of zooplankton on phytoplankton was calculated from measurement of the specific activities of the phytoplankton and zooplankton and a hypothetical, stationary, three-compartment model;

$$\text{Water} \xrightarrow{\lambda_1} \text{Algae} \xrightarrow{\lambda_2} \text{Zooplankton} \xrightarrow{\lambda_3}$$
$$\quad\; q_1 \qquad\qquad q_2 \qquad\qquad\quad q_3$$

where q_1, q_2, and q_3 are the specific activities in water, algae, and zooplankton, respectively. The term λ_1 is the ^{14}C uptake rate by algae, λ_2 is the uptake rate of ^{14}C-labeled algae by zooplankton (i.e., feeding rate), and λ_3 is the removal rate of ^{14}C from zooplankton (via defecation, excretion, and respiration). The changes in ^{14}C concentration in q_1, q_2, and q_3 are expressed as

$$\frac{dq_1}{dt} = -\lambda_1 q_1 \quad \text{or} \quad q_1 = q_{1,0} e^{-\lambda_1 t} \tag{7.14}$$

$$\frac{dq_2}{dt} = \lambda_1 q_1 - \lambda_2 q_2 \quad \text{or} \quad q_2 = q_{2,0} e^{-\lambda_2 t} + \frac{\lambda_1 q_{1,0}}{\lambda_2 - \lambda_1} (e^{-\lambda_1 t} - e^{-\lambda_2 t}) \tag{7.15}$$

$$\frac{dq_3}{dt} = \lambda_2 q_2 - \lambda_3 q_3 \quad (\doteqdot \lambda_2 q_2, \text{ as } \lambda_3 = 0 \text{ is assumed}) \quad \text{or}$$

$$q_3 = q_{1,0} + q_{2,0} (1 - e^{-\lambda_2 t}) + q_{3,0} + \frac{q_{1,0}}{\lambda_2 - \lambda_1} (\lambda_1 e^{-\lambda_2 t} - \lambda_2 e^{-\lambda_1 t}) \tag{7.16}$$

where $q_{1,0}$, $q_{2,0}$, and $q_{3,0}$ are q_1, q_2, and q_3 at time 0, respectively. The experiment is designed so that $q_{2,0} = 0$ and $q_{3,0} = 0$, and hence these equations simplify to

$$q_1 = q_{1,0}e^{-\lambda_1 t} \tag{7.17}$$

$$q_2 = \frac{\lambda_1 q_{1,0}}{\lambda_2 - \lambda_1} (e^{-\lambda_1 t} - e^{-\lambda_2 t}) \tag{7.18}$$

$$q_3 = q_{1,0} + \frac{q_{1,0}}{\lambda_2 - \lambda_1} (\lambda_1 e^{-\lambda_2 t} - \lambda_2 e^{-\lambda_1 t}) \tag{7.19}$$

Since $q_{1,0}$ is very high and the incubation time is very short, we can assume that $q_1 = q_{1,0}$, and $\lambda_1 q_1 \gg \lambda_2 q_2$. Then, Equations (7.15) and (7.16) are further simplified:

$$\frac{dq_2}{dt} = \lambda_1 q_{1,0} \quad \text{or} \quad q_2 = \lambda_1 q_{1,0} t \tag{7.20}$$

$$\frac{dq_3}{dt} = \lambda_2 q_2 \quad \text{or} \quad q_3 = \frac{1}{2} \lambda_1 q_{1,0} \lambda_2 t^2 \tag{7.21}$$

$$= \frac{1}{2} \lambda_2 q_2 t$$

Then

$$\lambda_2 = \frac{2q_3}{q_2 t} \tag{7.22}$$

In other words Equation (7.22) is derived assuming that ^{14}C uptake by algae is a linear function of time [Eq. (7.20)] and that ^{14}C-labeled algae uptake by zooplankton is a parabolic function of time [Eq. (7.21)]. While either Equation (7.18) or (7.19) can be solved for λ_2 by knowing all other parameters involved in each equation, equation (7.22) is an extremely simplified form with which to calculate λ_2 and will provide the same results if the assumptions are correct. Daro (1978) calculated λ_2 using these complex and simplified equations and found a good agreement between them.

As Daro's method was developed primarily to measure the rate of zooplankton feeding on phytoplankton in natural assemblages, possible differential ^{14}C uptake by the component algal species and incomplete separation of phytoplankton from zooplankton at the end of incubation are potential sources of error. These errors would be minimal in laboratory experiments with sorted zooplankton and specific algal foods. As noted by Daro (1978) this method is not applicable for measuring the feeding rate of zooplankton in the dark. Recently, Roman and Rublee (1981) adopted this method for the simultaneous measurement of zooplankton feeding on phytoplankton and bacteria-detritus in natural assemblages.

In addition to the use of $NaH^{14}CO_3$ for phytoplankton, (methyl-3H)-thymidine was used as a tracer of bacteria.

Besides ^{14}C and ^{32}P, radioisotopes used to study zooplankton feeding have been ^{131}I (Marshall and Orr, 1955c), ^{35}S (Zimmerman, 1973), and ^{65}Zn (Zillioux, 1973). Rigler (1971) suggested ^{59}Fe to be a good tracer. In light of the recycling problem mentioned above, ^{59}Fe, ^{65}Zn, and ^{68}Ge appear to be advantageous because these radioisotopes do not participate in the various metabolic processes of organisms. These isotopes have relatively long half-lives (^{59}Fe, 45 days; ^{65}Zn, 244 days; ^{68}Ge, 287 days) and all emit γ rays so that radioactivities are easily and directly measured with intact live specimens. Azam and Chisholm (1976) and Sullivan (1976) are good references regarding the labeling of algae with ^{68}Ge.

7.3.3 Marking Method

For filter-feeding zooplankton that do not discriminate unnatural from natural food particles, artificial particles can be mixed with natural ones and used as a marker to estimate the volume of water being swept clear by the animal.

Spittler (1973) and Heinbokel (1978) used corn starch to measure the clearance rate of Tintinnida. Corn starch was prepared in deionized water (range of particle diameter: 2.5–25μm) and added to glass bottles filled with 300–800 ml of natural seawater containing natural phytoplankton, tintinnids, and detritus assemblages. These bottles were attached to a "Ferris" wheel (Fig. 7.3A) for 5–30 min. At the end of the incubation the contents of the bottles were fixed with Lugol's iodine fixative and gently poured through 20-μm mesh nettings. Tintinnids retained on the nettings were washed into a settling chamber. Corn starch particles in individual tintinnids were stained black and measured and counted with an inverted microscope.

Assuming the largest corn starch particle ingested represents the maximum-sized particle a tintinnid could ingest and knowing the concentration of corn starch particles equal or less than this size (P_s, number of particles·ml^{-1}), the clearance rate (F, μl·tintinnid^{-1}·hr^{-1}) is calculated from the equation

$$F = \frac{I}{P_s t} \tag{7.23}$$

where I is the number of corn starch particles in the tintinnid and t is the incubation time (hr). This equation replaces the exponential form [Eq. (7.6)] considering that the incubation time is very short and only a small fraction of the total number of particles is ingested by the animal. To obtain accurate results with this method, Heinbokel (1978) noted that proper combinations of corn starch concentration and incubation time are essential to avoid low particle numbers being ingested and digestion or egestion of the particles.

Appendicularia feed using a house equipped with fine mucus nets (Section 7.1). Food particles in the surrounding water are pumped in, retained on the feeding nets, and eventually transported to the mouth. The first two steps of this feeding process appear purely mechanical. Alldredge (1981) captured individual appendicularians (*Oikopleura dioica* and *Stegasoma magnum*) in clear plastic containers (125- or 250-ml capacity) while scuba diving and injected a suspension of small red plastic beads (diameter: 12 μm) into the container through a syringe fitted to its lid. The concentration of beads in the container was 2–50 beads·ml^{-1} and did not significantly alter the mean concentration of particles in the water at the study site. Feeding of the animal was terminated after 5 min, before any beads were defecated, by injecting formalin into the container. These operations were conducted under depth, temperature, and light conditions similar to where the animal was captured. Then the containers were brought into the ship laboratory, the animals removed, and the number of beads in the house and gut of each animal determined under a microscope. The clearance rate of the appendicularia was calculated using Equation (7.23).

7.3.4 *Experimental Factors Affecting Feeding Rate*

Temperature and food supply both affect the feeding rate. Within the normal temperature ranges of animals, an increase in water temperature is known to accelerate the feeding rates of several copepod species (Conover, 1956; Anraku, 1964a).

In earlier feeding experiments on boreal copepods (*Calanus finmarchicus, C. glacialis, C. hyperboreus*), specimens collected from the field often were starved in filtered seawater for one or several days prior to the experiment to enhance the feeding rate (Conover, 1962; Marshall and Orr, 1955a; Mullin, 1963). Mullin (1963) and McAllister (1970) examined the change in the filtering rate of prestarved copepods at various intervals during incubation and found a progressive decrease in the rate. A gradual decrease in food concentration during incubation is not considered to cause this, because a low food concentration generally augments filtering rates of copepods (see Fig. 7.6). Runge (1980) compared the filtering rate of *Calanus pacificus* starved for 22–26 hr with the rate of nonstarved specimens of the same species and found that the rates of prestarved specimens were 1.5–3.1 times greater than those of nonstarved ones; he used food concentrations of 15–185 μg C·liter^{-1}. However, this "hunger" effect on filtering rates was not consistent with the results of feeding experiments with *C. pacificus* and other copepod species (Hargrave and Geen, 1970; Paffenhöfer, 1976), even when the food concentration was less than 300 μg C·liter^{-1} (Frost, 1972). In light of the probable intermittent nature of feeding during the vertical migration of some zooplankton, this "hunger" effect, if any, has to be considered in order to obtain realistic feeding rates from laboratory experiments and, thus, warrants further investigation (see McAllister, 1969, 1970).

Cushing (1958) measured the filtering rate of *Temora longicornis* using containers of 5–500 ml capacities and observed a proportionate increase in the rate with an increase in container size. However, experiments with *Calanus finmarchicus* and *Acartia tonsa* showed that this "volume" effect existed for containers of less than the 100-ml capacity but not for containers greater than this capacity (Marshall and Orr, 1955a; Anraku, 1964b). The predation rates of *Centropages typicus* and *Tortanus discandatus* again increased in the 500-ml container (Anraku, 1964b). As the volume effect has been examined for only a few zooplankton species, further examination is needed for other species.

Maintenance of a homogeneous suspension of food particles in the containers is essential in feeding experiments with filter-feeding and suspension-feeding zooplankton and accomplished by the use of appropriate mixing devices. Although mechanical disturbance may affect the feeding behavior of zooplankton and, subsequently, their feeding rate, little is known about this. According to Harbison (personal communication), most salps are easily damaged by mechanical mixing of water. Therefore, none of the mixing devices shown in Figure 7.3 nor the "plunger-jar" (Fig. 6.1A and B) are suitable for feeding experiments with salps and other delicate gelatinous zooplankton (see Harbison and Gilmer, 1976). For predation experiments certain mixing devices may adversely affect the feeding of animals. Omori (unpublished) measured the rate of predation by the copepods *Pleuromamma xiphias* and *Labidocera bipinnata* on *Artemia* nauplii in the dark and found that the rates obtained from jars on the "Ferris" wheel were 45–70% of those obtained in identical jars which were not agitated. These copepods appeared to detect prey by visual, chemosensory, or tactile senses and the mixing of water apparently interfered with the detection and location of the prey.

Feeding experiments conducted in the dark are needed to minimize the growth of food algae and/or to promote a random distribution of prey. One might believe that the growth of algal cells during an experiment can be accounted for by using control bottles and Equation (7.5). However, growth rates of algal cells in experimental and control bottles may be unequal due to nutrient excretion by animals in the experimental bottles (Roman and Rublee, 1980). This is especially true when nutrient-poor seawater is used. Marshall and Orr (1955a) reported more rapid feeding by *Calanus finmarchicus* on algal cells in the dark, but such a "light effect" was not seen for *Acartia tonsa* and *Calanus helgolandicus* fed algal cells (Conover, 1956; Richman and Rogers, 1969) and *Tortanus discaudatus* fed *Artemia* nauplii (Anraku, 1964b). A recent study of *Corycaeus anglicus* indicated that this cyclopoid copepod locates animal prey visually; pre-

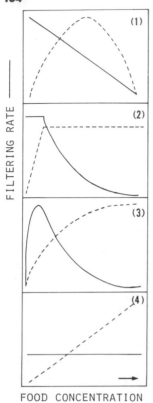

FILTERING RATE ——————

INGESTION RATE --------

FOOD CONCENTRATION

Figure 7.6. Response curves showing filtering and ingestion rates of zooplankton in relation to various concentrations of food particles. (1) *Calanus hyperboreus* feeding on several diatom species (Mullin, 1963); (2) *Calanus pacificus* feeding on several diatom species (Front, 1972); (3) *Calanus helgolandicus* feeding on a diatom species (*Biddulphia*) (Corner et al., 1972); (4) *Calanus hyperboreus, C. glacialis*, and *C. finmarchicus* feeding on natural phytoplankton assemblages (Huntley, 1981).

dation rates measured in the light were more than 10 times the rates determined in the dark (Gophen and Harris, 1981). The different feeding responses of zooplankton to light conditions may reflect their diversified mechanisms of sensing and capturing prey.

Size and concentration of food particles are two major intrinsic variables that affect the feeding rates of marine zooplankton. Many studies on this subject have been done with suspension-feeding copepods. With regard to the size of food, higher feeding rates are generally observed for larger foods, within a limited size range. Figure 7.6 illustrates four types of response curves for filtering and ingestion rates in relation to various food concentrations. Note that the response of feeding food concentration is ill-defined near the origin for each type of curve. It has been suggested that copepods may cease or slow filtering at a food concentration above zero. This threshold food concentration has been determined experimentally for different types of animals and foods (laboratory cultures of algal cells or natural particles) and varies considerably, that is, $7-150 \ \mu g \ C \cdot liter^{-1}$ (See Reeve and Walter, 1977). Above this threshold level of food concentration, filtering rates of types 1, 2, and 3 behave differently but eventually decrease with increasing food concentration. Experimental determinations of the response type for a given category of animal and food particle are usually difficult due to the large variability within the data, especially at low food concentrations (see Mullin et al., 1975b). Frost (1974) analyzed the type 2 response in detail. If food concentration and ingestion rate are p (μg C·liter^{-1}) and I (μg C·animal^{-1}·hr^{-1}), respectively, then the rectilinear response of I to p can be expressed as

$$I = I_{max} \left(\frac{p}{S_p} \right) \quad \text{when } p \leqq S_p \tag{7.24}$$

and

$$I = I_{max} \quad \text{when } p > S_p \tag{7.25}$$

where I_{max} is the maximum ingestion rate and S_p is the food concentration above which the ingestion rate does not increase any further. Both S_p and I_{max} are functions of the body weight for a given copepod and can be expressed as

$$S_p = aW^b \tag{7.26}$$

and

$$I_{max} = cW^d \tag{7.27}$$

where a, b, c, and d are all constants (Frost, 1974). Type 4 response is different from types 1, 2, and 3 in that no saturation of ingestion rate is seen. This nonsaturated ingestion rate has been confirmed up to a food concentration near 1000 μg C·liter^{-1} (Huntley, 1981). Note that the type 4 relation was generated from feeding experiments with natural assemblages of algae while the other three types were established with unialgal cultures in the laboratory. From the viewpoint of applying laboratory-measured feeding rate of zooplankton to field populations, the discrepancy between response type 4 and the other types is serious enough a warrant further examination.

Table 7.3 is a summary prepared by Marshall and Orr (1962) of the maximum filtering rates of several copepods (see also Conover, 1978 Tables 5-5, 5-6). As they noted, variations in the filtering rates of individual specimens of a given species or development stage are commonly observed even when similar experimental procedures are employed. Although such variations may be unimportant in feeding experiments with many individuals, they seriously affect measurements made with a single specimen. One might notice in Table 7.3 that mature female copepods were most extensively used. This is because the feeding activity of adult females is usually greater than that of adult males (Marshall and Orr, 1955a; Mullin, 1963; Ikeda, 1977c). In addition to sex, body size of copepods is an important intrinsic variable that affects feeding rates. Generally, smaller

Table 7.3. Filtering rates of copepods[a]

Species	Stage	Food[b]	Maximum filtering rate $(\text{ml·ind}^{-1}\text{·day}^{-1})$	Method
Calanus finmarchicus and	Adult	C	7	Cell count
Calanus helgolandicus	Adult	D	240	Cell count
	C_V	B	101	Cell count
	♀	A	4	[32]P
	♀	B	84.3	[32]P
	♀	D	43	[32]P
	C_V	D	4	[32]P
	♀	Natural sea water	36	Chemical analysis
Metridia lucens	♀	D	0.4	[32]P
	C_V ♀	B	1.3	[32]P
Centropages hamatus	?	B	15	Cell count
	♀	A	0.3	[32]P
	♀	B	15.0	[32]P
	♀	D	2.7	[32]P
Temora longicornis	?	B	11.6	Cell count
	♀	A	1.1	[32]P
	♀	B	15.5	[32]P
	♀	D	18.3	[32]P
	C_V ♀	D	6.6	[32]P
	?	D	150.0	Cell count
	♀	D	20.7	[32]P
Pseudocalanus minutus	?	B	8.6	Cell count
	♀	A	3.7	[32]P
	♀	B	9.1	[32]P
	♀	D	2.3	[32]P
	C_V	B	1.7	[32]P
	C_V	D	0.9	[32]P
Acartia clausi	?	D	10.4	Cell count
	♀	A	0.1	[32]P
	♀	B	8.6	[32]P
	♀	D	2.7	[32]P
Acartia tonsa	?	D	25.1	Cell count
Paracalanus parvus	C_V	D	0.6	[32]P
Oithona similis	♀	A	0	[32]P
	♀	B	0.02	[32]P
	♀	D	0	[32]P

[a] After Marshall and Orr, 1962.
[b] A = Flagellates less than 10 μm. B = Flagellates over 10 μm. C = Diatoms less than 10 μm, nonchain forming. D = Diatoms over 10 μm, or in chains.

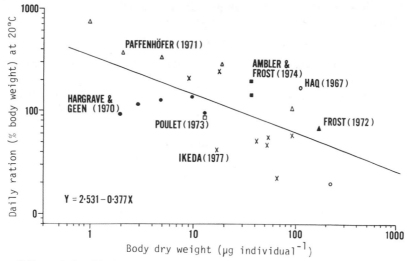

Figure 7.7.   Relationship between daily ration (in terms of percent body weight) and body weight in various marine planktonic copepods. The figure includes data on *Metridia longa* and *M. lucens* fed *Artemia* nauplii at 5°C (Haq, 1967), *Acartia tonsa, Oithona similis, Pseudocalanus minutus*, and *Temora longicornis* fed natural phytoplankton at 16°C (Hargrave and Geen, 1970), nauplii and copepodites of *Calanus helgolandicus* fed diatoms at about 15°C (Paffenhöfer, 1971), *Calanus pacificus* fed diatoms at 12.5°C (Frost, 1972), *Pseudocalanus minutus* fed natural particles at 4°C (Poulet, 1973), *Tortanus discaudatus* fed on nauplii of *Calanus pacificus* at 12.5°C (Ambler and Frost, 1974), and *Eucalanus subcrassus, Tortanus gracilis, Calanopia elliptica, Temora turbinata*, and *Paracalanus aculeatus* fed *Artemia* nauplii and diatoms at 22–25°C (Ikeda, 1977c). Daily rations cited from Haq (1967), Frost (1972), Poulet (1973), and Ambler and Frost (1974) are maximum values or close to the maximum given. All these data are standardized to 20°C, using a Q_{10} value established by Ikeda (1974) for respiration rates of zooplankton from boreal to tropical seas as a function of body weight.

species or earlier stages have higher weight-specific ingestion rates than larger species or later stages of the same species (Fig. 7.7).

7.3.5 *Conversion of Feeding Rate to Carbon or Calorific Units*

Although ingestion rates can be expressed in terms of the weight of various quantities, including wet, dry, and organic matter, carbon and calories have been used most commonly. If not measured directly (Chapter 5, Sections 5.5.2 and 5.7; see also Table 6.2), the available weight units of the food material have to be accurately converted to these units.

Strathmann (1967) proposed the following equations for the relationships between carbon content and volume of phytoplankton cells. For diatoms

$$\log C = -0.422 + 0.758 \log V \tag{7.28}$$

and for phytoplankton other than diatoms (e.g., Chlorophyceae and Dinophyceae)

$$\log C = -0.460 + 0.886 \log V \tag{7.29}$$

where C is the amount of carbon in a single cell (pg $C \cdot cell^{-1}$) and V is cell volume (μm^3).

Ingestion rates of algal cells measured in terms of chlorophyll a may be converted to carbon units by using a carbon/chlorophyll a ratio. However, it must be noted that the carbon/chlorophyll a ratio of algal cells both in laboratory cultures and in nature is easily changed by nutrient, light, and temperature conditions, and the ratio is known to range widely from 30 to 150 (Uno, 1971; Eppley, 1972; Uno and Ueno, 1981).

The caloric content of mixed phytoplankton from the sea is correlated with its carbon content. The relationship is expressed as

$$Ca = 0.632 + 0.086 \, C \quad or \quad Ca = -0.555 + 0.113 \, C + 0.054 \, CN$$
$$\tag{7.30}$$

where Ca is the caloric content (cal·mg dry weight^{-1}), C is the carbon content (% dry weight), and CN is the ratio of carbon to nitrogen of the natural, mixed phytoplankton. The latter equation gives a slightly better estimate of the caloric content than the former (Platt and Irwin, 1973). Similarly, Platt et al. (1969) give equations to estimate caloric content of mixed zooplankton from measurements of its organic matter, carbon, and ash:

$$Ca = -2.27 + 1.52 \, C$$

$$Ca = 13.51 + 1.06C - 0.212Ah \quad and$$

$$Ca = -33.7 + 1.36Og - 0.00514Og^2 \tag{7.31}$$

where Ah is the ash content (% dry weight) and Og is the organic matter content (% dry weight). Both Equations (7.30) and (7.31) were established for mixed phytoplankton and zooplankton, respectively, from St. Margaret's Bay, Nova Scotia, to study the energy flow in that area. Application of these equation sets to mixed phyto- and zooplankton from different areas may involve significant error and, if so, would require an independent determination of the relationships.

7.4 Field Measurement of Feeding Rate

Feeding rates of zooplankton in the sea can be estimated from gut contents of specimens. Morisita (1969) proposed an equation for this purpose. Let f and α be the instantaneous rates of feeding and food digestion by an animal. If we assume that f and α are constant, the change in the amount of food in the gut can be expressed as

$$\frac{dV}{dt} = f - \alpha V$$

where V is the amount of food in the gut. Integration of this equation yields

$$V = \frac{f}{\alpha} (1 - e^{-\alpha t}) + V_0 e^{-\alpha t} \tag{7.32}$$

where V_0 is the amount of food in the gut at $t = 0$. Then

$$f = \frac{\alpha(V - V_0 e^{-\alpha t})}{(1 - e^{-\alpha t})} \tag{7.33}$$

When field samples are collected at regular time intervals $(t_0, t_1, \ldots, t_n)$ and the changes in V are measured as $V_0, V_i, \ldots, V_n$, the amount of food eaten during each sampling interval $t_0 - t_1, t_1 - t_2, \ldots, t_{n-1} - t_n$ can be summed to estimate the total amount of food (F) eaten over the sampling period:

$$F = \alpha t \left\{ \sum_{i=1}^{n} V_{t,i} - \frac{(V_0 - V_n) e^{-\alpha t}}{1 - e^{-\alpha t}} \right\} \tag{7.34}$$

When samples are taken over 24 hr and the amount of food at the start and end of the sampling is found to be the same $(V_0 - V_n = 0)$, Equation (7.34) simplifies to

$$F = \alpha t \sum_{i=1}^{n} V_{t,i} \tag{7.35}$$

Application of Morisita's equation requires the value of α, which may be obtained from laboratory experiments (see Section 7.3.2 and Note 1). However, for the special conditions when values of both α and f are known to be constant over a long period, it is possible to estimate α from the amount of food in the gut measured at equally spaced sampling intervals. In such a case we can assume that $e^{-\alpha t}$ in Equation (7.32) is constant. Then Equation (7.32) is rewritten as

$$V = A + B V_0 \tag{7.36}$$

where $A [= f(1 - e^{-\alpha t})/\alpha]$ and $B (= e^{-\alpha t})$ are constants. Since this relation between V and V_0 is linear, α and f can be obtained from the slope and intercept of the linear regression of V on V_0.

Chlorophyll a is known to degrade to phaeophorbide a (Shuman and Lorenzen, 1975) or phaeophytin a (Hallegraeff, 1981) after passing through the gut of a zooplankton. It is possible to estimate the quantity of phytoplankton in the gut by measuring the amount of pigment with a fluorometer (Nemoto, 1968; Nemoto and Saijo, 1968). For the general procedures of fluorometric analysis of phytoplankton pigments, see Strickland and Parsons (1972). Mackas and Bohrer (1976) applied this sensitive method to intact zooplankton. Use of intact specimens simplifies the measurement but introduces a source of error, that is, extra fluorescence

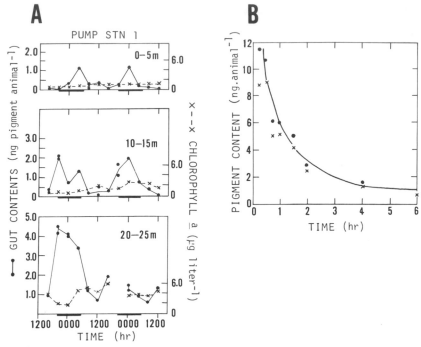

Figure 7.8. (A) Changes in gut fluorescence of *Centropages typicus* collected from three depths in the New York Bight. The readings of gut fluorescence were corrected by blank readings which were obtained from specimens starved for 24 hr in filtered seawater. (B) Change in gut fluorescence of fed *C. typicus* after placement in filtered seawater. Filled-in circles denote total pigments (chlorophyll plus phaeopigments); crosses are phaeopigments. (After Dagg and Grill, 1980.)

due to the carotinoid pigment astaxanthin commonly present in the cuticle of crustaceans. To correct for this fluorescence, Mackas and Bohrer (1976) subtracted the pigment value for test animals kept starved in filtered seawater from readings of animals just collected from the field. Using this procedure, Boyd et al. (1980) and Dagg and Grill (1980) calculated feeding rates of several copepod species in the field. Figure 7.8A illustrates the results of gut fluorescence analyses of *Centropages typicus* measured at 4-hr intervals over 48 hr in the New York Bight. It is seen that *C. typicus* feeds most actively at night at all three depths. The increase in the maximum gut fluorescence with increasing depth reflects higher ambient chlorophyll concentrations. Dagg and Grill (1980) fed *C. typicus* and measured the change in gut fluorescence after transfer of animals to filtered seawater (Fig. 7.8B). As chloroplastic pigments (chlorophyll and its degradation products) are not digested by zooplankton (Shuman and Lorenzen, 1975), the observed change in gut fluorescence in Figure 7.8B represents not time rate of digestion but of passage of ingested food.

When the animal is feeding continuously, a constant gut passage time can be expected, in contrast to the results of Fig. 7.8B in which passage is rapid initially but then slows, perhaps reflecting starvation of the animal. For this reason Dagg and Grill (1980) estimated the gut passage time to be 1.5 hr from the initial portion of the curve; that is, the gut fluorescence measured in the specimens is the result of feeding for 1.5 hr. Finally they converted the gut fluorescence readings to carbon ingestion rates using a carbon/phytoplankton pigment ratio of 50.

The above method enables one to make a simple estimate of the feeding rate in the field but has several problems at present which need to be solved. First, the absolute fluorescence values of the food are difficult to obtain due to uncertain corrections for the added fluorescence of astaxanthin in the animal which may vary with time (see Hairston, 1980). Second, the gut passage time of food has not been measured accurately and most certainly is not constant, that is, it varies with changes in the food concentration. Third, application of this method is limited to strict herbivores so that the food habits of the zooplankton of interest must be evaluated before the method is applied.

Among carnivorous zooplankton, attempts to estimate feeding rates from gut content measurements have been made almost exclusively on chaetognaths because recently ingested prey may be easily observed through their transparent bodies (Nagasawa and Marumo, 1973; Szyper, 1978; Pearre, 1981). In contrast to the gut fluorescence method, food in the gut is observed visually (directly) and can thus be quantified more accurately. Pearre (1981) used digestion time (t_d) instead of instantaneous digestion rate (α) to calculate the feeding rate of *Sagitta elegans*. The relationship between t_d and α is

$$t_d = \frac{1}{1 - e^{-\alpha}} \tag{7.37}$$

Digestion time in *S. elegans* is a function of water temperature ($T°C$):

$$t_d = 0.098e - 0.095T$$

As an alternative approach to estimating the feeding rate from field samples, the change in biomass of prey organisms due to feeding activity of predators can be quantified by knowing other processes that affect the biomass of prey organisms. An example of this approach can be seen in the study of Cushing (1958) in the North Sea. Cushing (1958) calculated the possible feeding rates of *Calanus finmarchicus*, the most dominant copepod in the North Sea, from field observations of the temporal change in phytoplankton biomass.

7.5 Selective Feeding

Zooplankton use foods limited in terms of size, shape, and quality due to their ability to detect different prey types and the functional adaptations of their feeding appendages. *Selective feeding* is defined as the ability of animals to differentially ingest or avoid some kinds of foods. Harvey (1937) first noted that *Calanus finmarchicus* feeds best on larger diatom cells such as *Lauderia borealis* than on smaller cells when diatoms of different sizes are available together. This preference for larger foods has been confirmed for several copepod species by many investigators (e.g., Mullin 1963, 1966; Richman and Rogers, 1969), but an exception was noted by Huntley (1981) in his experiments using natural algal particles.

The "electivity" index E proposed by Ivlev (1955) has been used as a quantitative expression of food selectivity:

$$E_i = \frac{r_i - n_i}{r_i + n_i} \tag{7.38}$$

where r_i denotes the proportion (often in terms of numbers) of a given food (type i) in the diet of the animal and n_i denotes the proportion of the same food (type i) available in the environment. When the animal eats all foods without selection, each $E_i = 0$. Active selection or rejection of a given food is described by the magnitude of E in positive and negative directions, respectively. The main problem with E is that it is not easily interpretable in biological terms. It compares food type i with "the rest," whatever "the rest" may be and depends only on the ratio $r_i : n_i$ (Chesson, 1978).

Chesson (1978) proposed a measure of preference alpha (α) which is derived from a stochastic model involving probabilities of encounter and ingestion upon encounter. Suppose there are m types of food and n_i ($i = 1, 2, \ldots, m$) individuals of food type i in the environment. Let r_i be the number of items of food type i in the diet ($\sum_{i=1}^{m} r_i = r$). When n_i is constant over time or very nearly so (i.e., the number of food items eaten is extremely small compared to the number available), α_i is expressed by the following equation:

$$\alpha_i = \frac{r_i}{n_i} \left(\sum_{i=1}^{m} \frac{r_i}{n_i} \right)^{-1} \qquad 1 \geq \alpha_i \geq 0 \tag{7.39}$$

Similarly, an index of food rejection beta (β) is estimated by the following equation (Checkley, 1982):

$$\beta_i = \frac{q_i}{n_i} \left(\sum_{i=1}^{m} \frac{q_i}{n_i} \right)^{-1} \qquad 1 \geq \beta_i \geq 0 \tag{7.40}$$

where q_i is the proportion of food type i in the assemblage of rejected foods. In another situation, when the values of n_i change significantly (i.e., a feeder consumes a substantial part of the available food), the changing number of food items must be taken into account (see Chesson, 1978).

Discriminate feeding by copepods on larger food particles appears not to be caused by active selection but by passive selection resulting from the structure of the feeding appendages. In copepods the maxillae and maxillipeds are believed to be the most important mouth parts involved in suspension feeding. Nival and Nival (1976) examined distances between setules on the maxillae of *Acartia clausi* and pointed out that the intersetule distance is quite variable. From these measurements, they predicted filtration efficiencies for various size food particles. Theoretically, the average size of the smallest food particle that *A. clausi* can retain increases with developmental stage, that is, 3 μm for copepodite I to 7 μm for adults. The particle size for which the retention efficiency is theoretically 100% also increases with developmental stages (5 μm for copepodite I and 12 μm for adults). These predicted efficiencies of food particle retention were in good agreement with the results of experiments in which various sizes of algal cells were fed to *A. clausi*. Boyd (1976) further generalized this size preference scheme in suspension-feeding copepods, comparing the filtering apparatus (maxillae) of copepods to a "leaky sieve." He noted that in addition to the retention of food particles on the leaky sieve (=setules), intersetule retention or raptorial feeding may be a supplemental mechanism that copepods use to feed on large particles with greater efficiency than hypothetical case which is assumed to be 100%. Frost (1977) tested this mechanistic feeding hypothesis experimentally with *Calanus pacificus* fed algal cells and plastic spheres of various sizes. He concluded that *C. pacificus* differentially filters food particles using a purely mechanical, or passive, mechanism. For a detailed account of the mechanistic models of filter feeding, see Lam and Frost (1976) and Lehman (1976).

While mechanical or passive filter feeding, instead of active selective feeding, has been hypothesized for copepods, Allan et al. (1977) and Richman et al. (1977) observed that *Eurytemora affinis*, *Acartia tonsa*, and *A. clausi* could shift their feeding perference to particles comprising the biomass peak of the natural food assemblage, nearly independent of particle size. They used a Coulter counter to measure the size and abundance of natural particles. As noted earlier, natural particles of various shapes may be sized inappropriately by particle counters and the results may involve serious errors. In addition, the potential breakage of fragile chain forms, often common in natural samples, adds another source of error to feeding rate calculations (O'Connors et al., 1976; Deason, 1980). Thus, suggestions of active particle selection by copepods based on electronic particle counts of natural particulates (e.g., Poulet, 1974, 1978; Cowles, 1979) must be reexamined.

Notwithstanding the technical difficulties that have been pointed out for some selective feeding experiments, there are several other experiments which suggest active selection of food particles by copepods. In early studies of copepod feeding, the age of cultured algal cells has been known to affect the filtering rate of animals (Conover, 1956; Mullin, 1963). Recently it was demonstrated that when *Acartia clausi* is provided two different size algal cells, together with plastic spheres (indigestable food) of an intermediate size, *A. clausi* could ingest both sizes of algal cells and avoid the plastic spheres (Donaghay and Small, 1979). Perhaps the plastic spheres were captured and rejected afterward without being ingested. In the same study *A. clausi* fed small algal cells prior to the experiments ate both sizes of algal cells, while those fed large algal cells prior to the experiment did not eat small algal cells, thus demonstrating the effect of preconditioning on the feeding of this animal. Poulet and Marsot (1978) fed a mixed population of *Acartia clausi* and *Eurytemora herdmani* artificial particles previously immersed in a homogenate of natural phytoplankton as well as the same particles without this treatment. Copepod did ingest the treated artificial particles, but the untreated particles were seldom or not at all ingested, suggesting the existence of a chemosensory mechanism to stimulate the feeding of these copepods. In fact, chemoreceptors have been detected on copepod mouth parts (Ong, 1969; Friedman and Strickler, 1975).

The functional mechanism of selective feeding by copepods is not yet fully understood. Inconsistent or contradictory results by previous workers may in part be due to experimental artifacts or may reflect diversified behavioral responses of different species of copepods. Recently, the direct observation of the behavior of filter-feeding copepods has become possible with the technique of high-speed cinematography (Alcaraz et al., 1980; Rosenberg, 1980; Koehl and Strickler, 1981; Paffenhöfer et al., 1982). At the present time these observations are limited, but they reveal complicated feeding behaviors in copepods. In the future it is desirable to combine such direct observation with feeding experiments for a better understanding of the feeding processes of zooplankton.

Experimental studies of selective feeding by carnivorous copepods are few. Landry (1978) showed that predation by *Labidocera trispinosa* on copepod nauplii increased with an increase in size of nauplii, despite increasing escape ability of the larger nauplii, and decreased suddenly when nauplii developed to copepodites. *Tortanus discaudatus* showed preferential predation on adult specimens when provided with mixed developmental stages, from nauplius to adult, of *Acartia clausi* (Mullin, 1979). This preference of *T. discaudatus* for adult *A. clausi* was not affected by the relative abundance of the adults in the mixture of developmental stages. Carnivorous zooplankters appear to detect prey with various senses—for example, mechanical, tactile, visual, or chemical. Results of predation experiments with *Sergestes similis* larvae (Omori, 1979) suggest

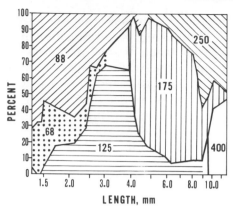

Figure 7.9. The relation between the size (μm) of zooplankton ingested by *Sagitta hispida* and the body length of *S. hispida*. Food size classes are specified as the mean of the retaining mesh size and that immediately above it. Unshaded areas indicate that a particular size of zooplankton was unavailable. (After Reeve and Walter, 1972.)

that while detection of prey by the predator increases with increasing prey size, the largest prey can, by its escape response, avoid becoming a prime food, whereas it becomes difficult or inefficient for predators to detect and ingest a sufficient number of prey below a certain size. Handling efficiencies, which probably decline for both large and small prey items, may also contribute to this apparent size-selecting feeding response.

For zooplankton other than copepods and larval fish, experimental demonstrations of selective feeding have rarely been made. Notable exceptions are the studies by Lalli (1970) and Conover and Lalli (1972) of the pteropod *Clione limacina* which feeds exclusively on the pteropods *Spiratella retroversa* and *S. helicina*.

The feeding habits of zooplankton change during the course of their growth (see Omori, 1971, 1979, and Chapter 6, Section 6.2.5). Generally, the size of prey increases during development of a predator. Sysoeva and Degtereva (1965) illustrated this phenomenon in their study of the food composition of larvae and juveniles of the Arctic-Norwegian cod. With an increase in the body length from 3–35 mm, the most important food items of the cod changed from nauplii of *Calanus* and *Oithona* to their copepodites and some other larger organisms (see also Checkley, 1982, for Atlantic herring). Figure 7.9 shows another example, the size of zooplankton ingested by *Sagitta hispida* of different sizes (Reeve and Walter, 1972).

Selective feeding may be evaluated by comparing the composition of organisms found in the gut of zooplankton with the composition of organisms in the habitat where zooplankton live. In this case the indices E (Ivlev, 1955) and α (Chesson, 1978) [Eq. (7.38) and (7.39)] can be applied. Sullivan (1980) compared the abundance of *Oithona similis* in the gut of *Sagitta elegans* with that in the same water column and obtained a negative value for E, which became increasingly negative as the body length of *S. elegans* increased. This means that as they grew older, the chaetognaths rejected *Oithona* more actively. Older stages of the *Sagitta* preferred prey larger than *Oithona*. For this kind of investigation it is critical to precisely sample the natural assemblage of food organisms at the predator's feeding site. The sampling strategy and analysis require particular care if the food species tends to aggregate and have a patchy distribution. For the same reason, the evaluation of selective feeding of vertically migrating animals requires special attention with regard to the field sampling program. For example, let us consider a species that feeds in the upper layers at night and then descends to deeper layers during the day when it does not feed. The apparent selective feeding of this species would be unnecessarily biased towards certain prey organisms in the upper layer if the evaluation was made on the gut contents of specimens collected from the deep layer during the day. Knowledge of the feeding behavior and migration pattern is prerequisite to field investigations of selective feeding by zooplankton.

7.6 Assimilation and Growth Efficiencies

7.6.1 Assimilation Efficiency

Ingested food is digested in the gut, assimilated, and utilized for growth, reproduction, or metabolsim. Undigested material is eliminated as feces. The assimilation efficiency ($=$ digestion efficiency) (A) is the percentage of ingested food which is digested:

$$A = \frac{F - E}{F} \times 100\% \qquad (7.41)$$

where F is the amount of food ingested and E is the amount of feces eliminated. When the amount of food utilized for growth plus reproduction (G) and metabolism (M) are known ($F - E = G + M$) this equation can be rewritten as

$$A = \frac{G + M}{F} \times 100\% \qquad (7.42)$$

In order to estimate A from either Equation (7.41) or (7.42), each term needs to be measured quantitatively in units of weight (wet, dry, carbon, nitrogen, etc.) or calories.

Equation (7.41) may be used with the tracer method. In this case F represents total radioactivity in the animal body and animal's products during incubation (feces, excreta, molts, eggs, etc.) and E represents the activity of feces (see Section 7.3.2 for potential errors in this method).

For the measurement of assimilation efficiency, Conover (1966a) developed the ratio method. The ratio method requires neither a quantitative measure of the ingestion rate of food nor the quantitative recovery of feces. The variables measured are the fraction of dry organic weight in

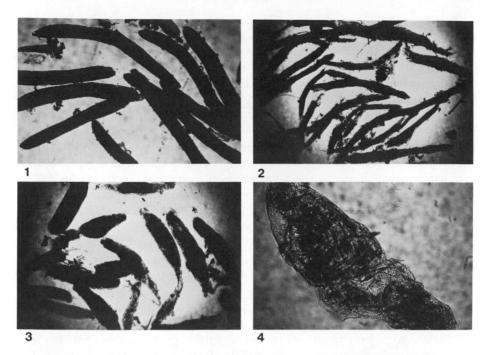

Figure 7.10. Fecal pellets of some zooplankton species. (1) *Calanus plumchrus* (length of a pellet, about 0.9 mm); (2) *Euphausia pacifica* (about 3 mm); (3) *Parathemisto japonica* (about 1.5 mm); (4) *Sagitta elegans* (about 1.5 mm).

the food (F') and feces (E'). A is then calculated from the following equation:

$$A = \frac{(F' - E')}{(1 - E')F'} \times 100\% \tag{7.43}$$

An important assumption of the ratio method is that inorganic matter (ash) in food is not assimilated by animals [for details of the derivation of Eq. (7.43), see Conover (1966a)]. For the procedure to estimate the fraction of dry organic weight in food and feces, refer to Chapter 5, Section 5.3.5.

Most zooplankton species produce feces of various shapes as well as sizes. Figure 7.10 shows the feces of some zooplankton. Most feces are pellet shaped and are covered with a "peritrophic membrane." Starved copepods often produce membrane-covered fecal pellets containing little or no solid contents. These fecal pellets are called *ghost pellets* (Marshall and Orr, 1955b). According to Gauld (1957), the peritrophic membrane consists of a chitinlike substance, but details of its chemical nature are not known. Feces of zooplankton are generally fragile and are often difficult to collect quantitatively. Hence, the ratio method, which requires no quantitative collection of feces, is advantageous over other methods.

It is thought that herbivorous zooplankton do not stop feeding at high concentrations of phytoplankton, as in the case of spring or summer blooms of phytoplankton at high latitudes. As a consequence, zooplankton eliminate largely undigested food as feces. The ecological significance of this phenomenon, which is called *superfluous feeding*, has been discussed by Beklemishev (1962). According to Beklemishev, superfluous feeding occurs when the phytoplankton concentration exceeds 390 μg C·liter^{-1}. Conover (1966b) tested the superfluous feeding hypothesis in laboratory experiments in which *Calanus hyperboreus* were fed algal cells in various concentrations. His results showed that the assimilation efficiency of *C. hyperboreus* determined by the ratio method, remained high (up to 70%), beyond the hypothesized critical concentration for the occurrence of superfluous feeding (390 μg C·liter^{-1}). The assimilation efficiency was not affected by either the concentration of diatom cells available or the number of diatom cells ingested. With these results Conover (1966b) concluded that superfluous feeding does not normally occur in nature. Butler et al. (1970) reached the same conclusion with *Calanus* spp.

Table 7.4 summarizes information on assimilation efficiencies for various food components by zooplankton species under various conditions and measured with various methods. Note that different components are assimilated at different efficiencies even in the same experiment (Butler et al., 1970; Conover and Lalli, 1974). Cosper and Reeve (1975), in measuring assimilation efficiency of *Sagitta hispida* feeding on small copepods, found that the ratio method gave considerably lower estimates (43%) than those obtained with the quantitative recovery method [up to 80%, Eq. (7.41), using dry weight as a basis of F and E]. They explained this discrepancy by the possible assimilation by *S. hispida* of inorganic matter in the food copepods.

7.6.2 Growth Efficiency

Conover (1964) separated the growth efficiency into the gross growth efficiency (K_1) and the net growth efficiency (K_2) and defined each as

$$K_1 = \frac{G}{F} \times 100\% \tag{7.44}$$

and

$$K_2 = \frac{G}{A'F} \times 100\% \tag{7.45}$$

where A' is the assimilation efficiency in decimal form ($=A/100$). Both growth (G) and food ingested (F) are expressed in units of weight (wet, dry, carbon, nitrogen, etc.) or calories.

Values of K_1 for various zooplankton measured by various workers are summarized in Table 7.4. Values of K_2 can be calculated if A' is known in the same experiment. Except in the studies of Butler et al. (1969, 1970), all values of K_1 are derived from laboratory growth experiments with zooplankton. Butler et al. (1969) calculated K_1 values in terms of nitrogen ($K_{1,N}$) and phosphorus ($K_{1,P}$) using the following equations:

$$K_{1,N} = \frac{62}{[1 + (a_2/a_3)(a_3 - 0.90a_1)/(0.90a_1 - a_2)]} \tag{7.46}$$

and

$$K_{1,P} = \frac{69}{(1 + (0.90a_1 - a_3)/(a_2 - 0.90a_1)} \tag{7.47}$$

where 62 and 69 are assumed assimilation efficiencies of N and P, respectively, and 0.90 is their ratio (62/69). The symbols a_1, a_2, and a_3 are the mass ratios of N to P (N/P ratio) in phytoplankton, soluble excretion products, and animal bodies, respectively. For details of the derivation of these equations, see Butler et al. (1969).

In addition to the unit of measurement several variables are known to affect the growth efficiency of zooplankton. Copepods of the genera *Calanus* and *Temora* have higher values of K_1 when grown at lower food concentrations (Mullin and Brooks, 1970b; Paffenhöfer, 1976). In other words a higher gross efficiency occurred at lower rates of ingestion, as is commonly observed for fish (see Paloheimo and Dickie, 1966). However, this inverse relationship between ingestion rate and K_1 was not the case for *Clione limacina* (Conover and Lalli, 1974). Conover and Lalli suggested that this inverse relationship might be due to the provision of foods of fixed size to growing animals; when larger foods are given to growing experimental animals, as in their study of *C. limacina*, the relationship is not seen. Mullin and Brooks (1970a) failed to find any systematic changes in K_1 in the course of development of *Rhincalanus nasutus* and *Calanus helgolandicus*, whereas Paffenhöfer (1976) found a maximum of K_1 between copepodite I and III stages in growing *C. helgolandicus*. Within the normal range, temperature appears to have an insignificant effect on K_1 (Mullin and Brooks, 1970a), although some temperature effects were noted for *Sagitta hispida* (Reeve, 1970).

Both K_1 and K_2 can be calculated with the egg produced by adult females not growing any further. In this case G is replaced by the amount of eggs released, using identical units. Gaudy (1974) and Checkley (1980) calculated K_1 for egg production by several neritic copepods. Checkley (1980) observed maximal K_1 for *Paracalanus parvus* at intermediate inges-tion rate and a dependence of K_1 on food type.

7.6.3 *Possible Sources of Error*

It is conceivable that during the handling and mastication of prey by a predator, some portion of the prey is lost to the surrounding water. If

Table 7.4. Summary of assimilation efficiencies (A) and gross growth efficiencies (K_1) for various zooplankton[a]

Species	Food	Temp. (°C)	A	Method	K_1	Method	Source
Copepods							
Calanus finmarchicus	P	10–20	26–99	^{32}P			Marshall and Orr (1955a)
Calanus finmarchicus	P	?	53–78	^{14}C			Marshall and Orr (1955b)
Calanus finmarchicus and *Calanus helgolandicus*	P	10			36	N	Corner et al. (1965)
Calanus finmarchicus and *Calanus helgolandicus*	P	10			14–34	N	Corner et al. (1967)
Calanus finmarchicus and *Calanus helgolandicus*	P	12			21–38	N	Butler et al. (1969)
					19–35	P	
Calanus finmarchicus and *Calanus helgolandicus*	P	6–18	77	P	17	P	Butler et al. (1970)
			62	N	27	N	
Calanus helgolandicus	P	10	74–91	DW			Corner (1961)
Calanus helgolandicus	P	10, 15			34–35	C	Mullin and Brooks (1970a)
Calanus helgolandicus	P	15			18.5–29	C	Paffenhöfer (1976)
Calanus helgolandicus	A	10	80–99.9	N			Corner et al. (1976)
Calanus hyperboreus, $C_{IV, V}$	P	2–5	44–65(55)	R	4–36(21)	OW	Conover (1964)
					5–50(30)	Cal	
Calanus hyperboreus	P	?	39–86(69)	R			Conover (1966a)
Calanus hyperboreus	P	2–11	40–87	R			Conover (1966b)
Rhincalanus nasutus	P	10, 15			30–45	C	Mullin and Brooks (1970a)
Paracalanus parvus	P	18			32–42(37)	N	Checkley (1980)
					15–41	C	
Pseudocalanus elongatus	P	12.5			14–18	C	Harris and Paffenhöfer (1976b)

Species	Food						Reference
Chiridius armatus	A	6	81–97	R			Alvarez and Matthews (1975)
Temora longicornis	P	10?	52–98	^{32}P			Berner (1962)
Temora longicornis	P	12.5			17–27	C	Harris and Paffenhöfer (1976a)
Metridia longa	A + P	5	54–57	R			Haq (1967)
Metridia lucens	A + P	5	35–94(70)	R			Haq (1967)
Mysids							
Metamysidopsis elongata	A	14–20	19–29	Cal	32	Cal	Clutter and Theilacker (1971)
Euphausiids							
Euphausia pacifica	P	10	>90	C	11–74(32)	^{14}C	Lasker (1960)
Euphausia pacifica	A	7–16	46–95(84)	^{14}C	6–46(26)	^{14}C	Lasker (1966)
Decapods							
Lucifer chacei	A	ca. 25	8–22	^{35}S	7–14	Cal	Zimmerman (1973)
Sergia lucens, early postlarva	A	22	63–96	C	53	C	Omori (unpublished)
Pteropods							
Clione limacina	A	15	83–98(97) 98–99(99)	C N	47–76(61)	DW	Conover and Lalli (1974)
Chaetognaths							
Sagitta hispida	A	16–26			19–50(36)	N	Reeve (1970)
Sagitta hispida	A	24–26	54–97(80)	DW			Cosper and Reeve (1975)

[a] Methods of measurements are with radioactive isotopes (^{14}C, ^{32}P, ^{35}S), elemental analyses (C, N, P), calories (Cal), dry weight (DW), organic matter (OW), and ratio method (R). Means in parentheses. Food: P, plant; A, animal.

this is the case, the resulting overestimation of the ingestion rate will affect the calculations of assimilation and growth efficiencies described above.

Dagg (1974) examined this loss for the amphipod *Callipius laeviusculus* fed ^{14}C-labeled *Calanus* spp. by measuring the specific activity of the seawater medium after feeding. The results show a significant loss in this prey–predator interaction, reaching a maximum of nearly 40% of the prey organic matter. The loss tends to decrease with an increase in the weight of the predator. With a similar ^{14}C technique Lampert (1978) estimated the loss of algal organic matter by a freshwater cladoceran *Daphnia pulex* provided with various unicellular algae. Up to 17% of the algal carbon "ingested" was lost as dissolved organic matter from algal cells damaged during feeding. Apparently, a smaller loss is associated with smaller algal cells.

Although the number of observations is limited, both the results of Dagg (1974) and Lampert (1978) suggest that the relative size of the prey to the predator is important in determining the magnitude of this initial loss. More information is required to establish the general importance of this initial loss in the feeding of various zooplankton.

It is known that zooplankton release dissolved organic nitrogen and phosphorus in the absence of food (see Section 7.4). Although little is known about the origin of this dissolved organic matter (e.g., whether from food in the gut or from an organic matter pool in the animal's body) and its chemical nature, it is another source of potential error in calculating assimilation and growth efficiencies using the equations described above.

Notes

[1] The most common technique for measuring the residence time of food in the gut is as follows. Carmine particles or vitally stained microorganisms are fed to animals and the ingestion and production of feces is observed microscopically. The duration between the first feeding and first discharge of the colored material is the residence time [see also the fluorescent method by Dagg and Grill (1980) in Section 7.4].

Chapter 8 _____

Respiration and Excretion

_____

8.1 Measurement of Respiration Rate

The respiration rate (= oxygen consumption rate) of animals is defined with respect to the activity of animals as follows (Prosser, 1973b): *basal metabolism* is the respiration rate for maintenance only, *routine metabolism* is the respiration rate measured with uncontrolled but minimal motor activity, and *active metabolism* is the respiration rate with enforced activity at a maximal level. When the respiration rate is measured at different activity levels, the rate extrapolated to zero activity is the *standard* or *resting metabolism*. When measuring the respiration rate of zooplankton as described below, activity level is not controlled. Therefore, the results can be taken as somewhere between routine and active metabolism.

8.1.1 *Water Bottle Method*

In the water bottle method bottles with and without experimental animals are prepared simultaneously and the difference in dissolved oxygen concentration after a certain period of incubation is attributed to respiration of the animals. We will describe this method in some detail because it has been the one most extensively used since the original respiration measurements on *Calanus finmarchicus* by Marshall et al. (1935).

BOD bottles are ideal, but ordinary glass reagent bottles fitted with airtight lids may also be used. The capacity of each bottle should be measured before the experiment. It is obtained from the difference in the weight of each bottle with lids before and after filling it with distilled water (1.00 g ≈ 1.00 ml).

Water sampled from the site from which the animals were obtained is used to wash and subsequently incubate the animals. In order to remove other organisms and particles, water should be filtered through a glass fiber filter or another filter of equivalent quality. Vacuum filtration usually reduces the dissolved oxygen content in water. Therefore, aeration of the water is necessary to adjust the oxygen content to near saturation level. Most experiments start with oxygen at saturation. However, special attention is required for experimentation on animals from oxygen-undersaturated habitats.

Three typical washing procedures are illustrated in Figure 8.1A–C. The best washing method depends on the character of the animals and the experimental design. Generally, the washing procedure is not so critical for respiration measurements, but it is a very critical source of error when excretion measurements are combined with respiration measurements. Care must be taken not to damage the specimens while washing them. The number of animals should be counted before incubation and again at the end of incubation. Control bottles without animals should be prepared using exactly the same procedure. In a typical experiment with 10 experimental bottles, two control bottles are prepared before the first experimental bottle and two after the last experimental bottle. An additional control bottle may be inserted between the fifth and sixth experimental bottles.

Upon completion of washing, lids should be fitted so as to avoid trapping any air bubbles. The lids are firmly wrapped with a plastic sheet and rubber band to reduce the risk of air bubbles in the bottles. The bottles are ready for incubation after wrapping with aluminum foil or black plastic (Fig. 8.1D). A water bath with a temperature control unit is most appropriate for incubating the bottles.

The incubation period will vary depending on the experimental temperature, density of animals relative to the volume of the bottle, and respiration rate of the animals. For medium size zooplankton 12 or 24 hr is recommended to eliminate the effect of diurnal rhythm, if any, in the respiration rate. The period may be shortened to only a few hours for very active animals that are less tolerant to prolonged starvation during the incubation period. For respiration measurements of very active animals, the oxygen electrode method is the technique of choice.

Antibiotics, such as streptomycin and chloromycetin, are often used to minimize bacterial respiration in water. The difficulty with using antibiotics is the determination of doses, which vary among different animals. Unfortunately, little is known about the appropriate doses of antibiotics for various zooplankton, and excess antibiotics seriously affects the activity of experimental animals (Marshall and Orr, 1958). Some antibiotics, such as penicillin, interfere with the dissolved oxygen analysis using the Winkler method. When the bacterial respiration is of the same magnitude in experimental and control bottles, this effect cancels out in the final calculation of the respiration rate. Bacterial respiration can be estimated from the difference between the dissolved oxygen content in control bottles before and after incubation, but according to our experience it usually is not significant in most experiments.

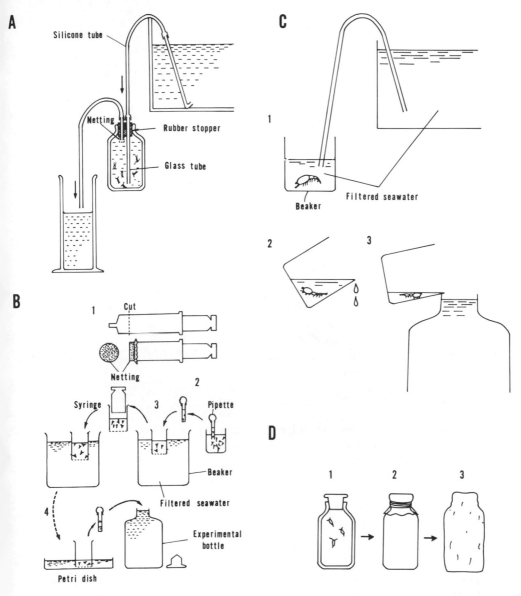

A

Silicone tube

Netting

Rubber stopper

Glass tube

B

1   Cut

Netting

Syringe   2   3   Pipette

Beaker

Filtered seawater

4

Experimental
bottle

Petri dish

C

1

Beaker

Filtered seawater

2   3

D

1   2   3

Figure 8.1. Three procedures for washing zooplankton for respiration and excretion experiments, and preparation of bottles for incubation after washing. (A) A siphon system, originally designed by Marshall et al. (1935) for respiration measurements. Seawater needed for flushing is 6–7 times the bottle volume. (B) Transparent plastic syringe. Place animals in the syringe with a pipette, and rinse them in several beakers. Then, place the syringe into a petri dish and introduce the animals into bottle with pipette. A convenient size of syringe is 20 ml (2 cm in diameter), but larger ones can be used when fine netting is placed over the tip. (C) Glass beaker for relatively large zooplankton. Gently add filtered seawater in the beaker and decant carefully. Repeat this treatment 3–4 times, and then place the animal into a bottle. (D) Preparation of bottles for incubation. See text.

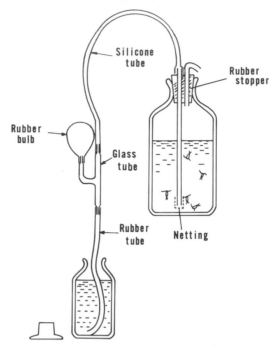

Figure 8.2. A siphon system to transfer sample water from incubation bottle into oxygen bottle for the Winkler titration.

In order to determine dissolved oxygen by the Winkler method, a certain amount of water is taken from the bottle after incubation. Figure 8.2 shows a siphon system we use to transfer water from incubation bottles into small oxygen bottles. Duplicate water samples are sufficient when the analyst is skilled in the technique. General cautions when sampling the water and in the analytical procedure of the Winkler method are described in the manual of Strickland and Parsons (1972).

1. **Analyzing Apparatus.** BOD bottle, 50-ml pipette, 1-ml pipette (dispensing type), 10-ml titration burette (with 0.05-ml scale divisions), 125–250-ml conical flasks or beakers (it is easier to discriminate between the colors during the titration if the entire bottom and half of the sides are painted white on the outside). As a titration burette, the Metrohm pipette[1] is accurate and easy to handle, especially for shipboard measurements.

2. **Reagents.**
   a. Manganous sulfate reagent: 480 g of manganous sulfate tetrahydrate $MnSO_4 \cdot 4H_2O$ (or 400 g of $MnSO_4 \cdot 2H_2O$ or 365 g of

$MnSO_4 \cdot H_2O$) is diluted in distilled water to make a volume of 1 liter.

b. Alkaline iodide solution: 500 ml of sodium hydroxide (analytical reagent grade) is dissolved in 500 ml of distilled water. 300 g of potassium iodide (analytical reagent grade) is dissolved in 450 ml of distilled water. Mix the two solutions together. A great deal of heat will be liberated.

c. Approximately $0.5N$ (or $0.01N$) sodium thiosulfate: To make a $0.5N$ solution, 145 g of high-grade sodium thiosulfate $Na_2S_2O_3 \cdot 5H_2O$ and 0.1 g of sodium carbonate Na_2CO_3 are dissolved in 1 liter of distilled water and one drop of carbon disulfide CS_2 per liter is added as a preservative. To make a $0.01N$ solution, prepare as above, using only 2.9 g of sodium thiosulfate per liter. Both solutions should sit 24 hr before use and are stabile over a long period if stored below 25°C in a dark well-stoppered bottle.

d. $0.100N$ (or $0.0100N$) iodate solution: To make $0.100N$ iodate solution, a small amount of potassium iodate KIO_3 (analytical reagent grade) is dried at 105°C for 1 hr and weighed out exactly 3.567 g after cooling. This is dissolved in 200–300 ml of distilled water by warming slightly. After cooling, distilled water is added to make a solution of 1 liter. To make up $0.0100N$ iodate solution, dilute the solution to 1/10. These solutions are stable indefinitely; keep in a brown bottle in a dark cool place.

e. Starch indicator solution: 0.1–0.2% of soluble starch solution is used. For long-term use preservation solution can be made as follows; 2 g of soluble starch is suspended in 300–400 ml of distilled water and a 20% solution of sodium hydroxide is gradually added with vigorous stirring until it becomes clear. After 1–2 hr concentrated hydrochloric acid is added until the solution becomes slightly more acidic than litmus paper. Then 2 ml of glacial acetic acid is added. Finally, distilled water is added to make a solution 1 liter.

3. Determination of the Factor for the Sodium Thiosulfate Solution. A 300-ml BOD bottle is filled with sample water; add 1.00 ml of concentrated sulfuric acid and 1.00 ml of alkaline iodide solution and mix thoroughly. Then 1 ml manganous sulfate solution is added and mixed well. 50-ml aliquots is poured into each titration flask. Use one or two flasks for blank determinations. Exactly 5.00 ml of iodate solution d ($0.100N$ for $0.5N$ sodium thiosulfate, and $0.0100N$ for $0.01N$ sodium thiosulfate) is added with a 5-ml pipette. Allow iodine liberation to proceed for 2–5 min at a temperature below 25°C and out of direct sunlight, then titrate the iodine with the thiosulfate solution. The starch indicator is added when the yellow color of the liberated iodine fades; the first moment of the disappearance of blue color is the end point. If V is the titration volume of sodium thiosulfate solution in milliliters, the calibration factor f is

$$f = \frac{1.00}{V} \quad \text{(in case of } 0.100N \text{ iodate and } 0.5N \text{ sodium thiosulfate)}$$

or

$$f = \frac{5.00}{V} \quad \text{(in case of } 0.0100N \text{ iodate and } 0.01N \text{ sodium thiosulfate)}$$

The titration should be performed with at least three flasks and the average value obtained becomes f. A blank determination is made by titrating without adding iodate solution. When the blank correction exceeds 0.1 ml of $0.01N$ sodium thiosulfate, the reagents are suspect and should be made afresh.

4. **Analytical Procedure.**

a. Remove the stopper from the 300-ml BOD bottle containing the sample water, and by placing the tip of a pipette a little below the water surface, add 1.0 ml of the manganous sulfate solution a followed by 1.0 ml of alkaline iodine solution b^2. Then restopper the bottle immediately and carefully so that bubbles are not trapped in it and shake the bottle well. (When small reagent bottles are used as oxygen bottles, immerse them in water after adding manganous sulfate reagent and alkaline iodide solution to prevent any introduction of air bubbles prior to the titration.) After 2–3 min, shake the bottle again. Then allow the samples to stand quietly for several hours at room temperature until the precipitate has settled at least one-third of the way down the bottle, leaving a clear supernatant solution.

b. The titration is performed within several hours to one day after fixation. Remove the stopper from the bottle and add 1.0 ml of concentrated sulfuric acid by placing the tip of a pipette just below the water surface. Then restopper the bottle and shake the bottle well in order to dissolve the precipitate.

c. Transfer 50.0 ml of this solution into the specially painted flask or beaker by means of a pipette. Titrate immediately with standard $0.5N$ (or $0.01N$) thiosulfate solution c. During titration the water should be stirred by a magnetic stirrer. When the straw color of the iodine becomes very pale, add 5 ml of starch indicator. The titration is continued until the blue color disappears; this moment is the end point. Read the volume of sodium thiosulfate used up to the end point. When an oxygen bottle of less than 100 ml is used, the titration of its entire contents in a titration flask is recommended rather than taking a 50-ml aliquot. In this case X = Y in Equation (8.1).

5. Calculations. If the volume of BOD bottle is Y (ml) and that of sample water used for titration is X (ml) and the titration volume of $0.01N$ sodium thiosulfate solution is expressed as V (ml) and the blank correction as V' (ml), the dissolved oxygen concentration in the sample water can be obtained by the following formula:

$$\text{mg-at } O_2 \cdot \text{liter}^{-1} = \frac{Y}{Y-2} \times \frac{5.00}{X} \times f \times (V - V') \qquad (8.1)$$

Thus, when a 50-ml aliquot is taken from a 300-ml BOD bottle, mg-at $O_2 \cdot \text{liter}^{-1} = 0.1006 \times f \times (V - V')$.

The milliliters of oxygen in a liter of water can be calculated from the expression

$$\text{ml } O_2 \cdot \text{liter}^{-1} = \text{mg-at } O_2 \cdot \text{liter}^{-1} \times 11.20 \qquad (8.2)$$

1 ml O_2 is equivalent to 1.43 mg O_2 or 0.089 mg-at O_2. When expressed by mg $O_2 \cdot \text{liter}^{-1}$

$$\text{mg } O_2 \cdot \text{liter}^{-1} = \text{mg-at } O_2 \cdot \text{liter}^{-1} \times 16.00 \qquad (8.3)$$

The activity of animals remaining in the experimental bottles should be checked. These animals are then transferred to a petri dish (large animals) or directly onto a piece of mesh (small animals). Counting of animals can be done unaided or under a dissecting microscope for small animals. Animals should then be dried and weighed. Water adhering to the body of an animal should be removed by placing the animal on filter paper. For small animals that are difficult to handle individually, this can be achieved by placing the filter paper on the other side of the mesh. Although a brief rinse with distilled water is sometimes a routine procedure to remove salts, we do not recommend this because of simultaneous loss in body organic matter (see Omori, 1978). The animals on the mesh are then transferred by means of a dissecting needle or fine forceps to tared containers (such as aluminum pans) for later determination of dry weight (see Chapter 5, Section 5.3.2).

Respiration rate (R) is calculated using the following equation:

$$R = \frac{(C_{ox} - E_{ox})1000V}{t(N \text{ or } W)} \qquad (8.4)$$

where C_{ox} and E_{ox} are the dissolved oxygen contents (ml $O_2 \cdot \text{liter}^{-1}$) in the control and experimental bottles, respectively; V is the volume of the experimental bottles (liter); t is the incubation time (hr); N is the number of animals; and W is dry weight of animals (mg). R is expressed in two ways: (1) total respiration rate (μl $O_2 \cdot \text{animal}^{-1} \cdot \text{hr}^{-1}$) and (2) weight-specific respiration rate (μl $O_2 \cdot \text{mg dry wt}^{-1} \cdot \text{hr}^{-1}$). Note that the body dry weight is often not mentioned in publications in which the weight-specific expression is used. In either expression of respiration rate, however, the body weight of the zooplankton should be provided with the respiration rate so that other researchers can convert the results into both expressions.

8.1.2 *Oxygen Electrode (or Polarographic) Method*

In the oxygen electrode method changes in dissolved oxygen content in water are measured with an oxygen electrode, as for example the work of Teal and Halcrow (1962) and Halcrow (1963) who measured the respiration rate of *Calanus finmarchicus*. Measurements with this method are not so accurate as with the Winkler method, but the electrode method has the advantage of simple operation and the continuous recording of the changing dissolved oxygen content in experimental containers.

There are two types of oxygen electrodes, membrane-covered solid electrodes and wide-bore dropping-mercury electrodes. For general characteristics of these two types, see Teal (1971). Membrane-covered solid electrodes are in common use since they can be set directly into an animal chamber. Many kinds of membrane-covered solid electrodes are available from commercial sources. In the use of electrodes of this type, efficient mixing of water is essential to prevent a local oxygen deficiency around the electrodes, and because of this problem, the size of the animal chamber is limited. Although some electrodes with a temperature-compensating circuit give a stable reading despite temperature fluctuations during the measurements, other electrodes without this circuit are very sensitive to

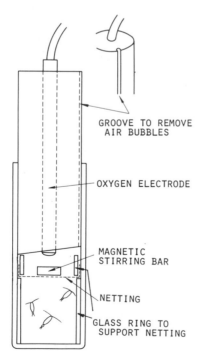

GROOVE TO REMOVE
AIR BUBBLES

OXYGEN ELECTRODE

MAGNETIC
STIRRING BAR

NETTING

GLASS RING TO
SUPPORT NETTING

Figure 8.3. Diagram showing an incubation cell for respiration measurement with the oxygen electrode.

temperature changes and thereby need to be linked with suitable temperature control units. The effect of temperature on the electrode is due to the change in permeability to dissolved oxygen of the membrane. A detailed diagram of an incubation cell is illustrated in Figure 8.3.

Prior to and after the measurement, the oxygen electrode needs to be calibrated in water with dissolved oxygen concentrations of 0% and 100% air saturation. A dissolved oxygen concentration of 0% can be obtained by bubbling the water with nitrogen gas, sodium dithionite solution, or simply by electrical adjustment. The absolute amount of dissolved oxygen in 100% air-saturated water can be determined directly with a Winkler titration or, by knowing salinity and temperature, it can be found in oxygen solubility tables. General calibration procedures should be mentioned in the operation manual of each brand of oxygen electrode and the apparatus associated with it.

A measurement starts by filling the incubation cell with air-saturated water and then placing animal(s) washed with the same air-saturated water in this container (Fig. 8.1). Then the oxygen electrode is positioned, the magnetic stirrer switched on, and the dissolved oxygen concentration in the animal chamber recorded. An example record and calculation of the respiration rate is shown in Figure 8.4. In this experiment two individuals of *Parathemisto pacifica* were placed in two incubation cells and the change in dissolved oxygen content was measured twice with two electrodes (E1 and E2). Usually, a rapid decrease of the dissolved oxygen content is observed followed by gentle linear decrease. The initial rapid decrease is considered to be due to increased activity of the animal(s) after introduction into a new environment. The respiration rate is calculated from the linear decrease phase. A blank run without the animal is necessary to correct for oxygen consumption by the electrode. In the example the calculation was made using the records of E1 in the first (experimental) and third (blank) runs. The oxygen electrode method may not be suitable for fragile animals that are easily damaged by the mixing of water during the measurement.

8.1.3 *Manometric Method*

Animals are placed in a respiration chamber with a small amount of water. The dissolved oxygen consumed by the animals is measured directly by a manometer as the decrease in oxygen gas volume, while CO_2 produced by the animals is absorbed by NaOH in the chamber. In order to achieve this measurement, the whole system must be closed and vigorous shaking of the respiration chamber is needed to maintain gas equilibrium between the air and water in the system. Warburg and Gilson respirometers are typical apparatus for this method. General description of the manometric method can be found in Dixon (1934). Advantages of this method are the

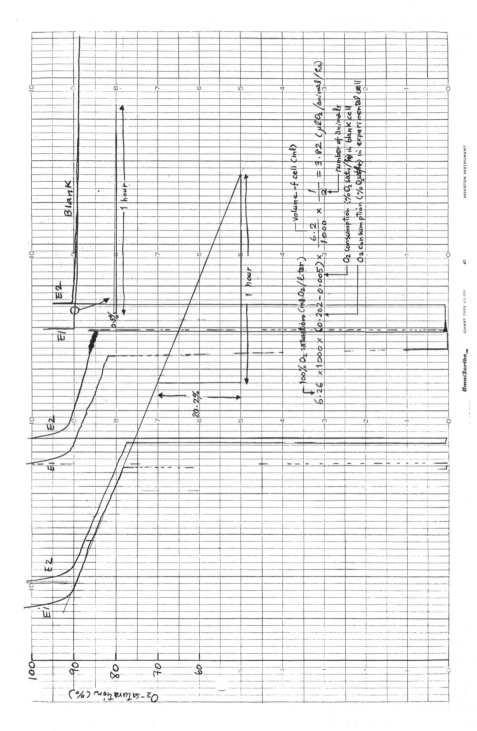

Figure 8.4. A chart record of respiration measurements with an oxygen electrode (YSI O₂ monitor Model 54).

continuous recording of the respiration rate of animals under constant oxygen partial pressure and the possible simultaneous measurement of CO_2 production. Early applications of this method to respiration measurements of marine zooplankton are found in Clarke and Bonnet (1939), Raymont and Gauld (1951), and Gauld and Raymont (1953). At present, however, the manometric method is not commonly used for zooplankton because of the following reasons: (1) injury of or damage to animals is possible by shaking the respiratory chamber, (2) simultaneous measurement of CO_2 production is difficult for marine zooplankton because of the strong buffering action of seawater, and (3) the apparatus required is generally delicate and difficult to use on shipboard experiments.

8.1.4 *Cartesian Diver Method*

The experimental animal(s) is placed in a small, stoppered diver enclosing a gas bubble, which is buoyant in the flotation medium ($0.1N$ NaOH) (Fig. 8.5). The animal's respiration changes the oxygen partial pressure in the diver and subsequently causes it to move up and down in the medium in a vessel held in a temperature-regulated water bath. This movement is measured and is highly sensitive, capable of detecting a change in oxygen volume in the order of $0.001\ \mu l\ O_2 \cdot hr^{-1}$ (Klekowski, 1971b). This method is suitable for the respiration measurements of smaller animals weighing less than 0.5 mg of wet weight per individual, such as microzooplankton. However, the preparation and operation of divers requires considerable skill so that this method has rarely been used to study zooplankton, except by Zeuthen (1947) and Epp and Lewis (1980). Cartesian divers and associated equipment are not available on the commercial market yet, so that all equipment must be homemade. For a detailed description of this method, see Klekowski (1971a, 1971b) and Hamburger (1981). Løvtrup

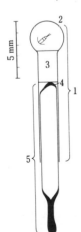

Figure 8.5. Section of Cartesian diver. (1) Diver chamber; (2) Chamber head; (3) Gas bubble; (4) $0.1N$ NaOH solution; (5) Stopper. (After Klekowski, 1971b.)

and Larsson (1965) devised an automatic recording system for respiration measurements with the diver method.

The Cartesian diver method is not applicable on shipboard as the essential prerequisite for the use of this method is complete stability of the apparatus during measurement. Klekowski (1977) devised a microrespirometer (constant-volume manometric gasometer similar to a Cartesian diver) for the shipboard measurement of respiration of oceanic microzooplankton.

8.1.5 *Flow-through System*

Except for the manometric method, all previous methods for measuring the respiration rate of zooplankton are closed systems in which oxygen concentration decreases proportionally with time. Measurements with open systems, such as the flow-through system, have several advantages compared with closed systems. With an open system we can maintain the oxygen concentration at a constant level during the experiment and eliminate the effect of accumulation of metabolic wastes. Measurements of time series and detection of any intrinsic variation in respiration under constant environmental conditions is possible. A disadvantage of the flow-through system is its limited application to only agile and active zooplankton

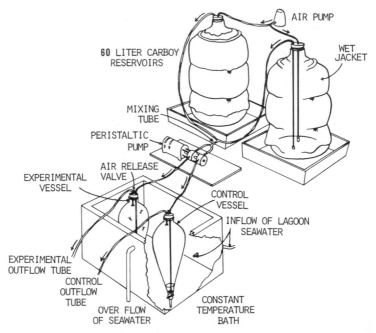

Figure 8.6. Flow-through system used to measure ingestion, respiration, and assimilation by lagoon zooplankton. (After Gerber and Gerber, 1979.)

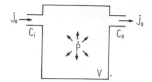

Figure 8.7. Schematic of flow system. (After Northby, 1976.)

species. Figure 8.6 shows a flow-through system used by Gerber and Gerber (1979) for the simultaneous measurement of ingestion and respiration rates of tropical lagoon zooplankton (*Undinula vulgaris* and *Creseis acicula*).

Analyses of the results from measurements with an open system are sometimes misleading, as noted by Northby (1976). As an example, consider the open system illustrated in Figure 8.7, in which j_o is the flow rate, V is the volume of the container, C_i and C_o are the concentrations of some substance p at the inflow and outflow, respectively. The symbol $\dot{p}$ is the rate of production or consumption of p in the container which we wish to estimate. In the simplest case C_i and $\dot{p}$ are independent of time t, and we have

$$\dot{p} = j_o(C_o - C_i) \tag{8.5}$$

In fact, this equation has been used commonly in studies of open systems. When C_i or $\dot{p}$, or both, are a function of time, the calculation of $\dot{p}$ becomes more complicated. However, if the mixing time in the container is short in comparison to all other characteristic times in the problem, solution of this time-dependent case becomes simple. As the concentration of p in the container becomes spatially uniform and equal to C_o, the mass conservation equation will be

$$[j_o C_i(t) + \dot{p}\,(t)] - [j_o C_o(t)] = \frac{d}{dt}[V C_o(t)] \tag{8.6}$$

In other words the rate at which $\dot{p}$ enters plus the production rate minus the rate at which $\dot{p}$ leaves equals the rate of change of $\dot{p}$ in the container. If we define the flushing time $\tau \equiv (V/j_o)$, then we have

$$\dot{p}(t) = j_o[C_o(t) - C_i(t) + \tau(dC_o/dt)] \tag{8.7}$$

An approximation of dC_o/dt is $\frac{1}{2}\Delta t\,[C_o(t + \Delta t) - C_o(t - \Delta t)]$. Therefore, it is apparent that for the calculation of $\dot{p}(t)$, the four measurements $C_o(t)$, $C_i(t)$, $C_o(t + \Delta t)$, and $C_o(t - \Delta t)$ are required. Choice of Δt is arbitrary but needs to be short in comparison to $C_i(t)$ and also well beyond sampling and measurement errors in C_o. An example of the varied effects originating from different Δt values in the same sets of data is discussed by Northby (1976). Equation (8.7) can be applied not only to the flow-through measurement of respiration rate but also to excretion and feeding rates.

8.1.6   *Assay of the Respiratory Electron-Transport System*

The enzyme assay of the respiratory electron-transport system (ETS) is an indirect method of estimating potential respiration in zooplankton. Packard (1971) first used the ETS assay to estimate the respiration rate of phytoplankton. Since then this method has been modified by Packard and his co-workers (Packard et al., 1975; King and Packard, 1975a, 1975b; Owens and King, 1975). At present several ETS assay procedures differing slightly in the compositions of the buffer, substrate, and quench solutions are in use. For the intercomparison of results from different assay procedures, see Christensen and Packard (1979). The ETS assay procedure described below is from Owens and King (1975).

1.   Sample Preparation.   Ahmed et al. (1976) observed that although a cell-free extract of *Calanus finmarchicus* frozen at $-2°C$ lost considerable ETS activity in 24 hr, no measurable loss was found when intact specimens were stored at $\leqslant -20°C$ for at least one week. On the other hand Båmstedt (1980) found that frozen intact specimens of *Acartia tonsa* lost 50% of their ETS activity within 30 hr at $-20°C$ and virtually no activity was detected after 74 hr. Until detailed information becomes available, it is only safe to use fresh zooplankton samples for the determination.

2.   Reagents.

a.   Phosphate buffer: $0.1M$, pH 8.5

b.   ETS-B solution: 75 μM $MgSO_4$, 1.5 mg·ml^{-1} polyvinyl pyrrolidone, and 0.2% (V/V) Triton X-100 in $0.1M$ phosphate buffer, pH 8.5

c.   Substrate solution: 1.7mM nicotinamide adenine dinucleotide (NADH), 0.25 mM nicotinamide adenine dinucleotide phosphate (NADPH), and 0.2% (V/V) Triton X-100 in $0.1 M$ phosphate buffer

d.   2-(*p*-iodophenyl)-3-(*p*-nitrophenyl)-5-phenyl   tetrazoliumchloride (INT) solution: dissolve 2 mg INT per ml of double distilled water

e.   Quench solution: 50% formalin plus 50% $1M$ H_3PO_4, pH 2.5

3.   Assay Procedure.

a.   A zooplankton sample on a GF/C filter is homogenized in ETS-B solution with a teflon–glass tissue grinder and test tubes with ground samples are kept in crushed ice. The amount of zooplankton sample must be less than 5 mg wet wt·ml^{-1} homogenate. The homogenate is centrifuged at 6000*g* for 5 min in a refrigerated centrifuge (0–4°C).

b.   1 ml of supernatant (cell-free extract) is incubated with 3 ml of substrate solution and 1 ml of INT solution for 20 min at desired temperature (*in situ* temperature of zooplankton, in most cases).

c.   1 ml of quench solution is added immediately at the end of incubation to stop the reaction (final volume of the homogenate is 6 ml at this step).

Table 8.1. **The ratio of electron-transport system (ETS) activity to respiration rate for 13 species of tropical zooplankton from inshore waters of the Great Barrier Reef[a]**

Zooplankton	Number of individuals	ETS/R Mean	1 SD	CV(%)
Ctenophora				
Pleurobrachia sp.	6	4.11	1.44	35.0
Siphonophora				
Diphyes sp.	3	2.40	0.16	6.7
Pteropoda				
Cavolinia longirostris longirostris	7	1.54	0.30	19.5
Creseis acicula	8	4.66	0.78	16.7
Copepoda				
Acartia pacifica	5	3.82	0.57	14.9
Acartia australis	4	6.21	1.04	16.7
Calanopia elliptica	3	6.72	0.65	9.7
Tortanus gracilis	4	3.65	0.54	14.8
Eucalanus subcrassus	8	2.67	0.18	6.7
Labidocera acuta	5	2.49	0.29	11.6
Decapoda				
Acetes sibogae australis	6	2.99	0.39	13.0
Mysis larva	3	4.03	1.13	28.0
Chaetognatha				
Sagitta enflata	10	1.28	0.42	32.8
Average		3.58	1.63	45.4

[a] ETS activity was measured by the method of Owens and King (1975). SD, standard deviation of the mean (Skjoldal and Ikeda, unpublished).

d. Absorbance at 490 nm is read with a spectrophotometer; correction should be made with a turbidity blank (1 ml of supernatant from step a, plus 4 ml of $0.1M$ phosphate buffer and 1 ml of quench solution) and a substrate INT blank (1 ml of ETS-B solution plus 3 ml of substrate solution and 1 ml of INT solution; incubate for 20 min at the same temperature, and add 1 ml of quench solution).

4. **Calculation.** ETS activity is calculated as

$$0.188 \times H \times S \times COD \quad (\mu\text{g-at } O_2 \cdot \text{hr}^{-1})$$

or

$$11.2 \times 0.188 \times H \times S \times COD \quad (\mu\text{l } O_2 \cdot \text{hr}^{-1}) \tag{8.8}$$

where H is the total homogenate volume, S is the final reaction volume (6 ml), and COD is the corrected absorbance. The factor 0.188 is derived from the conversion of incubation time to hours and the extinction coefficient of INT-formozan in 0.133% Triton X-100 (E 490 nm = 15.9 $mM^{-1} \cdot cm^{-1}$), and 11.2 is the conversion of μg-at O_2 to μl O_2. If assays are run at other than *in situ* temperatures, corrections are made to the final results, using the Arrhenius equation:

$$\text{ETS }_{in\ situ} = \text{ETS}_{incu} \times \exp\left[E_a\left(\frac{1}{T_{incu}} - \frac{1}{T_{in\ situ}}\right)/R \right] \qquad (8.9)$$

The value of the Arrhenius activation energy (E_a) has been determined to be about 15 kcal·mole^{-1} in *Calanus pacificus* and in the range of 13–16 kcal·mole^{-1} for a variety of plankton. The symbol R is the gas constant (1.987 × 10^{-3} kcal·mole^{-1}).

Convert ETS activity to respiration rate using the correlation between empirical ETS activity and respiration rate (ETS/R ratio). According to Owens and King (1975), the ratio for *Calanus pacificus* is 2.02 ± 0.29 ($\bar{x}$ ± 1 SD). Table 8.1 shows this ratio for 13 zooplankton species from the Great Barrier Reef (Skjoldal and Ikeda, unpublished). The ETS/R ratio appears to be species-specific and caution is needed when applying the ratio to different species.

8.2 Experimental Factors Affecting Respiration Rate

The respiration rate of animals is affected by the oxygen content of the surrounding water. For some animals (metabolic conformers) respiration rate is proportional to the concentration of oxygen, while for others (metabolic regulators) a constant respiration rate is observed as the oxygen concentration is reduced down to some critical level (Pc), below which the respiration rate declines rapidly (see Prosser, 1973b). However, intermediate types are also seen. In general, animals that inhabit oxygen-saturated environments are metabolic conformers.

Marshall et al. (1935) observed a rapid decrease in the respiration rate of *Calanus finmarchicus* when the oxygen concentration decreased to below 3 ml O_2·liter^{-1} (about 50% saturation). On the other hand respiration rates of two mysid species (*Archaeomysis grebnitzkii* and *Neomysis awatschensis*) were not affected until the oxygen concentration decreased to 30% saturation, below which the rates decreased (Jawed, 1973). An extreme capacity to respire at low oxygen concentrations is reported for some midwater crustaceans (e.g., *Gnathophausia ingens*) from the oxygen minimum layer in the eastern Pacific (Teal and Carey, 1967; Childress, 1971, 1975). Ikeda (1977a) studied the relationship between respiration rate and oxygen saturation level for seven zooplankton species from Saanich Inlet, British Columbia (Fig. 8.8). He observed no distinct Pc concentration for the seven species, but the effect of lowered oxygen concentration was less pronounced in *Calanus plumchrus* and *Holmesiella anomala*, collected from oxygen-deficient bottom water, than in species from oxygen-rich surface

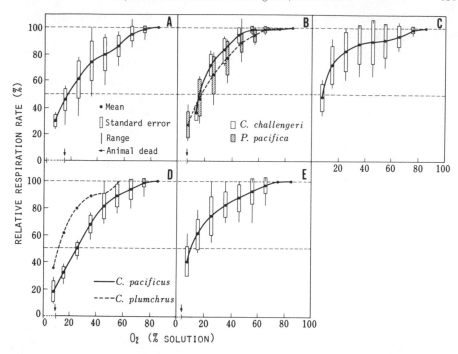

Figure 8.8. Respiration-ambient oxygen concentration (% saturation) relations for boreal zooplankton species. (A) *Euphausia pacifica*; (B) *Cyphocaris challengeri* and *Parathemisto pacifica*; (C) *Holmesiella anomala*; (D) *Calanus pacificus* and *C. plumchrus*; (E) *Sagitta elegans*. Respiration rates are expressed as relative rates, taking the rate as 100% at 80–90% oxygen saturation. (After Ikeda, 1977a.)

water. In comparing his respiration measurements on *Euphausia pacifica* with those of earlier workers, Paranjape (1967) found that his results were considerably low, although he used oxygen-saturated water, and concluded that they reflected the low oxygen concentration in the natural habitat of the *E. pacifica* used in his experiment. For the best estimate of the respiration rate of zooplankton in the sea, it is important to take into account not only the oxygen concentration in the experimental containers but also in the water where the specimens live. In most respiration experiments with zooplankton from oxygen-saturated environments, a reduction of the oxygen concentration to no less than approximately 80% saturation by the end of the incubation will safely avoid the effects of low oxygen.

Illumination is a factor that may affect the respiration rate of zooplankton. The effect has been investigated for in some copepods and euphausiids (Marshall et al., 1935; Pearcy et al., 1969; Kils, 1979). Fernandez (1977) revealed that the respiration rate of several species of copepods increases in proportion to the amount of light above a certain threshold. This threshold was different in different species, depending on the illumination at the depth where they were captured.

In order to avoid oxygen production or consumption by other organisms, filtered seawater is commonly used for incubations. While this procedure provides a common basis of interspecific comparison of various zooplankton from the sea, about which little is known of their natural foods and nutritional history at the time of capture, respiration rate measured under nonfeeding of zooplankton may underestimate the rate in the field. According to Gaudy (1974), the respiration rates of three copepod species incubated with food algae increased in proportion to their ingestion rates. Conover and Lalli (1974) measured the respiration rate of *Clione limacina* in the presence of the prey (*Spiratella retroversa*) and found the rate to increase 2.0–3.5 times during feeding. Vidal (1980), however, failed to find any significant effect of the presence of food on the respiration rate of *Calanus pacificus*. As a different approach to this problem, Ikeda (1976) compared the respiration rates of several zooplankton species before and after a short feeding period (3–4 hr). Higher rates after feeding were observed in *Pleurobrachia pileus* and *Parathemisto pacifica*, but this was not the case in five copepod species. These contradictory results may be partly due to the specific differences in the food requirements of zooplankton. Clearly, the functional relationship between feeding activity and respiration rate of zooplankton need to be established for better application of the laboratory data to field populations of zooplankton.

In most respiration experiments the density of zooplankton in the experimental containers is kept higher than that in the field. Although crowding is necessary to cause a significant difference in the dissolved oxygen contents of the experimental and control containers, it may also affect the respiration rate (Marshall and Orr, 1958; Zeiss, 1963). Razouls (1972) measured respiration rates of *Temora stylifera* and *Centropages typicus* at densities from 0.1 ind·ml^{-1} to 2 ind·ml^{-1}. The highest respiration rates were obtained at 0.5 ind·ml^{-1} and these rates decreased at lower and higher densities in both species.

A diel rhythm of respiration, if it exists, is a critical problem, especially for short-term experiments. Although Pearcy et al. (1969) did not find such a rhythm for *Euphausia pacifica* in their laboratory experiments, positive evidence for a diel rhythm was obtained for some freshwater zooplankton (Duval and Geen, 1976); the diel rhythm in respiration rate was bimodal, with maximum values at dawn and dusk.

Molting in euphausiids is known to accelerate the respiration rate (Paranjape, 1967; Ikeda and Mitchell, 1982). Bulnheim (1972) made detailed measurements on respiration acceleration in the course of molting in benthic gammarid amphipods. The respiration rates increased 2.2–3.9 times during molting.

With regard to the comparison of the results obtained from different methods, Conover (1959) and Pearcy et al. (1969) observed no significant difference between the respiration rates of copepods and a euphausiid measured using the water bottle and manometric methods. According to Lawton and Richards (1970), measurement of the respiration rates of the larvae of the damselfly (*Pyrrhosoma nymphula*) by the Cartesian diver and water bottle methods and the rates of a snail (*Polamopyrgus jenkinsi*) by the Cartesian diver and manometric methods are essentially the same.

With respect to the several experimental factors mentioned above, inconsistent results between workers are present in most cases. These inconsistencies may result from the use of different zooplankton species. Effects caused by the interaction of two or more factors must also be studied. However, we still sometimes see considerable variation in the results of researchers for a given experiment with a single species. Further improvement and possible standardization of handling techniques should be sought. Experimental factors such as temperature and salinity are not mentioned here since these are adjusted to field conditions quite easily by using appropriate temperature control systems and water collected from the site of zooplankton collection, respectively. If the temperature changes before or during the experiment, interpretation of the results is often difficult since the mode and duration of temperature change affects the results (see Halcrow, 1963). Figure 8.9 shows the relationship between the weight-specific respiration rate and the body dry weight of marine zooplankton as a function of temperature. The relationship can be used to roughly estimate the respiration rate of zooplankton by knowing their body dry weight and temperature prior to the experiment, and the figures may help to determine the appropriate number of animals, size of incubation containers, and incubation period in conjunction with the analytical sensitivity of the dissolved oxygen technique adopted.

8.3. Conversion of Respiration Rate to Carbon and Calorific Units

The molar ratio of carbon dioxide produced to oxygen consumed as a result of respiration is called the *respiratory quotient* (RQ). The RQ varies from 0.7 to 1.0 depending on metabolic substrate (lipid, 0.71; protein, 0.80; carbohydrate, 1.0; from Prosser, 1973a). Recently, Gnaiger (1983) pointed out that the RQ value for lipid is 0.72 and for protein it changes as a fuction of the excretory product (i.e., 0.97 for ammonia and 0.84 for urea). Since marine zooplankton are considered to be primarily ammonotelic (see Section 8.4), RQ = 0.97 should be more appropriate to convert respiration data of marine zooplankton to carbon unit. The respiration rate (R) expressed as $\mu l\ O_2 \cdot animal^{-1} \cdot hr^{-1}$ can be converted to a carbon equivalent (C: $\mu g\ C \cdot animal^{-1} \cdot hr^{-1}$) by

$$C = R \times RQ \times \frac{12}{22.4} \qquad (8.10)$$

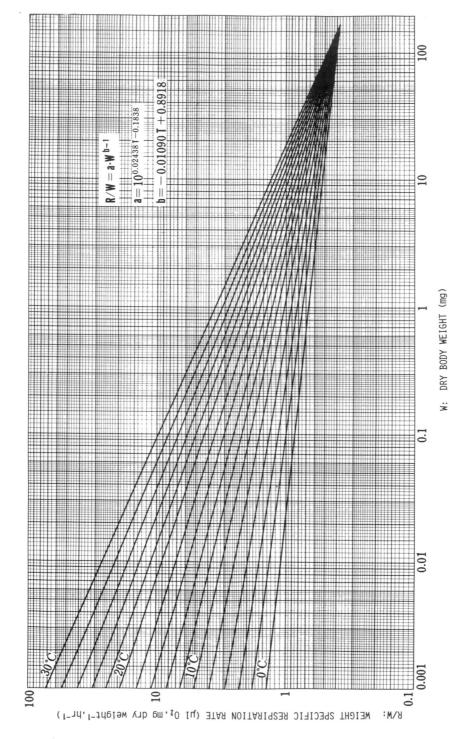

Figure 8.9. Relationship between respiration rate and body dry weight of marine zooplankton as a function of habitat temperature. (Modified after Ikeda, 1974.)

where 12/22.4 is the weight (12 g) of carbon in 1 mole (22.4 liters) of carbon dioxide. When the metabolic substrate of zooplankton is unknown, minimal and maximal C values can be calculated by using RQ values of 0.71 and 1.0, respectively. In nonfeeding zooplankton or zooplankton being starved, the major metabolic substrates are considered to be protein and lipid because the amount of carbohydrate is usually very small in zooplankton (see Table 5.1). Information on the protein and lipid contents of zooplankton is needed to determine which is the major metabolite. Protein-dominated metabolism is often identifiable from the low ratio ($<$ 24, by atoms) of the respiration rate to the ammonia excretion rate, the O/N ratio (see Section 8.6).

Respiration rate expressed in units of carbon represents the carbon requirement of zooplankton for metabolism. It can be used as an index of the minimum food requirement when digestion efficiency and growth are not taken into account.

The caloric equivalent to a unit volume of oxygen respired varies depending on the metabolic substrate utilized. According to Prosser (1973a), there are 5.0, 4.7, and 4.5 kcal liberated per 1-liter of O_2 respired for carbohydrate, lipid, and protein metabolism, respectively. Winberg (1971) proposed an average value of 4.86 for ecological studies. A detail conversion table between respiration rate and calories, joules, watts, and other biochemical units is provided by Gnaiger (1983).

8.4 Measurement of Excretion Rate

Although the excreta of zooplankton may include not only liquid forms but also solid forms, such as fecal pellets, we consider here only liquid forms. The excretion of zooplankton is usually measured as dissolved nitrogen (total N, amino-N, urea-N, and ammonia-N) and dissolved phosphorus (total P, organic-P, and inorganic-P) compounds. Among these, special interest has been paid to ammonia and inorganic phosphorus because of their importance as nutrients for phytoplankton (see Corner and Davies, 1971).

It is well documented that ammonia is the major form of dissolved nitrogen excreted by zooplankton.[3] For *Neomysis rayii* and *Euphausia pacifica*, ammonia accounted for 75–85% of the total nitrogen excreted, followed by 10–25% as amino acids and 1% as urea (Jawed, 1969). Corner and Newell (1967) reported that 60–100% of the total nitrogen excreted by *Calanus helgolandicus* is ammonia and some of the rest is urea. Corner et al. (1976) confirmed this result for the same species fed barnacle nauplii; only 9–10% of total nitrogen excreted was urea. Ammonia is also the dominant form of nitrogen excreted by the ctenophore *Mnemiosis leidyi* (Kremer, 1977). Mitamura and Saijo (1980) analyzed nitrogenous compounds in the excreta of *Calanus helgolandicus* and *Acartia pacifica* and reported that urea and ammonia contribute 7–8% and 38–44% of total nitrogen excreted, respectively.

Regarding dissolved phosphorus compounds in zooplankton excreta, inorganic and organic fractions have been separated (total phosphate = inorganic plus organic phosphates). Pomeroy et al. (1963) reported that 33–50% of the total phosphate excreted by mixed zooplankton is in organic forms. Later, Hargrave and Geen (1968) found a higher percentage in the organic fraction (up to 74% of the total phosphate excreted) for three copepod species. According to Butler et al. (1969, 1970), the percentage in the organic fraction of the total phosphate excreted by *Calanus helgolandicus* is the highest (70%) in spring when phytoplankton food is most abundant, but the inorganic fraction dominates during food impoverishment in winter. It should be noted that this seasonal change in the relative amounts of organic and inorganic phosphate was not seen for nitrogen excretion in which ammonia comprised a constant 88% of the total nitrogen excreted.

To measure the excretion rate of zooplankton, the water bottle method described in Section 8.1.1 can be used. If the simultaneous measurement of respiration is not required, complete exclusion of air in the bottles is not necessary. In the normal range of water temperature (0–30°C) and pH (7.0–8.5), the loss of ammonia from seawater into the air is negligible (see Bower and Bidwell, 1978). Compared with respiration measurements, however, extra caution is necessary for excretion measurements to avoid contamination from various sources. All glass bottles must be rinsed with $6N$ HCl. Careful and complete washing of test animals, as shown in Figure 8.1, is essential. In the case of small animals, or when washing is difficult, it is necessary to measure the initial concentration of nitrogen and phosphate by taking a sample of seawater at the beginning of the incubation of animals in the experimental bottle.

Analytical methods to determine dissolved nitrogen and phosphorus compounds can be found in Strickland and Parsons (1972). Here, we refer to the quantitative analysis of ammonia-N and phosphate-P (inorganic-P). The methods are spectrophotometric and the final extinction readings for samples from experimental and control bottles are converted to concentrations of each compound by use of the respective standard curves.

A. Ammonia-N

1. Reagents.

a.   Deionized water: Remove the ammonia from distilled water by passing it through a small column of cation exchange resin in the hydrogen form just before use and store the water in a tightly stoppered glass flask.

b.   Phenol solution: Dissolve 20 g of crystalline phenol (analytical reagent grade) in 200 ml of 95% (V/V) ethanol.

c.   Sodium nitroprusside solution: Dissolve 1.0 g of sodium nitroprusside $Na_2Fe(CN)_5NO\cdot2H_2O$ in 200 ml of deionized water. The solution is stored in a brown bottle. It is stable for about a month.

d.   Alkaline solution: Dissolve 100 g of trisodium citrate and 5 g of sodium hydroxide (analytical reagent grade) in 500 ml of deionized water. This solution is stable for a long period.

e.   Sodium hypochlorite solution: Use a solution of commercial hypochlorite (e.g., Chlorix) which should be at least $1.5N$.

f.   Oxidizing solution: Mix 100 ml of reagent d and 25 ml of reagent e. It is best to prepare this solution immediately before the analysis.

2.   Analytical Procedure.   Add 50 ml of the sample water to an Erlenmeyer flask with accompanying stopper and then add 2 ml, 2 ml, and 5 ml of solutions b, c, and f, respectively. Mix well after each addition. Restopper the flask in order to avoid contamination from ammonia in the air and allow to stand at a temperature between 20–27°C for 1 hr. Then measure the extinction at 640 nm relative to distilled water in a spectrophotometer using 10-cm cells. It is best to conduct the reaction in a constant-temperature water bath. The reaction requires a full 60 min for completion. During that time the samples should never be placed in direct sunlight or near a window. The detrimental effect of sunlight on the reaction has been pointed out often in recent years (Liddicoat et al., 1975).

3.   Calibration.   Standard solution and procedure: Dissolve 0.6607 g of ammonium sulfate (analytical reagent quality) in 1 liter of deionized water (1 ml $\equiv$ 10 µg-at N). Add 1 ml of chloroform and store in a dark place with stopper. At the time of use dilute the standard solution 100 times with distilled water to make a secondary solution and dilute this solution with filtered seawater containing as little ammonium as possible to make the standard solution. If seawater is added to 10 ml of the secondary solution to make 1 liter of the standard solution, the resulting ammonium concentration is equivalent to 1.0 µg-at $N\cdot liter^{-1}$ of ammonia-N (A serial dilution of the standard solution should be made once during the experiment for linearity check of extinctions against ammonia concentration. The range of extinctions should cover the expected range of readings of experimental values. When any reagents of this analysis are renewed, the linearity should be reexamined).

4.   Calculation.   If the average extinctions of filtered seawater and the standard solution are expressed by A_b and A_{b+1}, respectively, the concentration of ammonia-N in the sample water can be obtained from the following equation:

$$\mu\text{g-at N}\cdot\text{liter}^{-1} = \frac{A - r}{A_{b+1} - A_b} \tag{8.11}$$

where A is the extinction of the sample water and r is the value of a reagent blank. The reagent blank value is obtained exactly as described in the above analytical procedure but using deionized water. Blank extinctions using a 10-cm cell should not exceed 0.075.

B. Phosphate-P

   1. Reagents.
   a. Ammonium molybdate solution; Dissolve 15 g of ammonium paramolybdate (NH_4) $Mo_7O_{24} \cdot 4H_2O$ (analytical reagent grade, preferably fine crystals) in 500 ml of distilled water. It should be protected from direct sunlight and be preserved in a polyethylene bottle. This solution is stable indefinitely.
   b. Sulfuric acid solution: Add 140 ml of concentrated sulfuric acid (analytical reagent grade; specific gravity of 1.8) to 900 ml of distilled water. Allow the solution to cool and preserve in a glass bottle.
   c. Ascorbic acid solution: Dissolve 27 g of good quality ascorbic acid in 500 ml of distilled water. Store the solution frozen in a polyethylene bottle. Thaw for use and refreeze at once. This solution can be kept only about one week at room temperature but it is stable for several months if frozen.
   d. Potassium antimonyl-tartrate solution: Dissolve 0.34 g of good quality potassium antimonyl-tartrate $C_2H_2(OH)_2COOKCOO$ $(SbO) \cdot \frac{1}{2}H_2O$ in 250 ml of warm distilled water. The solution should be preserved either in a glass or polyethylene bottle. It is stable for several months.

   2. Analytical Procedure. Immediately before the analysis, the above solutions (a, b, c, and d) are mixed in the ratio of $2:5:2:1$ (V/V), respectively. Use this reagent for one-batch samples and discard any excess; it should not be kept for more than 6 hr. To 50 ml of sample water in a measuring cylinder or 100-ml flask, add the mixed reagent in the ratio of $1:10$ (V/V) and mix immediately. After 5 min or at most within 1–2 hr, measure the extinction of the solution relative to distilled water in a 10-cm cell at a wavelength of 885 nm.
   3. Calibration. Standard solution and procedure: Dissolve 0.816 g of anhydrous potassium dihydrogen phosphate KH_2PO_4 in 1 liter of distilled water (1 ml $\equiv$ 1.0 µg-at P) and store in a dark bottle with 1 ml of chloroform. At the time of use dilute the standard solution with distilled water or sodium chloride solution, close to the chlorinity of seawater, in the ratio of $1:10$ to make substandard liquid a (1 ml $\equiv$ 0.10 µg-at P) and in the ratio of $1:40$ to make substandard liquid b (1 ml $\equiv$ 0.25 µg-at P). Using these substandard solutions, make serial dilutions (Table 8.2) and carry out the phosphate determination exactly as described in the above analytical procedure. Obtain mean values of the extinction for each of

Table 8.2.   Preparation of standard solutions of phosphate-P

Volume of two substandard solutions (ml)		Concentration of standard solution when diluted to 50 ml (μg-at·liter^{-1})
a	b	
—	0.40	0.2
—	0.80	0.4
—	1.20	0.6
—	1.60	0.8
—	2.00	1.0
0.75	—	1.5
1.00	—	2.0
1.25	—	2.5
1.50	—	3.0
1.75	—	3.5
2.00	—	4.0

different concentrations and subtract the mean extinction of the blanks from these values; draw a standard curve.

Depending on the size of the animals, a reduction of the volume of the sample water may be required. We use 10 ml of sample water and glass test tubes instead of 50 ml for the determinations of ammonia and inorganic phosphate excreted by macrozooplankton. The amounts of reagents added to the sample water are reduced to one-fifth, accordingly. As the storage of sample water for these analyses is always associated with uncertain errors (see Gilmartin, 1967; Degobbis, 1973), immediate analyses are recommended. A simple method of ammonia determination with an ammonia electrode has come into use recently. However, the general sensitivity of this electrode is around 0.1 mg NH_3-N·liter^{-1} (Barica, 1973; Merks, 1975), which is still too low for zooplankton excretion measurements, especially for oceanic species.

The excretion rate is calculated from the difference in the concentration of the compounds between experimental and control bottles divided by the number of animals (or total dry weight) and incubation time, analogous to Equation 8.4. Excretion rates thus obtained are (1) total excretion rate = μg N or P·animal^{-1}·hr^{-1} or (2) weight specific excretion rate = μg N or P·mg dry weight^{-1}·hr^{-1}. μg N or P can be converted to μg-at N or P by dividing by 14 or 31, respectively.

8.5   Experimental Factors Affecting the Excretion Rate

Ideally, excretion experiments should satisfy the following conditions:

1.   Incubation should be long enough to show significant differences in the concentration of excreta between experimental and control bottles.

2. The effect of the starvation of zooplankton during incubation should be minimal.

3. The accumulation of organic wastes in the experimental bottles should not affect the excretory activity of the zooplankton.

Although the effects of accumulated excreta on the activity of zooplankton is little known, it may be significant for ammonia.[4]

The results of ammonia-N and inorganic phosphate excretion measurements on *Calanus pacificus* for various incubation times indicate that both excretion rates are rather stable for incubation times up to 4 hr (Ikeda, unpublished). The ratio of ammonia excretion to phosphate excretion also remained constant during this period. A stable ammonia excretion rate for *Temora stylifera* over 1–7 hr was reported by Nival et al. (1974), but a progressive decline in the phosphate excretion rate of *Acartia tonsa* incubated for 4–24 hr was observed by Hargrave and Geen (1968). The timing and length of the incubation period in excretion measurements should be determined in relation to a possible diel rhythm in the excretion rate of the zooplankton. According to Hargrave and Geen (1968), *Acartia tonsa* excretes phosphate at higher rate at night, when it is actively feeding.

There is little information on the effects of density of animals in the experimental bottles, but the effects of accumulation of excreta are apparent. In *Acartia tonsa* the phosphate excretion rate decreased when the density of individuals exceeded 400 ind·liter^{-1} (Hargrave and Geen, 1968). In contrast, Nival et al. (1974) obtained abnormally high ammonia excretion rates for *Temora stylifera* incubated at densities greater than 200 ind·liter^{-1}.

Hargrave and Geen (1968) incubated several zooplankton species separately, with and without antibiotics (penicillin and streptomycin sulfate), to measure phosphate (inorganic and organic) excretion. In all species the excretion rates were lower for specimens incubated without antibiotics, thus showing bacterial uptake of phosphate during incubation. Jawed (1969) maintained *Neomysis rayii* and *Euphausia pacifica* with antibiotics prior to nitrogen excretion measurements in autoclaved seawater and bottles. The nitrogen excretion rates of these treated specimens were not significantly different than the rates of untreated specimens. As mentioned previously, the use of antibiotics requires caution concerning its appropriate dosage, and more importantly, some antibiotics seem to interfere in the chemical analyses of zooplankton excreta. In most previous excretion experiments with zooplankton, no antibiotics have been used.

The use of unfiltered seawater should be avoided unless a correction for the simultaneous uptake by phytoplankton of zooplankton excreta can be made. The magnitude of error if this correction is not made will vary, depending on the concentrations of phytoplankton and excreta. Figure 8.10 shows the results of excretion experiments in which *Metridia pacifica* was incubated in seawater containing various amounts of phytoplankton

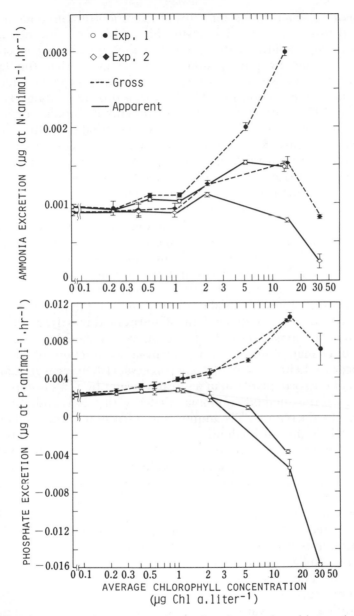

Figure 8.10. Apparent and gross rates of excretion of ammonia and inorganic phosphate by *Metridia pacifica* incubated in bottles (900 ml) with different concentrations of phytoplankton (*Skeletonema costatum*) for 24 hr at 8°C. About 100 specimens of *M. pacifica* were used in each bottle. Concentration of phytoplankton is expressed as chlorophyll *a*. Gross excretion rates are calculated by correcting for the uptake by phytoplankton of ammonia and inorganic phosphate during the incubation period. (After Takahashi and Ikeda, 1975.)

(*Skeletonema costatum*). Apparent rates of excretion of ammonia and phosphate tend to increase with increasing phytoplankton concentration, but both rates drop rapidly at the maximum phytoplankton concentration. Phosphate uptake by phytoplankton was so great that the apparent excretion rate of phosphate by zooplankton became negative. In order to estimate the real excretion rate of *M. pacifica* during feeding, Takahashi and Ikeda (1975) corrected for the phytoplankton uptake of ammonia and phosphate using kinetic relationships. Excretion rates thus corrected are shown in Figure 8.10 as gross excretion rates. The increase in the gross excretion rate in relation to phytoplankton concentration reflects a proportional increase of zooplankton feeding on phytoplankton. In a similar study Lehman (1980) estimated ammonia and phosphate excretion rates of *Daphnia pulex* feeding on *Chlamydomonas reinhardtii*. To correct for nutrient uptake by *Chlamydomonas*, he added both nutrients to the incubation medium of *Daphnia* so as to achieve a constant nutrient uptake by *Chlamydomonas*. [Note: the pattern of nutrient uptake by phytoplankton is known to follow Michaelis–Menten kinetics when only nutrients limit phytoplankton metabolism. If nutrient concentrations are adjusted to saturating levels, rates of uptake of nutrients by phytoplankton will be relatively constant and independent of nutrient fluxes from the animals during short periods. Rates of regeneration can then be assessed from net rates of change of nutrient concentrations as functions of zooplankton abundance (see Lehman, 1980)]. Caperon et al. (1979) measured ammonia excretion of microzooplankton in natural seawater by a ^{15}N isotope dilution technique. In this method ammonia labeled with ^{15}N is added to natural seawater. The decrease in ^{15}N ammonia as a fraction of the total ammonia (^{15}N + ^{14}N) is due to its dilution by microzooplankton excretion. This method is an alternative approach to the analysis of zooplankton–phytoplankton interactions in zooplankton excretion studies.

The use of filtered seawater instead of unfiltered seawater eliminates the complication of phytoplankton uptake of zooplankton excreta, but the results may underestimate the excretion rate of zooplankton in the field where they feed. Although measurements of excretion of herbivorous zooplankton in the presence of phytoplankton require rather complicated corrections, similar measurements are not so difficult for carnivorous zooplankton. The correction can be made by preparing extra control bottles to measure the excretion by prey animals (Corner et al., 1976).

In addition to the various nitrogen and phosphorus compounds in zooplankton excreta, the rapid release of phosphate from injured and dead specimens often becomes a great source of error in excretion measurements. For example, respiration and excretion were measured simultaneously for *Euphausia pacifica* whose physiological condition were altered artificially by adding $CuSO_4$ to the seawater (Table 8.3). An increase in phosphate excretion was significant for moribund and dead specimens. Ikeda et al. (1982) compared the phosphate excretion rates of

Table 8.3. Simultaneous measurements of respiration and excretion of *Euphausia pacifica* whose activity was altered by addition of $CuSO_4$ to the seawater medium[a]

Cu added (ppb)	Activity at end of experiment	Respiration rate (μl $O_2 \cdot ind^{-1} \cdot hr^{-1}$)	Ammonia excretion rate (μg NH_3-N $\cdot ind^{-1} \cdot hr^{-1}$)	Phosphate excretion rate (μg PO_4-P $\cdot ind^{-1} \cdot hr^{-1}$)	O/N	N/P	O/P
0	Active swimming	6.70	0.407	0.051	20.6	17.6	364
0		5.79	0.272	0.037	26.5	16.1	427
76	Moribund	6.26	0.411	0.132	19.1	6.9	132
76		5.70	0.299	0.159	23.8	4.1	99
305	Dead	5.26^b	0.419^b	0.325^b	15.7	2.8	44
305		5.39^b	0.431^b	0.350^b	15.6	2.7	42

[a] Two specimens of *E. pacifica* were kept in each 300-ml bottle for 24 hr at 13°C (Ikeda, unpublished).
[b] Assuming the specimens were alive during experiment.

artificially injured and uninjured tropical copepods and found the former rates elevated by as much as one order of magnitude. In order to eliminate the possible damage of specimens during sorting, phosphate excretion rates are sometimes measured on unsorted mixed zooplankton. However, Mullin et al. (1975a) pointed out that the phosphate excretion rate of mixed zooplankton always overestimates the sum of excretion rates for healthy individuals. Death and injury of specimens are not only caused during sampling but also by possible predation or cannibalism during the incubation of mixed species.

Recently, Ikeda and Mitchell (1982) observed an increase in phosphate excretion by *Euphausia superba* which had molted during the incubation, but this was not the case for ammonia. In our experience the results of phosphate excretion measurements are much more variable than those of ammonia excretion.

8.6 Relationship Between Respiration and Excretion: O/N, N/P, and O/P Ratios

From simultaneous measurements of respiration and excretion, the ratios of respiration to ammonia excretion (O/N), ammonia excretion to phosphate excretion (N/P), and respiration to phosphate excretion (O/P) are calculated. Let the rates of respiration, ammonia excretion, and phosphate excretion measured simultaneously be a μl $O_2 \cdot$animal$^{-1} \cdot$hr^{-1}, b μg N$\cdot$animal$^{-1} \cdot$hr^{-1}, and c μg P$\cdot$animal$^{-1} \cdot$hr^{-1}, respectively, the ratios are calculated on an atomic basis in the following manner:

$$\text{O/N} \quad \frac{a/11.2}{b/14} = \frac{a}{b} \times 1.25$$

$$\text{N/P} \quad \frac{b/14}{c/31} = \frac{b}{c} \times 2.21$$

and

$$\text{O/P} \quad \frac{a/11.2}{c/31} = \frac{a}{c} \times 2.77$$

2N/O is called the ammonia quotient[5] and is often used in fish physiology (e.g., Kutty, 1978). These ratios change depending on the metabolic substrates of the organism. The nitrogen and phosphorus contents of animal protein, carbohydrate, and lipid, together with the amount of oxygen[6] required for complete combustion of each type of organic matter, are summarized in Table 8.4. The O/N, N/P, and O/P ratios for each class of organic matter are also calculated. It must be noted that these ratios will change with chemical variations, particularly for protein and lipid, and no precise information is available for zooplankton material. Even so, remarkable variations in these ratios are seen among protein, carbohydrate,

Table 8.4.   **Average nitrogen and phosphorus composition of organic matter and the oxygen required to oxydize each class of organic matter in an animal body**[a]

	Carbohydrate	Lipid	Protein
N (g/g)[b]	0	0.0061	0.178
P (g/g)[b]	0	0.0213	0.007
O (liter/g)[c]	0.82	2.03	0.97
O/N	∞	415	6.8
N/P	—	0.63	56
O/P	∞	263	383

[a] Ratios are by atoms.
[b] From Rogers (1927).
[c] From Prosser (1973a).

and lipid. It is apparent from Table 8.4 that a carbohydrate-dominated metabolism causes high O/N and O/P ratios. Low N/P and O/N ratios are characteristic of lipid- and protein-dominated metabolisms, respectively.

Raymont and Krishnaswamy (1960) and Raymont and Conover (1961) observed little change in the carbohydrate content of starved zooplankton. In conjunction with these observations, the carbohydrate content of zooplankton was found to be only a few percent of the dry weight (5% at most). Even if all the carbohydrate in a zooplankton were metabolized, it would not be large enough to support the organism's respiration for 24 hr. Therefore, protein and lipid are considered the major metabolic substrates of starved zooplankton.[7] Ikeda (1974) calculated an O/N ratio of 24 when protein and lipid are metabolized in equal quantities at the same time; hence, an O/N ratio less than 24 indicates protein-dominated metabolism and a ratio greater than 24 indicates lipid-dominated metabolism. This method was used to compare the metabolic substrates of zooplankton inhabiting various regions (Fig. 8.11). Zooplankton in tropical, subtropical, and temperate seas are characterized by protein metabolism, while those from boreal seas showed a wide range in the O/N ratio. According to Conover and Corner (1968), the O/N ratio of some boreal zooplankton is high at the end of summer but declines throughout winter, with a corresponding decrease in lipid. During the spring phytoplankton bloom, zooplankton feed actively on phytoplankton and deposit lipid in their bodies, but the O/N ratio remains low during this period. The O/N ratio increases only after the zooplankton have deposited a large amount of lipid. Thus, the O/N ratio is closely related to the lipid content of zooplankton.

The N/P and O/P ratios are not so sensitive as the O/N ratio for assessing metabolic substrates of zooplankton, and in earlier studies the major

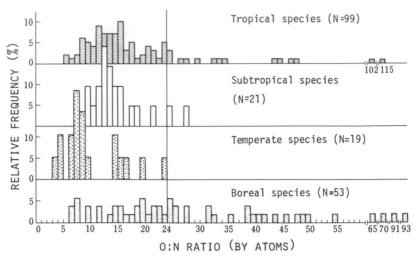

Figure 8.11. Relative frequency (%) of the O:N ratio by atoms, calculated from respiration and ammonia-nitrogen excretion, for tropical (46 species), subtropical (13 species), temperate (7 species), and boreal (19 species) zooplankton (After Ikeda, 1974.)

interest centered around the constancy of N/P and O/P values. From a seasonal study of total nitrogen and total phosphorus excretion by *Calanus finmarchicus* and *C. helgolandicus*, Butler et al. (1970) found that the atomic ratio of N/P is quite stable through spring (11.0) and winter (14.6). Satomi and Pomeroy (1965) compared the O/P ratio for excretion by mixed zooplankton from three different sites and obtained quite similar values (range; 63–75; mean, 72, by atoms). If the N/P and O/P ratios are constant, estimates of respiration, nitrogen excretion, and phosphorus excretion can be made by knowing only one of these variables. Table 8.5 summarizes ratios reported by various workers. There is large variation in these ratios and the N/P and O/P ratios are not constant. Alternatively, N/P and O/P ratios can be used as indicators of feeding history and magnitude of physical damage of test zooplankton, as phosphate excretion slow down more rapidly than respiration and ammonia excretion in food-depleted zooplankton (Ikeda, 1977b; Ikeda and Skjoldal, 1980), and abnormally high phosphate liberation is a characteristic of damaged zooplankton (Mullin et al., 1975a; Ikeda et al., 1982).

Redfield et al. (1963) proposed the average C:N:P composition of marine phytoplankton and zooplankton is 106:16:1 (by atoms). Richards (1965) constructed a model of the organic composition of phytoplankton and zooplankton based on Redfield's ratio and a sequence of decomposition:

$$(CH_2O)_{106}(NH_3)_{16}H_3PO_4 + 106O_2 = 106CO_2 + 16NH_3$$
$$+ H_3PO_4 + 106H_2O$$

$$16NH_3 + 32O_2 = 16HNO_3 + 16H_2O$$

Table 8.5. Oxygen:nitrogen (O/N), nitrogen:phosphorus (N/P), and oxygen:phosphorus (O/P) ratios from measurements of respiration, ammonia–nitrogen excretion and inorganic phosphate–phosphorus excretion rates reported by previous workers[a]

Zooplankton	O/N	N/P	O/P	Source
Mixed zooplankton (UFW)	7.7	7.0	54	Harris (1959)
Mixed zooplankton (UFW)	41	9.98	222	Martin (1968)
Calanus cristatus (UFW)	5.7	19	110	Taguchi and Ishii (1972)
Calanus plumchrus (UFW)	6.8	13	89	Taguchi and Ishii (1972)
Mixed zooplankton (UFW)	13.48	10.33	142.4	Le Borgne (1973)
Sagitta hispida (FW)	—	11.3	—	Beers (1964)
Mixed zooplankton (UFW)	—	—	72	Satomi and Pomeroy (1965)
Calanus helgolandicus (FW)	9.8–15.6[b]	—	—	Corner et al. (1965)
Boreal zooplankton (10 species) (UFW)	6–200[b]	—	—	Conover and Corner (1968)
Calanus finmarchicus (FW)	—	10.8	—	Butler et al. (1969)
Calanus finmarchicus (FW)	—	11.0[c]	—	Butler et al. (1970)
Sagitta hispida (FW)	6.8	—	—	Reeve et al. (1970)
Calanus helgolandicus (FW)	—	16.5[c]	—	Corner et al. (1972)
Temora stylifera (FW)	7–15[d]	—	—	Nival et al. (1974)
Phronima sedentalia	4.27	—	—	Mayzaud (1973)
Meganyctiphanes norvegica (FW)	4.77–12.13	—	—	Mayzaud (1973)
Acartia clausi (FW)	1.61	—	—	Mayzaud (1973)
Sagitta setosa (FW)	1.75	—	—	Mayzaud (1973)
Boreal, temperate, subtropical and tropical zooplankton (81 species) (FW)	4–115	—	—	Ikeda (1974)
Mixed zooplankton (UFW)	—	6.8	—	Mullin et al. (1975a)

[a] Use of unfiltered (UFW) and filtered (FW) seawater for measurement is noted. Ratios are by atoms. — means no data (after Ikeda, 1977b).
[b] Ninhydrin N.
[c] Total N/total P.
[d] Taken from Nival et al. (1974, Fig. 10).

Hence

$$(CH_2O)_{106}(NH_3)_{16}H_3PO_4 + 138O_2 = 106CO_2 + 122H_2O$$
$$+ 16HNO_3 + H_3PO_4$$

From this scheme, the O/N, N/P, and O/P ratios are predicted to be 17, 16, and 276, respectively. When biological oxidation by zooplankton is considered, the nitrogenous end product is not HNO_3 but NH_3. Thus, the appropriate ratios would be 13, 16, and 212. Application of these ratios to individual zooplankton species is cautioned, since a large departure of the $C:N:P$ ratio from that of Redfield has been noted (see Corner and Davies, 1971, and Chapter 5, Section 5.5.2).

8.7 Further Remarks on Handling Techniques for Zooplankton Respiration and Excretion Experiments

It has been pointed out that the rates of respiration and excretion are high in freshly caught zooplankton and decrease with time in laboratory conditions (Marshall et al., 1935; Ikeda, 1974; Roger, 1978), although in some observations a peak respiration rate is recorded not just after capture but sometime later (4–5 hr) (Roff, 1973; Holeton, 1974). Two possible explanations can be offered for the phenomenon: (1) higher rates for freshly caught specimens are due to the increased activity of specimens during sampling and (2) lowered rates after prolonged laboratory maintenance of specimens reflect laboratory effects, such as depletion of food or overcrowding. The former indicates that the lower rates of specimens maintained over long periods would better estimate the rates under natural conditions, while the latter explanation indicates that the higher rates of freshly caught specimens are closer to the natural activity of the zooplankton. In vertebrates initial high respiration rates are known to be abnormal (Holeton, 1974) and this has been verified for some fishes, which have been shown to have high concentration of lactic acid in the blood (see Love, 1970). Skjoldal and Båmstedt (1977) compared ATP, ADP, AMP, and adenylate energy charge (EC)[8] values in zooplankton (*Euchaeta norvegica* and *Meganyctiphanes norvegica*) immediately after capture and subsequently after maintenance in the laboratory. EC values were low for freshly caught specimens but increased to normal levels during subsequent maintenance, indicating severe capture stress.

Figure 8.12 shows the results of simultaneous measurements of respiration and excretion rates, together with biochemical components, including EC, for *Acartia australis* captured and maintained in natural seawater for 2 days. A rapid decrease was observed in respiration and excretion rates at the start of the experiment, followed by moderate changes thereafter. Ammonia excretion increased after the initial decrease. The changes in these rates are moderated when the rates are expressed on a

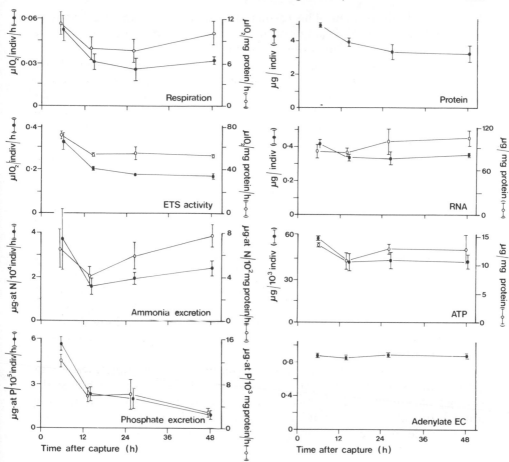

Figure 8.12. Changes with time after capture in respiration rate, ETS activity, ammonia and phosphate excretion rates, protein, RNA and ATP contents, and adenylate energy charge (EC) for *Acartia australis*. Filled circles, left ordinates, show rates or contents per individuals, excepting adenylate EC which is a quantity with no dimension. Open circles, right ordinates, show rates or contents on mg protein basis. Vertical bars, standard deviation. *Acartia* was maintained in unfiltered natural seawater at 24–25°C for 2 days. For the measurements of respiration and excretion, 120–200 specimens were transferred from the aquarium into glass bottles (capacity, 270 ml) filled with filtered seawater. Experiments lasted for 3.5–4.5 hr. (After Ikeda and Skjoldal, 1980.)

per mg of protein basis. Despite these changes in respiration and excretion rates, the EC value was stable (0.85–0.89). From these results we can infer that changes in respiration and excretion were not due to capture stress but to food shortage during maintenance of these specimens (see Ikeda and Skjoldal, 1980). This inference is substantiated by the progressive decrease in body protein. The results also indicate that the ETS activity and the RNA and ATP contents in zooplankton are easily altered by prolonged maintenance of zooplankton after capture.

Since *Acartia australis* is a tropical neritic species with a small body weight (9 μg dry wt·ind^{-1}), its weight-specific metabolic rates are considerably higher than those rates for other zooplankton species living in cold and/or deep waters. The acute effects of food depletion may be typical among tropical and subtropical coastal zooplankton in which body size is small and life span is relatively short. There are many examples showing decreases in the respiration and excretion rates of various zooplankton species after prolonged starvation in the laboratory (Conover and Corner, 1968; Omori, 1970; Ikeda, 1974, 1977b; Mayzaud, 1976). When animals are kept for a long time before an experiment, appropriate food must be provided to minimize the reduction of metabolic rates. When several boreal zooplankton species were fed, their respiration rates were stable, but the excretion rates of ammonia and phosphate changed in an irregular manner for unknown reasons (Ikeda, 1977b). One obvious problem in providing food is the lack of information about appropriate food for various zooplankton species. To reduce these problems, experiments should be started with freshly caught specimens without a long delay. In this case great care must be taken in sampling and handling specimens so as not to cause undue stress to the animals.

For measurement of respiration and ammonia excretion rates, Biggs (1977) enclosed gelantinous zooplankton directly in glass jars while scuba diving. This method is advantageous for it reduces the risk of damaging animals by net collection, but its application is limited to large zooplankton recognizable by divers in phytoplankton-poor, oligotrophic oceans.

Notes

[1] Manufactured by Metrohm AG. (Switzerland).

[2] For the use of oxygen bottles smaller than 300 ml, the amounts of reagents may be reduced proportionally.

[3] Although Webb and Johannes (1967) reported abnormally high excretion of dissolved free amino acids by zooplankton, this is considered to be an artifact caused by an extremely high density of zooplankton in their experimental containers (i.e., dissolved free amino acids leaked from the bodies of injured specimens). Webb and Johannes (1969), however, suggested that the contradicting results of other workers were due to the bacterial uptake of amino acids that occurred during the incubations in their studies.

[4] Ammonia is present in two forms in aqueous solution, i.e. un-ionized (NH_3) and ionized (NH_4^+). The equilibrium equation of these two forms is

$$NH_4^+ + H_2O \rightleftarrows NH_3 + H_3O^+$$

Un-ionized ammonia is known to be very toxic to aquatic animals, while ionized ammonia is less toxic. Zooplankton excrete NH_3. The relative proportions of NH_3

and NH_4^+ in solution depend mainly on temperature and pH. Salinity affects the relative proportions to a lesser extent. The proportion of NH_3 increases with increasing temperature and pH and decreases with increasing salinity. Bower and Bidwell (1978) tabulated the relative fraction of NH_3 in total ammonia ($NH_3 + NH_4^+$) at various temperatures, pH's, and salinities. In seawater at temperature 0–25°C, pH 7.5–8.5, and salinity 18–40‰, the percent of total ammonia as NH_3 is 0.2–14.0%.

[5] Ammonia quotient (AQ) is the molar ratio of ammonia excreted to oxygen consumed:

$$AQ = \frac{b/14}{a/(2 \times 11.4)} = \frac{b}{a} \times 1.60 \quad \left(= \frac{2}{O/N} \right)$$

[6] The values of oxygen (liter·g^{-1}) cited from Prosser (1973a) in Table 8.4 are similar to those calculated recently by Gnaiger (1983) (0.82 for carbohydrate, 2.01 for lipid, and 1.02 for protein, respectively).

[7] In phytoplankton carbohydrate comprises 20–42% of the total organic matter (Parsons and Takahashi, 1973). This suggests that carbohydrate may be a chief metabolic substrate of herbivorous zooplankton when they are feeding. To examine this hypothesis, simultaneous measurements of respiration and excretion should be done. However, such measurements have not yet been made because of the difficulty in accurately correcting for phytoplankton respiration and its uptake of zooplankton excreta.

[8] Adenylate energy charge (Atkinson, 1971) is calculated as

$$EC = \frac{(ATP) + \frac{1}{2}(ADP)}{(ATP) + (ADP) + (AMP)}$$

EC is a measure of the balance between energy-yielding and energy-requiring metabolic processes. EC ranges between 0 and 1. Under normal physiological conditions, the EC of most organisms is stable within the range 0.8–0.9 (Chapman et al., 1971; Niven et al., 1977).

CHAPTER 9 _____

Productivity

_____

9.1 Production and Terminology

Although many terms have been used to describe production, it is a process whose definition is not well agreed upon. We will now review terms most commonly used to describe the production of zooplankton.

The lives of individual zooplankton begin with eggs, larval release from females, or cell fission (in protozoans). Individuals that grow successfully to the adult stage die after producing the next generation. In general, the majority of zooplankton die without completing their life cycle. The production of a population over one generation is the sum of the body masses of individual animals at the time of their death.

The amount of food ingested (F) less egestion (E) is the gross assimilation of food (A_g). The total amount of organic matter assimilated by animals during a certain period of time is called gross production (P_g). Animals lose a great deal of assimilated organic matter to metabolism. Thus net production (P_n) is obtained by subtracting the metabolic expenditure (R) from A_g:

$$A_g = F - E \tag{9.1}$$

$$P_n = A_g - (R + U) \tag{9.2}$$

where U is the excretion of soluble organic matter.[1] For most aquatic animals U is considered to be very small relative to R. The term P_n is used for both individual growth (G) and reproduction. In a growing population total loss (L) is the sum of physiological mortality (L_m), predation (L_p), and other losses such as ecdysis (L_{im}) in crustaceans. The difference between P_n and L is the net increase of the population (I_n):

$$L = L_m + L_p + L_{im} \tag{9.3}$$

$$I_n = P_n - L \tag{9.4}$$

On an individual basis (P_g/F) × 100% is called the *assimilation efficiency*; (P_n/F) × 100%, the gross growth efficiency (K_1); and (P_n/P_g) × 100%, the net growth efficiency (K_2). Trophically, the ratio of net production of predators to that of prey organisms (P_{nf}), i.e., (P_n/P_{nf}) × 100%, is termed *ecological transfer efficiency.*

Production of zooplankton is usually expressed as P_n, but there are cases where P_g is used instead. Production in the following refers to P_n. The production (P) of a zooplankton population during a certain period of time (t) is often easiest to estimate with the following equation:

$$P = L + (B_t - B_o) \qquad (9.5)$$

where $B_t - B_0$ indicates the difference in biomass at the start and finish of period (t) and L is the mortality within that period. Production is commonly expressed as biomass per unit surface area (m^2) and unit time. With regard to time, it is considered that one year (annual production) is appropriate when the production is highly seasonal, as in high latitudes, while one day or one month (daily or monthly production) can be used when the production is relatively stable throughout a year, as in low-latitude waters. Biomass and production may be expressed as wet weight, dry weight, ash-free dry weight, and caloric content; in recent years, however, they are commonly expressed as carbon.

9.2　Life History Patterns

Life histories of zooplankton vary from one group to the other. The typical life history of Rotifera consists of parthenogenetic reproduction in favorable circumstances and sexual reproduction in conjunction with, or in anticipation of, unfavorable conditions. Parthenogenetic females often appear in the spring with the hatching of resting eggs, although some populations overwinter. Several generations are produced in the summer by diploid female parthenogenesis as each female produces a series of eggs with soft shells which develop without fertilization. After some period of time, a few females morphologically indistinguishable from those which are parthenogenetic appear and produce small haploid eggs by ordinary meiosis. Males develop from these small, unfertilized eggs. The fertilization of a haploid egg by a male produces the large resting egg capable of undergoing a prolonged diapause and carrying the population through overwintering periods.

The life history of Cladocera consists of sexual and parthenogenetic generations. A typical cycle commences in the spring with overwintering females or from resting eggs which develop into mature females. Summer eggs are produced parthenogenetically and develop rapidly and directly into miniature versions of the parent. After a variable number of instars, but typically five or six, these individuals mature into reproductive females, thus completing the parthenogenetic stage of the cycle. A substantial number of such asexual generations are produced through the spring and summer. Then, in response to various environmental stimuli, eggs are laid which develop into males while other eggs develop into sexually reproducing females. Sexual reproduction results in the production of resting eggs. In the waters around Japan where environmental variation is regular, the alternation of generations is normally observed once a year, and resting eggs appear at the end of autumn (Onbé, 1972). Although this reproductive pattern is analogous to a rotifer's life history, the underlying cytological and development mechanisms differ substantially, testifying to its dependent derivation.

Life cycles of Copepoda differ from those of the Rotifera and Cladocera in their obligate sexuality. Fertilized eggs hatch into a larval stage, the nauplius. Six naupliar stages and six copepodite stages are recognized, of which the last is the adult. Each stage is easily distinguished by morphological characteristics. Mating follows sexual maturity and the female may store sperm in a spermathecal sac, thus minimizing the need for subsequent mating. Spawning is divided into two types, the release of eggs freely into the water (e.g., *Calanus* and *Eucalanus*) or their envelopment in single (e.g., *Euchaeta* and *Corycaeus*) or paired (e.g., *Oithona* and *Oncaea*) egg sacs attached to the sides of the genital segment until hatching. Eggs develop into males or females, completing the life cycle. Generally, males tend to mature faster than females. Spawning of resting eggs is frequently seen in freshwater and neritic copepods (Kasahara et al., 1974; Uye and Kasahara, 1978; Grice and Marcus, 1981), but they are dioeceous.

Various rules have been formulated with regard to the development of marine copepods. One of these is isochronal development which is characterized, in principle, by the equal duration of each intermolt phase of the life cycle (Miller et al., 1977). This molting pattern was originally demonstrated in the genus *Acartia*, but McLaren (1978) suggested that it might apply equally well to all neritic copepods. This rule is in contrast to the development rule for oceanic copepods for which naupliar instars are generally of shorter duration than copepodite instars. However, according to Landry (1983), results of his study on rearing various neritic copepods do not support a general tendency toward isochronal development; no development is strictly isochronal. He notes that the assumption of isochronality in population dynamics studies can severely bias stage-specific mortality rate estimates.

Amphipoda and Mysidacea have brood pouches in which young larvae develop before they emerge. Euphausiacea and Decapoda are divided into those that shed free eggs (e.g., *Euphausia* and *Sergestes*) and those that attach eggs to the posterior pairs of thoracic legs (thoracopods) by membranes (e.g., *Nematoscelis*) or to pleopods by specialized setae (e.g., *Acanthephyra*). Eggs hatch into nauplii or more advanced stages and develop into adults after a series of molts. Typically the number of larval stages is constant.

Opisthobranchia, Chaetognatha, and Tunicata are mostly hermaphroditic, but the former appear to shed fertilized eggs into the water after

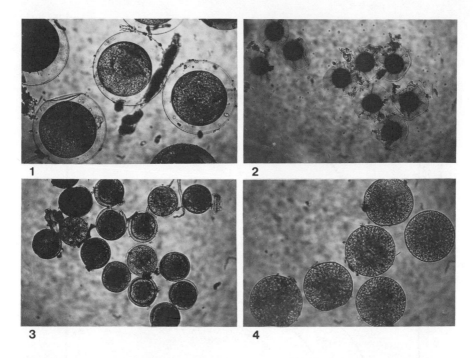

Figure 9.1. Eggs of representative zooplankton. (1) *Euphausia pacifica* (diameter, about 0.45 mm); (2) *Eucalanus bungii bungii* (diameter, about 0.15 mm); (3) *Tortanus discaudatus* (diameter, about 0.10 mm); (4) *Sagitta elegans* (diameter, about 0.3 mm).

copulation with other individuals. Several kinds of eggs of marine zooplankton are shown for reference in Figure 9.1.

Growth and reproduction of *Sergia lucens* in Suruga Bay and *Acetes japonicus* in Ariake Sea are shown in Figure 9.2. By comparison with seasonal changes in water temperature at the spawning ground of each shrimp, spawning clearly occurs during the time when the ambient temperature is increasing. Life history studies of several zooplankton species in temperate waters around Japan suggest that there are two typical growth patterns represented by these two species. One type (*Sergia*) is characterized by a life span of one year or more. It reproduces during the warm period of the year and larvae grow, but the growth rate decreases in the cold period. The juveniles grow again with increasing, ambient temperature and reach sexual maturity. Many oceanic plankton exhibit the similar life history pattern. The other type (*Acetes*) is characterized by a life span of only a few weeks or months. Many neritic plankton are included in this type. Eggs hatch during the warm period and the larvae grow rapidly to maturity and reproduce. Sequential generations with cohorts are produced in the summer and autumn before entering winter, when the population size decreases. Late generations either survive the

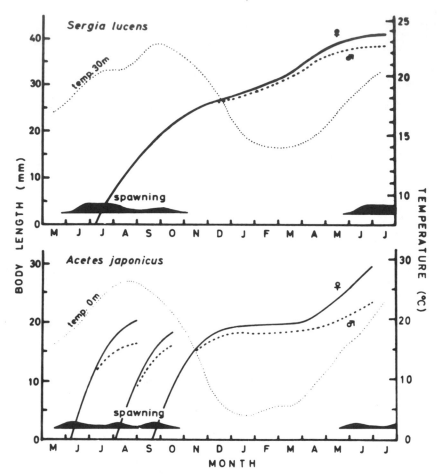

Figure 9.2. Growth trends of *Sergia lucens* in Suruga Bay and *Acetes japonicus* in the Ariake Sea and variation of the environmental temperature. (After Omori, 1974.)

cold months without producing eggs, die after releasing resting eggs, or possibly hibernate until the next spring. In the spring and summer overwintering individuals grow again and reproduce, while the resting eggs hatch and reconstitute the population. Generally, overwintering populations live much longer than summer populations. Individuals of the former populations are larger in size and produce more eggs than those of the latter populations.

9.2.1 Increment of Population

Needless to say, the different patterns of life history are amenable to quantitative analysis. Different species have different capabilities for

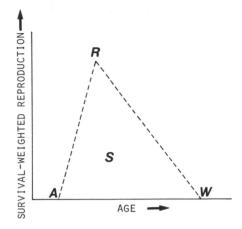

Figure 9.3. Hypothetical relationship of survivorship-weighted reproduction to age. (Modified after Allan, 1976.)

increasing their number in a given environment. Parthenogenetic zooplankton such as rotifers and cladocerans have short life cycles and develop large, transitory populations. Isochronal development or a short generation time with a shortened adult stage (e.g., *Acartia*) may be, like parthenogenesis, an adaptation for surviving unfavorable conditions found by species in highly variable neritic and estuarine environments. Shortening of the last life cycle stage could lead to a shorter total development time and hence to faster population increase.

For a population with a stable age distribution and constant survivorship and fecundity functions, the intrinsic rate of natural increase in number of individuals (r) can be estimated from

$$\int_0^\infty l_t m_t e^{-rt}\, dt = 1 \tag{9.6}$$

where l_t is the probability of surviving to age t and m_t is the average number of female offspring per female between ages t and $t + 1$. Lewontin (cited by Allan, 1976) noted that the relationship between age (t) and survivorship-weighted reproduction ($l_t m_t$) for a particular population tends to have a triangular shape, from which r can be derived from simple geometrical considerations. That is, if the $l_t m_t$ curve looks sufficiently like the graph in Figure 9.3, r may be estimated using the following equation:

$$\frac{W - A}{2S} = \frac{1}{r^2}\left[\frac{1}{R - A}(e^{-rA} - e^{-rR}) + \frac{1}{W - R}(e^{-rW} - e^{-eR})\right] \tag{9.7}$$

where A is the time to first reproduction (egg to egg), W is longevity, R is the age at which survival-weighted reproduction peaks, and S is total female offspring born to a female in her lifetime. If food is always in surplus and predation does not occur, these parameters are highly temperature dependent. The equation is most sensitive to changes in A, less so to S and R, and least of all to W (Lewontin, cited by Allan, 1976), in accord with the general conclusion of life history analyses that r is affected most by changes in prereproductive development time, less by changes in brood size, and least of all by changes in longevity.

9.2.2 Egg Ratio Method

The egg ratio method (Edmondson, 1960, 1968) is a technique for assessing the birth rate of clutch-bearing zooplankton. Only females are considered in the case of sexual generations. If N_e is the number of eggs per female ($= C_t/N_t$, C_t and N_t being the total number of eggs and individuals in the population at time t) and D is the development time (days) of eggs, the number of offspring produced daily per female (E) is given as

$$E = \frac{N_e}{D} \tag{9.8}$$

If the number of all females, including larvae, juveniles, and adults, of a population on a particular day is designated as 1, the number will be $1 + E$ on the next day if there is no population mortality. Then the instantaneous birth rate (b) can be obtained from the following equation:

$$b = \ln(1 + E) - \ln(1) = \ln(1 + E) \tag{9.9}$$

Caswell (1972) showed that when population mortality does exist, b is calculated using the following equation:

$$b = \frac{rE}{e^r - 1} \tag{9.10}$$

Then

$$N_t = N_0 e^{rt} \qquad r = \frac{\ln N_t - \ln N_0}{t} \tag{9.11}$$

where N_0 and N_t are the numbers of individuals at times 0 and t, respectively. The symbol r represents the instantaneous rate of population change. The relationship among b, r, and m (instantaneous mortality rate) is

$$r = b - m \tag{9.12}$$

Paloheimo (1974) further modified the egg ratio method, taking into consideration mortality during population growth and egg development (including the death of egg-carrying females). The modified equation is

$$b = \frac{\ln(N_e + 1)}{D} \tag{9.13}$$

Both the equations of Edmondson and Caswell assume a uniform age distribution of eggs. Leigth (in Edmondson, 1968) gives a correction factor (F_1) to calculate E for an exponentially growing population:

$$F_1 = \frac{e^{-rD} - e^{-r(D-1)}}{e^{-rD} - 1} \tag{9.14}$$

Then
$$E = F_1 N_e \tag{9.15}$$

Smith (in Cooper, 1965) derived a more generalized correction factor (F_2) taking into account the instantaneous, per capita mortality rate of eggs (m) during development:

$$F_2 = \frac{b}{e^{bD} - 1} \left(\frac{e^{b-m} - 1}{b - m} \right)$$

$$= \frac{b(e^r - 1)}{r(e^{bD} - 1)} \tag{9.16}$$

Again
$$E = F_2 N_e \tag{9.17}$$

Equation (9.16) is equivalent to Equation (9.14) when $m = 0$. Taylor and Slatkin (1981) showed that the Caswell's equation [Eq. (9.10)] reduces to Equation (9.13) when E of Equation (9.10) is calculated by using Equation (9.17).

Threlkeld (1979) proposed an age distribution analysis of eggs to directly and accurately estimate E. For this purpose, he divided the development of parthenogenetic eggs of *Daphnia* into five stages and estimated the duration of each stage relative to total developmental time (D). From this, the birth rate of the field population can be calculated by analyzing the age distribution of eggs. With field data on *Daphnia pulicaria* from Lake Tahoe, on the California–Nevada border, and *D. galeata mendotae* from Wintergreen Lake, Michigan, he compared E calculated from the age distribution of eggs with that estimated from F_1 and F_2 mentioned above. The result indicated that the equations using F_1 and F_2 overestimate the real number of offspring produced per day by 19–63%.

The egg ratio method has been applied to the study of the population dynamics of rotifers which have a simple development (Edmondson, 1960) and later to *Daphnia* (Hall, 1964) and *Diaptomus* (Edmondson et al., 1962). All these animals carry eggs in their bodies or egg sacs where they are easily counted. The method assumes a stable population characterized by constant values of $b, m, D,$ and E and a constant portion of mature females.

Note that the egg ratio method is related to the increase and decrease of population numbers but not to production, which is a function of biomass. However, rotifers are exceptions since their body size as adults does not differ greatly from that of larvae.

9.3 Growth

The growth rate of individuals is usually measured as the increase in size (in terms of body length or weight) during discrete time intervals. For animals such as copepods which undergo discrete morphological changes as they grow, their developmental stage can be regarded as an index of their biological age (nondirect development).[2] In this case growth rate is obtained by measuring the time required for this biological aging. On the other hand larvae of some organisms, such as gastropods and chaetognaths, grow to adulthood without significant morphological changes (direct development). In this case biological age is obtained mainly from body size or gonadal development, which are less accurate indices of age than morphological indices. In the case of either direct or indirect development, the life span of the animals is divided into several "age classes" for the calculation of the growth rate in the field.[3] "Age class" is synonymous with "size class" as the size of an individual is usually related to its age.

When a synchronized breeding stock (cohort) can be distinguished in a population by means of morphological characters and/or a size frequency distribution and the body weight determined, the finite growth rate (G) of animals in that cohort can be estimated with the following equation:

$$G = \frac{W_i - W_{i-1}}{W_{i-1}t} \tag{9.18}$$

where $W_i - W_{i-1}$ is the increase or body weight of an individual between adjacent age classes $i - 1$ and i and t is the time required to develop from class $i - 1$ to class i. The growth rate can also be expressed in exponential form:

$$W_i = W_{i-1}e^{gt}$$

Hence

$$g = \frac{\ln W_i - \ln W_{i-1}}{t} \tag{9.19}$$

where g is the instantaneous growth rate.

For animals which reproduce seasonally and grow at a variable rate during the year, age frequency distributions are generally broad due to age-dependent birth and mortality rates. Figure 9.4 illustrates an age class analysis (based on carapace length) for the shrimp *Metapenaeus joyneri* in the Ariake Sea, Japan.[4] It is seen that their growth is rapid from July to September and slows down thereafter. The average growth curve for the life span of this shrimp can be drawn by either connecting the peak frequency value of each month or fitting a smooth curve by eye to the observed data. In order to calculate the growth rate in terms of weight, the carapace length must be converted to body weight using an allometric relationship between carapace length and body weight. It is recommended

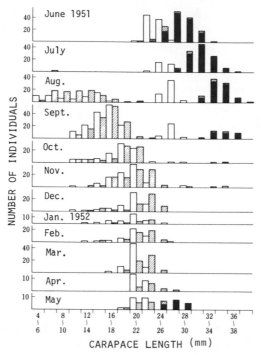

Figure 9.4. Size (carapace length) distribution and its seasonal change in *Metapenaeus joyneri* in the Ariake Sea, Japan. Histograms: open is male; slashed is virgin female; solid is copulated female. (After Ikematsu, 1963.)

that observations of maturity of the animals be made concurrent with body size measurements. Morphology of the copulatory organs and the conditions of the ovary are commonly used as indices of the degree of maturation.

If the length of an animal is plotted against age, the result is usually a curve of continuously decreasing slope with increasing age, which approaches an upper asymptote parallel to the X axis (Fig. 9.5A). Curves of weight at age also approach an upper asymptote but generally form an asymmetrical sigmoid,[5] the inflection occurring at a weight of about one-third of the asymptotic weight (Fig. 9.5B). When a growth curve is fit mathematically as accurately as possible to observe mean or modal size data, one can estimate growth with a limited number of observations. The method also permits a more meaningful comparison between cohorts as well as between populations. The mathematical expression must give a size (length or weight) at any given age which agrees with observed data.

There are various growth equations, but none seemed to be entirely satisfactory in all situations. "In fact it is most unlikely that a simple formula would always be able to describe the growth of even a single fish

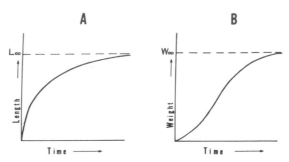

Figure 9.5. Hypothetical growth curve for length (A) and weight (B).

through most of its life" (Gulland, 1969, p. 34). Among the equations, however, one deduced by von Bertalanffy (1938) seems to particularly useful:

$$L_t = L_\infty (1 - e^{-k(t - t_0)}) \qquad (9.20)$$

where L_t is length at age t, t_0 is the theoretical age at which $L = 0$, L_∞ is the hypothetical maximum value of L (i.e., the asymptote of the curve in Fig. 9.5A), and $-k$ is the growth constant. The weight of many zooplankton species is usually closely proportional to the cube of their length:

$$W_t = W_\infty (1 - e^{-k(t - t_0)})^3 \qquad (9.21)$$

where W_∞ is the asymptotic weight corresponding to the asymptotic length L_∞.

In general, tracing the growth of a cohort by consecutive sampling is possible only for zooplankton species which have short spawning seasons relative to the length of the life spans, such as those inhabiting high-latitude seas. Note that, even in such regions, a single population of a zooplankton species may not always be found for a given area and time. For example, Sameoto (1971) found four to five subpopulations of *Sagitta elegans* in St. Margaret's Bay, Nova Scotia, and he used a technique of polymodal analysis to trace the growth rate of each subpopulation.

With a previous knowledge of synchronized reproductive patterns of zooplankton in the subtropical pacific near Hawaii, Newbury (1978) attempted to estimate the growth rates of *Pterosagitta draco* and *Scolecithrix danae* from temporal variations in their size and developmental stage. In a unique study to estimate growth rate, Heron (1972) followed a swarm of *Thalia democratica* using buoys attached to drogues and obtained samples every 1–4 hr. Using the size frequency of specimens in sequential samples, he calculated a remarkably high growth rate for *T. democratica*, which was confirmed by laboratory experiments with live specimens.

For animals with a long life span and short spawning season, a gross estimate of the growth rate may be obtained from infrequent sampling

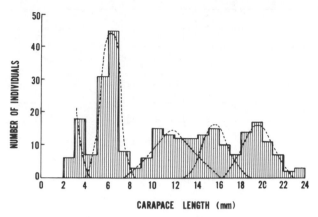

Figure 9.6. Size (carapace length) distribution of *Pasiphaea japonica* collected from Toyama Bay, Japan, on May 30, 1975. (After Omori, 1976b.)

for body size distribution and maturation. For example, consider *Pasiphaea japonica*, a shrimp in Toyama Bay (Fig. 9.6). It is known that eggs attach to the pleopods of females of this species hatch twice a year, at the end of May and in early October. First spawning occurs when females reach 14–16 mm in carapace length. The frequency distribution of carapace length in Figure 9.6 was obtained during the spring season of egg hatching. Since the mesh size of the net used for sampling was not sufficiently small, recently hatched larvae were not sampled adequately. However, the five distinct modes correspond to individuals hatched in May of the collection year (0 yr-class) for the smallest size mode, in October of previous year (0.5 yr-class) for the next smallest size mode, and so on, suggesting a life span of 2.0–2.5 years for this shrimp. The average net weight of larvae at hatching is 1.94 mg·ind^{-1} and that of females with carapace length of 15.5 mm is 907.90 mg. Assuming the time required for the larvae to grow to female size (15.5 mm) is 17 months, finite and instantaneous growth rates are calculated from Equations (9.18) and (9.19) as 0.92 day^{-1} and 0.012 day^{-1}, respectively.

9.3.1 *Probability Paper Method for Cohort Analysis*

As shown in Figure 9.6 a mixture of several age classes is commonly observed in a given sample. The separation of each age class can be made easily when age groups differ morphologically, but this is not the case for animals which develop directly. The probability paper method is commonly used for size distribution analysis in the latter case. This method assumes that the size distribution of a single age (size) group is normally distributed and hence can be represented by a straight line on probability paper.

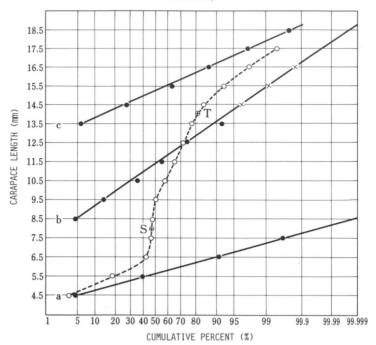

Figure 9.7. Polymodal analysis of carapace length of *Pasiphaea japonica* using the probability paper method.

Table 9.1 shows the size distribution analysis of *Pasiphaea japonica* data (Fig. 9.6). A range of carapace length from 4.1 to 19.0 mm was arbitrarily chosen for this analysis and divided equally into 1.0-mm increments. The median value and number of individuals in each size class are shown in columns 2 and 3. Column 4 indicates the cumulative number of individuals. The cumulative percentage is shown in column 5 and plotted on probability paper in Figure 9.7 (open circles). When the size distribution is unimodal, all open circles fall onto a straight line; however, the line curves with several inflection points in the case of a polymodal size distribution. In Figure 9.7 two inflection points are obvious (S and T) indicating the presence of three size groups. The symbol S corresponds to a carapace length of 8.0 mm and cumulative percentage of 46.0%. Assuming S to be a boundary between two groups (i.e., 0.5 and 1.0 yr-class), we recalculate the cumulative percentage for carapace lengths of 4.1–8.0 mm and adjust the last size interval to 100% (column 6). The adjustment factor for the cumulative percentage for the length interval from 4.1 to 5.0 mm, for example, is

$$2.08 \times 100/46.0 = 4.52\%$$

The adjusted cumulative percentages are shown in Figure 9.7 as dots. A

Table 9.1. Summary table for polymodal analysis of carapace length of *Pasiphaea japonica* using the probability paper method[a]

1	2	3	4	5	6	7	8	9	10	11	12	13	14	15	16
4.1– 5.0	4.5	4	4	2.08	4.52	4.3	4	4							
5.1– 6.0	5.5	31	35	18.22	39.61	39.5	34	30							
6.1– 7.0	6.5	45	80	41.67	90.59	89.0	78	44							
7.1– 8.0	7.5	8	88	45.83	99.63	99.7	88	10							
8.1– 9.0	8.5	3	91	47.40					4.06	4.5	3	3			
9.1–10.0	9.5	6	97	50.52					13.10	14.0	6	6			
10.1–11.0	10.5	15	112	58.33					35.74	31.0	19	10			
11.1–12.0	11.5	13	125	65.10					55.36	53.5	33	14			
12.1–13.0	12.5	12	137	71.35					73.47	73.5	45	12			
13.1–14.0	13.5	12	149	77.60					91.59	88.5	54	9	5.64	5.6	2
14.1–15.0	14.5	13	162	84.38						96.4	59	5	26.25	27.0	10
15.1–16.0	15.5	15	177	92.19						99.1	61	2	61.53	57.0	13
16.1–17.0	16.5	10	187	97.40						99.9	61	0	86.67	86.0	12
17.1–18.0	17.5	4	191	99.48									97.44	97.3	5
18.1–19.0	18.5	1	192	100.00									100.00	99.8	1
Total		192						88				61			43

[a] Column designations: (1) carapace length (mm), (2) median value (mm), (3) number of individuals, (4) cumulative number of individuals, (5) cumulative percentage, (6) first group (%), (7) corrected cumulative percentage, (8) cumulative number of individuals, (9) corrected number of individuals, (10) second group (%), (11) corrected cumulative percentage, (12) cumulative number of individuals, (13) corrected number of individuals, (14) third group (%), (15) corrected cumulative percentage, (16) corrected number of individuals.

223

straight line (line *a*) can be readily fitted to the points (dots), indicating there is little or no overlap between the 0.5 and 1.0 year-class. Using this linear regression, corrected cumulative percentages (column 7) are converted to cumulative numbers of individuals by multiplying by the total cumulative number of this group (88; column 8) and finally expressed as numbers of individuals in each size interval (column 9).

The analysis is similar for the second group (1.0 yr-class) between *S* and *T*. Carapace length at *T* is 14.0 mm and the cumulative percentage is 80.5%. Calculation of the cumulative percentage for the first size interval (8.1–9.0-mm carapace length) of this second group is

$$(47.4 - 46.0) \times 100/(80.5 - 46.0) = 4.06\%$$

The same calculation is repeated for the following intervals (column 10). The cumulative percentages thus obtained are plotted on a straight line (line *b*). In this case the last of the new points lies somewhat to the right of the fitted line. This is due to overlap with the third group (1.5 yr-class) and is a normal occurrence. The greater the degree of overlap, the greater the number of such points. This implies, first, that the fitting of the third group should commence below *T* and, second, that a correction must be made to the first few points. Several points (cross marks) occur beyond the upper size limit *T* of the second group, indicating a number of individuals of 1.0 year-class overlapping with the third group. The corrected cumulative percentages taken from line *b* are in column 11. Cumulative numbers of individuals are computed up to 17.0 mm (column 12), and numbers of individuals in each size interval (column 13) are obtained.

The remaining 43 individuals form the third group (1.5 yr-class). Assuming that carapace lengths of individuals in this group are greater than 13 mm, this group's cumulative percentages are calculated beginning at the same level at the point just below inflection *T*. The percentage of the total represented by the 1.0-year-class component will be

$$88.5(80.5 - 46.0)/100 = 30.5\%$$

To this the 46.0% represented by the 0.5 year-class is added:

$$30.5 + 46.0 = 76.5\%$$

The new point of the first size interval (13.1–14.0 mm) of the 1.5 year-class will then be

$$(77.6 - 76.5)100/(100.0 - 80.5) = 5.64\%$$

This process continues until at the third point above *T* the percentage on line *b* is nearly 99.9, and the correction need no longer be applied. The point for the size interval of 17.1–18.0 mm of 1.5 year-class is

$$(99.5 - 80.5)100/(100.0 - 80.5) = 97.44\% \quad \text{(column 14)}$$

The calculation procedures for correcting the cumulative percentage (column 15) and corrected numbers of individuals (column 16) are the same as those for the previous two groups.

The normal distribution curve is hypothetically symmetrical, but the curve obtained from the probability paper analysis is often not exactly symmetrical because of errors in reading from the probability paper and rounding values. The standard deviation (SD) of carapace length in each group is estimated from respective normal distribution curve by measuring the difference between the mean and the 15.87% or 84.13% level on the graph of the corrected cumulative percentages. Alternatively, the standard deviation can be estimated by dividing the difference between the values at 50% and 99.999% by 3.72. For example, for the second group (50%, 11.4; 99.999%, 18.7).

$$(18.7 - 11.4)/3.72 = 1.96$$

The chi-square test is often used to rigorously examine the fit of the data to the normal distribution curve.

Harding (1949) and Cassie (1954) describe in detail the method of size distribution analysis using probability paper.

9.3.2 *Instar Analysis*

For animals with distinctive morphological characters related to their stage of development, the rate of development can be estimated from the mean time between the peaks (pulses) of numbers of individuals appearing in successive developmental stages or instars in successive samples (e.g., McLaren, 1969; Rigler and Cooley, 1974). Absolute counts or relative abundances of certain developmental stages are used to trace cohorts. If development of the population is reasonably synchronous, the relative abundance of a given stage in field samples should show well-defined pulses through time. Growth rate is calculated with a knowledge of the average body weight of individuals at each stage.

However, this approach is sometimes unsatisfactory for the analysis of species at lower latitudes, especially in the open sea where several more or less recognizable cohorts coexist. The investigator must be aware of two problems with this method of analysis that may cause misinterpretation or unsuccessful tracing of cohorts. First, when stages/instars have widely different development rates or when the development time of a stage is not constant throughout the pulse, the accumulation and sustained abundance of long-lived stages may dwarf changes in the abundance of short-lived stages. Second, when stages have widely different mortality patterns or when the mortality rate of each stage does not remain constant, this method may give false cohorts. In certain cases, therefore, population data from field sampling must be analyzed with some subjectivity and/or undemonstrative general rules. There have been some attempts to theoretically derive development times and generation lengths for copepods

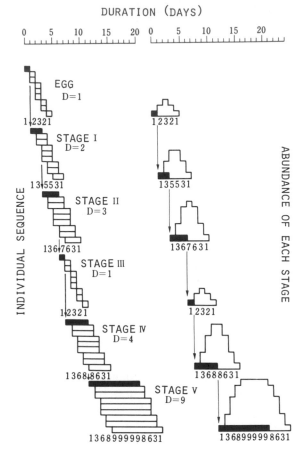

Figure 9.8. Diagram of the hypothesized relationship between the duration (*D*: in days) and the number of individuals expected in each developmental stage. Boxes on the left are the expected periods of appearance of each individual on the designated day. Histograms on the right are the expected abundances of individuals in each stage. Solid bar denotes the development of the first individual resulting from the egg laid at the time. (After Edmondson, 1971.)

in nature from simple generalizations based on physiological relationships for a species (McLaren, 1978). However, at present, we believe that the use of more detailed field data, together with information from laboratory experiments, is best for interpreting population dynamics and production of zooplankton.

We would like to consider the relationship of the time between the maximal abundances of two consecutive developmental stages and the residence time of each stage with a diagram presented by Edmondson (1971) (Fig. 9.8). In this diagram the initial population begins with one egg laid on day 1, two eggs laid on day 2, three eggs laid on day 3, two eggs laid on day 4, and one egg laid on day 5, a total of nine eggs (top left-hand diagram). Based upon the designated duration (D) for developmental stages I to V, the expected abundance of each stage was calculated. The development of one egg laid on day 1 is marked by a black square. The egg hatched to stage I on day 2. Two eggs laid on day 2 become stage I on day 3. The number of individuals of each stage on consecutive days is shown on the bottom of the rectangular graphs. As the duration of stage I is 2 days, the number of stage I individuals is five on day 4 when one individual of stage II appears. The number belonging to each stage increases in proportion to the length of residence in that stage. All nine individuals are in stage V for 5 days because of the longer duration of this stage (9 days). Clearly, the number of individuals in each stage is greatly influenced by the different durations of development. (Note that the total number of individuals is constant from egg through stage V because no mortality is assumed.) Thus, the time between peaks is not an accurate representation of the duration of the stages and therefore can be used only as an approximation of duration time. Ideally, the stage duration can be better estimated from the dates of the first appearance of individuals of one stage and that of the next stage. But such a sensitive determination is nearly impossible with field data and is limited to laboratory experiments.

In laboratory experiments[6] the growth of all individuals is usually not the same. A common criteria for the advancement of a cohort from one development stage to the next is the appearance of more than 50% of the individuals in the new stage. To examine this, subsamples are taken randomly from a batch culture.

9.4 Mortality

Mortality is the sum of physiological death and predation. Both physiological death and predation are usually difficult to estimate separately in field studies. The loss of exuviae of crustaceans may be included in mortality but is usually expressed as a separate term in laboratory experiments. Mortality rate is defined as the fraction of individuals dying during a unit time interval. For an initial number N_{i-1} decreasing to N_i during time t, the finite mortality rate (M) is

$$M = \frac{N_{i-1} - N_i}{N_{i-1}t} \tag{9.23}$$

When decrease in the population abundance is not linear but exponential, the mortality rate is better described by the instantaneous mortality (m):

$$N_i = N_{i-1}e^{-mt}$$

$$m = \frac{\ln N_{i-1} - \ln N_i}{t} \tag{9.24}$$

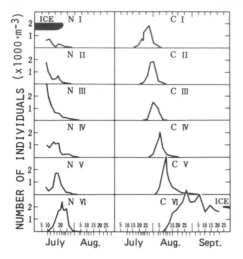

Figure 9.9. Occurrence of naupliar stage (NI–NVI) and copepodite stages (CI–CVI) of *Limnocalanus johanseni* in an arctic lake (Imikpuk) during summer. Sampling period: July 10 to September 21, 1952. (After Comita. 1956.)

The average abundance $(\overline{N})$ per unit time t is

$$\overline{N} = \frac{\int_0^t N_{i-1}e^{-mt}\,dt}{t} = \frac{N_{i-1}(1 - e^{-mt})}{mt} \tag{9.25}$$

For calculation, this "exponential mean" may be approximated more closely by the geometric mean $\sqrt{N_{i-1}N_i}$ than by the arithmetic mean $(N_{i-1} + N_i)/2$.

General procedures in the separation of age (size) classes to calculate the mortality rate of a given zooplankton population are similar to those described for estimating the growth rate. Note, however, that growth rate is related to the average body weight of individuals and the mortality rate to population abundance.

As an example, because the copepod *Limnocalanus johanseni* is distributed in a small enclosed arctic lake and its reproduction period is limited to only a few weeks each year, one may estimate its mortality rate by comparing the abundance of individuals of each developmental stage in samples from a few stations (Fig. 9.9). With short intervals between sampling (1–5 days) the succession of developmental stages from nauplius stage I to copepodite stage VI is well illustrated.

Except for this special situation of *L. johanseni* for which there is a cohort of limited age and instar composition at any one moment, the estimate of mortality from field samplings is often biased by sampling efficiency and patchiness of animals. In addition, special attention to emigration and immigration is needed when the mortality of a population is estimated in a large, open area because its distribution extends as individuals grow older. One of the attempts to determine the mortality rates by plankton populations in such cases was the study by Omori of

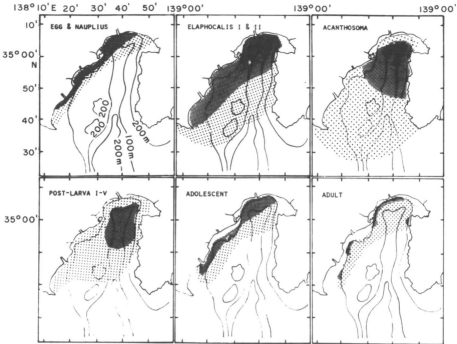

Figure 9.10. Schematic illustration of the distribution of various developmental stages of *Sergia lucens* in Suruga Bay. Density of dotted areas denotes relative abundance of animals. (After Omori, 1974.)

the population of *Sergia lucens* in Suruga Bay. The shrimp reaches sexual maturity after one year and intense spawning occurs in the innermost portion of the bay. Larvae disperse gradually with the flow of water in the bay, and one stage eventually distributed in the entire bay is the protozoea (elaphocaris) stage (Fig. 9.10). Knowing the pattern and range of the distribution, periodical samples were taken with a vertical net haul from 150 m to the surface at each of 11 stations (Fig. 2.1). The total number of larval stage i (N_i) produced in the survey area during the period (T_j) is determined from:

$$N_i = \sum_{j=1}^{m} [(N_{i,j} w_j T_j) D_i^{-1}] \tag{9.26}$$

where $N_{i,j}$ is the estimated number of larval stage i under 1 m^2 at the jth station, w_j is the area (m^2) represented by that station, T_j is the time weighting given to station j (equal to one-half the time elapsing since the preceding occupancy plus one-half the time elapsing before the succeeding occupancy), m is number of stations, and D_i is the duration of stage i. The term D_i was estimated from the average water temperature during the

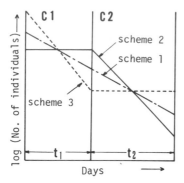

Figure 9.11. A diagram showing three possible schemes of mortality for *Calanus* copepodite stages I (C1) and II (C2). Scheme 1, equal mortality rate for both stages; scheme 2, mortality occurs in C1 but not in C2; scheme 3, mortality occurs in C2 but not in C1. (Modified after Fager, 1973.)

When $m_2 = 0$ the observed mortality occurred during stage I. The average abundances for stages I and II are $N_0(1 - e^{-m_1 t_1})/m_1 t_1$ and $N_0 e^{-m_1 t_1}$, respectively. Therefore

$$\frac{\overline{N}_2}{\overline{N}_1} = \frac{m_1 t_1 e^{-m_1 t_1}}{1 - e^{-m_1 t_1}} = \frac{m_1 t_1}{e^{m_1 t_1} - 1} \tag{9.31}$$

It is not usually possible to know which mortality scheme, or some intermediate scheme, is correct in a particular field study. This is mainly due to the difficulty of knowing the number of organisms which just entered each developmental stage (N_0, in this example); application of a single number for m should be made only after careful consideration.

Apart from this problem on the mortality schemes, let us consider ecological longevity of the adult stage (Petrusewics and MacFadyen, 1970). The number of individuals passing from one stage to the next between two consecutive sampling dates can be estimated by

$$N_i = \frac{\overline{N}_{i-1}}{D_{i-1}} T \tag{9.32}$$

where N_i is the recruitment into stage i, $\overline{N}_{i-1}$ is the average number of stage $i - 1$ on two consecutive sampling dates, 0 and t, D_{i-1} is the duration of stage $i - 1$ in days, and T is the time interval between the sampling dates. For a given cohort the numbers leaving stage i (n_i) is the sum of the "efflux" to the next stage and mortality (or loss L_i) during the time interval. Thus

$$n_i = \frac{\overline{N}_i}{D_i} T + L_i \tag{9.33}$$

If the change in stock from 0 to t is designated $N_{i,t} - N_{i,0}$, then

$$N_{i,t} - N_{i,0} = \frac{\overline{N}_{i-1}}{D_{i-1}} T - \frac{\overline{N}_i}{D_i} T - L_i \tag{9.34}$$

that is, Equation (9.32) minus (9.33); consequently

$$L_i = N_{i,0} - N_{i,t} + \frac{\overline{N}_{i-1}}{D_{i-1}} T - \frac{\overline{N}_i}{D_i} T$$

Mortality during the adult stage (L_a) can be calculated by a slightly different equation, as there is no "efflux" to the following stage:

$$L_a = N_{a,0} - N_{a,t} + \frac{\overline{N}_{a-1}}{D_{a-1}} T \tag{9.35}$$

where $\overline{N}_{a-1}$ and D_{a-1} are the average number and duration time of the stage just prior to the adult during T. In field studies it is generally difficult to estimate the longevity of animals after sexual maturity. However, if the population is stationary, that is, constant age structure and population size, and once the mortality in the adult stage is calculated, it is possible to estimate the average ecological longevity ($\overline{t}$) of the adults:

$$\overline{t} = \frac{\overline{N}_a}{L_a} T \tag{9.36}$$

It is not a surprising fact that the average ecological longevity of a species is much shorter than the average physiological longevity under severe predation pressure. According to Uye (1982), the average physiological longevity of adult *Acartia clausi* under laboratory conditions was 27–68 days, whereas the ecological longevity was only 1.4–9.8 days in Onagawa Bay, Japan.

9.5 Estimation of Production

At present there is no single method that provides the best estimate of production. In fact, the methods which have been used by various workers on different zooplankton species or populations vary considerably in terms of the kind of data collected and their analyses. For convenience of description, we divide the existing methods arbitrarily into three categories: the field-oriented population dynamics method, laboratory-oriented budgetary method, and other methods.

9.5.1 Field-oriented Population Dynamic Method

Consider a hypothetical zooplankton population with a generation time T and initial number and body weight of an average newborn N_0 and W_0, respectively ($t = t_0$). At consecutive sampling times $t_1, t_2, \ldots, t_n$ ($t_n - t_0 = T$), the numbers and mean body weights of this cohort are $N_1, N_2, \ldots, N_n$ and $W_1, W_2, \ldots, W_n$, respectively. In most cases the numbers decrease toward the end of the generation due to the death of individuals, while the body weights increase as a result of growth.

We will calculate the production of this hypothetical population four different ways.

Calculations

Removal Summation

Production (P_1) during times t_0 to t_1 is given by

$$P_1 = \tfrac{1}{2}(W_0 + W_1)(N_0 - N_1) + (W_1N_1 - W_0N_0)$$

where $N_0 - N_1$ is the number of specimens lost and $\tfrac{1}{2}(W_0 + W_1)$ is the average body weight between times t_0 and t_1. Therefore, $\tfrac{1}{2}(W_0 + W_1)(N_0 - N_1)$ represents the biomass lost during time $t_0 - t_1$ and $(W_1N_1 - W_0N_0)$ is the change in biomass over the same period. Similarly, the production during times t_{i-1} and t_1 can be written in a general form:

$$P_i = \tfrac{1}{2}(W_{i-1} + W_i)(N_{i-1} - N_i) + (W_iN_i - W_{i-1}N_{i-1}) \quad (9.37)$$

Therefore, total production over sampling period T is

$$P = P_1 + P_2 + \cdots + P_n$$

$$= \frac{1}{2} \sum_{i=1}^{n} (W_{i-1} + W_i)(N_{i-1} - N_i) + (W_nN_n - W_0N_0) \quad (9.38)$$

On the other hand, if the population of a species is the same from year to year, that is, $W_nN_n - W_0N_0 = 0$ in Equation (9.38), and the residence time in each size class is similar (i.e., when the number of classes is n, the residence time in each size class is $365/n$ days), the annual total production (P) can be determined using the size frequency (Hynes) equation modified by Hamilton (1969) and Menzie (1980):

$$P = \sum_{i=1}^{n} (N_{i-1} - N_i)(W_{i-1}\,W_i)^{1/2}$$

$$N_i = (n\bar{n}_i)\frac{\text{Pe}}{\text{Pa}}\frac{365}{\text{CPI}}$$

where n is the number of size categories in which the animals are classified, $\bar{n}_i$ is the mean number of individuals in a particular size class i, Pe is the estimated proportion of the life cycle spent in a particular size class, Pa is the actual proportion of the life cycle spent in a particular size class (i.e., Pe/Pa is a correction factor of the deviation from assumed linear growth rate), and CPI is the cohort production interval in days from hatching to the attainment of the largest size class, that is, the correction factor for deviation from an assumed 1-year life cycle. Obviously, the equation sets are analogous to Equation (9.38) except for the use of the geometric mean ($\sqrt{W_{i-1}W_i}$) rather than the arithmetic mean $[(W_{i-1} + W_i)/2]$ for the body weight. This method can be applied to several species when they are of the same trophic level, all univoltine (one generation per year), having similar maximum sizes and growing at similar rates.

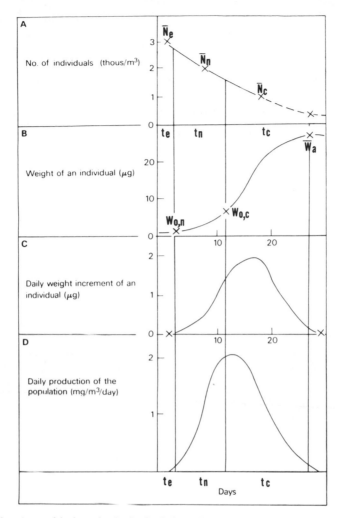

Figure 9.12. A graphical method of calculation of copepod production. (After Winberg et al., 1971.)

Increment Summation

Equation (9.37) can be reduced to

$$P_i = \tfrac{1}{2}(W_i - W_{i-1})(N_i + N_{i-1}) \tag{9.39}$$

where $\tfrac{1}{2}(N_i + N_{i-1})$ is the average number during times t_{i-1} and t_i and $W_i - W_{i-1}$ is the net increase in body weight during the same time interval. Production over the entire sampling period T is

$$P = \frac{1}{2} \sum_{i=1}^{n} (W_i - W_{i-1})(N_i + N_{i-1}) \tag{9.40}$$

Equations (9.38) and (9.40) are similar, but the latter is the summation of increments of biomass while the former is the summation of removed biomass. For this reason, Equation (9.40) does not need a second term to account for the change in biomass.[7]

The increment summation method is amenable to graphical analysis (Winberg et al., 1971). As shown in Figure 9.12, four graphs with a common X axis (development time of each stage: t_e, t_n, and t_c) are needed for this analysis. The stages are egg, nauplius, and copepodite in this case. In the first graph the average number of individuals of each stage for each day of life is plotted on the Y axis. The value of $\overline{N}$ is obtained by the division of the total number of individuals of a given stage (i.e., the area beneath the upper curve) by the development time and is marked in the middle of the segment on the X axis, which corresponds to the length of development of the given stage. The points are fitted with a smooth curve. The second graph is the growth curve of an individual. The initial weight of each successive stage ($W_{0,n}$, $W_{0,c}$) and the average weight of the adult (W_a) are marked on the Y axis, and assuming the increase in weight with time for each organism is described by a sigmoid curve (see Section 9.3), the required curve is plotted. In the third graph the daily weight ($\Delta W/T$) increment of an individual, computed from the growth curve in the second graph, is represented on the Y axis. The daily production curve in the fourth graph is obtained by multiplying, for each day, the value in the third graph by the value of the first graph. The area under the fourth curve represents the production of the entire population as a result of the growth of individuals.

Allen Curve

The Allen curve (Allen, 1950) is a graphical method in which the changes in the number and individual body weight of one cohort are plotted on the Y and X axes, respectively, as shown in Figure 9.13. Production over one generation is given by the area under the curve. As pointed out by Gillespie and Benke (1979), the method is in fact an integral version of the increment method with respect to the X axis and an integral version of the removal method with respect to the Y axis. In order to demonstrate this method, the production during times t_{i-1} and t_i is considered from the hypothetical curve in Figure 9.13. The biomass at t_{i-1} and t_i is $B_{i-1} = N_{i-1}W_{i-1} = X + V$ and $B_i = N_iW_i = V + Z$, respectively, where X is the amount of original biomass lost during t_{i-1} and t_i, Y is the amount of new production lost, and Z is the new production during the same period which has been retained.

From the removal summation [Eq. (9.37)], the production during times t_{i-1} and t_i is

$$P_i = \tfrac{1}{2}(W_{i-1} + W_i)(N_{i-1} - N_i) + (W_iN_i - W_{i-1}N_{i-1})$$

$$= \tfrac{1}{2}(X + X + 2Y) + (Z - X) = Y + Z$$

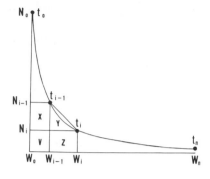

Figure 9.13. Hypothetical Allen curve.

From the increment summation [Eq. (9.39)], the production is

$$P_i = \tfrac{1}{2}(W_i - W_{i-1})(N_i + N_{i-1})$$
$$= \tfrac{1}{2}(Z + 2Y + Z) = Y + Z$$

Thus, it is apparent that the production estimate from the Allen curve gives results similar to those of the removal summation and increment summation methods when the time interval $t_i - t_{i-1}$ is short enough to assume linear changes in N and W. Gillespie and Benke (1979) suggested that when few samples are available over one generation for a cohort, or if the sampling error is large, a smoothed curve fitted by hand to data points may provide better results than would the arithmetic methods mentioned above.

Exponential Equations

Although the methods of removal summation and increment summation are valid when the sampling interval is short enough to assume linear growth and mortality, on most occasions an exponential equation is a much more accurate expression for both processes. In the above example of a hypothetical zooplankton population, the instantaneous growth (g_i) and mortality (m_i) rates during the time interval (t_i) are given by Equations (9.19) and (9.24):

$$W_i = W_{i-1}e^{g_i t_i} \quad \text{and} \quad N_i = N_{i-1}e^{-m_i t_i}$$

The change in biomass from class $i - 1$ ($B_{i-1} = N_{i-1}W_{i-1}$) to class i ($B_i = N_i W_i$) then becomes

$$B_i = B_{i-1}e^{(g_i - m_i)t_i} \tag{9.41}$$

The average biomass ($\bar{B}$) between classes $i - 1$ and i is

$$\bar{B} = \frac{\int_0^{t_i} B_{i-1}e^{(g_i - m_i)t_i}\, dt}{t_i} = \frac{B_{i-1}(e^{(g_i - m_i)t_i} - 1)}{(g_i - m_i)t_i} \tag{9.42}$$

As $B_i = B_{i-1}e^{(g_i - m_i)t_i}$

$$\overline{B} = \frac{B_i - B_{i-1}}{(g_i - m_i)t_i} \tag{9.43}$$

Then production (P_i) during time t_i is given by

$$P_i = g_i\overline{B}_it_i = \frac{g_i}{(g_i - m_i)}(B_i - B_{i-1}) \tag{9.44}$$

Equation (9.44) is valid when $B_i - B_{i-1} \neq 0$ and $g_i \neq m_i$ (Gillespie and Benke, 1979). When $B_i - B_{i-1} = 0$ or $g_i = m_i$, that is, there is no change in biomass during time t_i, Equation (9.44) becomes

$$P_i = g_iB_it_i \quad \text{or} \quad m_iB_it_i \tag{9.45}$$

Above we considered production of two adjacent size classes. A generalized equation for production (P) over one generation is

$$P = \sum_{i=1}^{n} \frac{g_i}{(g_i - m_i)}(B_i - B_{i-1}) \quad \text{or} \quad \sum_{i=1}^{n} g_iB_it_i \tag{9.46}$$

in which the former expression is used for intervals when a net change in biomass occurs and the latter is used when biomass remains constant.

In connection with Equation (9.44) LeBlond and Parsons (1977) suggested a simplified equation for production:

$$P_i = \left(1 + \frac{m_i'}{g_i'}\right)\Delta B_i \tag{9.47}$$

where $m_i' = m_i$ and $\Delta B_i = B_i - B_{i-1}$ in Equation (9.44). The term g_i' is derived not from $W_i = W_{i-1}e^{g_it_i}$ but from $B_i = B_{i-1}e^{g_i't_i}$ in our notation. Compared with Equation (9.41) ($B_i = B_{i-1}e^{(g_i - m_i)t_i}$), it is apparent that $g_i' = g_i - m_i$. Substituting these in Equation (9.47), we obtain

$$P_i = \left(1 + \frac{m_i}{g_i - m_i}\right)(B_i - B_{i-1}) = \frac{g_i}{g_i - m_i}(B_i - B_{i-1})$$

which is exactly the same as Equation (9.44). Although Equation (9.47) has some advantage because it uses g_i' directly from biomass data, numbers still need to be counted to obtain m_i'. Since g_i in Equation (9.44) is derived from $W_i(=B_i/N_i)$ and $W_{i-1}(=B_{i-1}/N_{i-1})$, the advantage of using LeBlond and Parsons' equation appears small, except that it removes the necessity of obtaining individual weights for animals of each age. One can determine B_i directly (dry weight of zooplankton sample) and numbers only.

It is now clear that all of the methods mentioned above (removal summation, increment summation, Allen curve, and exponential equations) are based on the same principle but involve different mathematical expressions and calculation procedures.

Examples

Yablonskaya (1962) calculated the annual production of the copepod *Diaptomus salina* in the Aral Sea. *Diaptomus salina* breeds once a year in April–May and then dies. Most individuals of the spring cohort reach the copepodite IV stage by July and 90% become adult in October of the same year. The adults hibernate until the next spawning season. Table 9.3 shows a summary of data from field samples of this copepod over 3 years. The number of individuals (N) of each developmental stage (the six nauplius stages are combined) was converted to biomass (B) by multiplying by the average body weight (W) of each stage, and annual production was calculated by the removal summation method. Annual production of this copepod is given by the sum of the biomass lost during one generation and the difference between the initial and final biomasses of one generation.

Another example of the application of the removal summation method may be found in the study of Heinrich (1962), who estimated the annual production of several copepod species in the Bering Sea, and that of Berkes (1977), who calculated production of the euphausiid *Thysanoëssa raschii* in the Gulf of St. Lawrence.

Field samplings provide data on size or developmental stage on each sampling date. However, when the spawning period is long and irregular, it is extremely difficult to estimate the growth rate of a cohort by connecting one size class or developmental stage to the next between discrete samples. In this case laboratory experiments to estimate the growth rate become necessary. Newbury and Bartholomew (1976) estimated the production of copepods (Paracalanidae) in tropical inshore waters near Hawaii. To determine growth rates, they reared the copepods in the laboratory and *in situ* by suspending culture containers in the field. Their calculation of production was based on the increment summation method, but they used the geometric mean instead of arithmetic mean to obtain the mean number between stages in Equation (9.40). The result of this calculation, summarized in Table 9.4, indicates that the average production of copepods was 15.18 mg dry wt·m^{-3}·day^{-1}. Mullin and Brooks (1970a) estimated the production of *Calanus helgolandicus* off La Jolla, California, with exponential equations [Eq. (9.44)] using stage abundances from field samples and stage durations from laboratory growth experiments with this copepod under simulated field concentrations of food. Similar production estimates were made by Hirota (1974) for *Pleurobrachia bachei* off La Jolla, California, and Reeve and Baker (1975) for *Mnemiopsis mccradyi* and *Sagitta hispida* in south Florida inshore waters.

Heinle (1966) estimated the production of *Acartia tonsa* in the Patuxent River estuary from mortality data for this copepod. This estimate is valid because the biomass of *A. tonsa* was shown to be quite constant over the period of investigation (i.e., $B_i - B_{i-1} = 0$, or $g_i = m_i$). Heinle classified

Table 9.3. **Seasonal abundance and biomass dynamics of *Diaptomus salinus* in the Aral Sea**[a]

Stages of development	W (mg·ind⁻¹)	May		August		October	
		N (ind·m⁻³)	B (mg·m⁻³)	N (ind·m⁻³)	B (mg·m⁻³)	N (ind·m⁻³)	B (mg·m⁻³)
Nauplii	0.0027	5720	15.444	291	0.786	130	0.351
C_I	0.0063	437	2.753	6	0.038	13	0.082
C_{II}	0.0110	317	3.487	11	0.121	7	0.077
C_{III}	0.0220	214	4.708	384	8.448	2	0.044
C_{IV}	0.0320	48	1.536	2244	71.808	38	1.216
C_V	0.0540	41	2.214	355	19.170	1629	87.966
♀	0.1020	446	45.492	25	2.550	208	21.216
♂	0.0610	742	45.262	67	4.087	351	21.411
Total	—	7965	120.896	3383	107.008	2378	132.363

Summary of production calculations using above data

Index	May — All stages except adults	August — All stages	New brood (nauplii) $\sim C_{III}$	Decrease during May–August	October — All stages	New brood (nauplii) $\sim C_{III}$	Decrease during August–October	May — Adults ($\female + \male$)	Decrease during October–May
N (ind·m^{-3})	6777 (n_1)	3383 (n_2)	692 (n_3)	4086 $[n_1 - (n_2 - n_3)]$	2378 (n_4)	152 (n_5)	1157 $[n_2 - (n_4 - n_5)]$	1188 (n_6)	1190 $(n_4 - n_6)$
B (mg·m^{-3})	30.142 (b_1)	107.008 (b_2)	9.393 (b_3)	83.354 (b_{d1})	132.363 (b_4)	0.554 (b_5)	52.528 (b_{d2})	90.754 (b_6)	78.540 (b_{d3})
W (mg·ind^{-1})	0.0045 $\left(\dfrac{b_1}{n_1}\right)$	0.0316 $\left(\dfrac{b_2}{n_2}\right)$	0.0136 $\left(\dfrac{b_3}{n_3}\right)$	0.0204 $\dfrac{1}{2}\left(\dfrac{b_1}{n_1} + \dfrac{b_2}{n_2} - \dfrac{b_3}{n_3}\right)$	0.0557 $\left(\dfrac{b_4}{n_4}\right)$	0.0036 $\left(\dfrac{b_5}{n_5}\right)$	0.0454 $\dfrac{1}{2}\left(\dfrac{b_2}{n_2} + \dfrac{b_4}{n_4} - \dfrac{b_5}{n_5}\right)$	0.0764 $\left(\dfrac{b_6}{n_6}\right)$	0.0660 $\dfrac{1}{2}\left(\dfrac{b_4}{n_4} + \dfrac{b_6}{n_6}\right)$

Decrease biomass for a year $(B_d = b_{d1} + b_{d2} + b_{d3})$ 214.422 mg·m^{-3}

Residual biomass $(B_r = b_6)$ 90.754 mg·m^{-3}

Production $(= B_d + B_r)$ 305.176 mg·m^{-3}

[a] Values represent averages for three years (after Yablonskaya, 1962). N, B, and W are number of individuals, biomass, and mean weight of one individual, respectively. See Yablonskaya (1962) for details.

Table 9.4. Calculation of the daily population production of Paracalanidae in southern Kaneohe Bay during August 1968[a]

Stage	N (ind·m^{-3})	W (μg dry wt·ind^{-1})	B (mg dry wt·m^{-3})	Stage duration (days)	Geometric mean density ($\sqrt{N_{i-1}N_i}$) (ind·m^{-3})	Weight increase (μg dry wt·ind^{-1})	Production rate (mg dry wt·m^{-3}·day^{-1})
Nauplius							
I[b]	(33,463)	0.004	0.13	—	—	—	—
II	(25,346)	0.006	0.15	0.5	22,058	0.014	0.62
III	(19,198)	0.02	0.38	0.5	16,708	0.03	1.00
IV	(14,542)	0.05	0.73	0.5	12,656	0.02	0.51
V	(11,015)	0.07	0.77	0.5	9,586	0.02	0.38
VI	(8,343)	0.09	0.75	0.5	7,261	0.04	0.58
Copepodid							
I	6,320	0.13	0.82	0.8	5,060	0.14	0.88
II	4,052	0.27	1.09	0.8	3,199	0.42	1.68
III	2,526	0.69	1.74	0.8	2,459	0.31	0.96
IV	2,394	1.00	2.39	0.9	2,243	0.82	2.04
V	2,103	1.82	3.83	0.9	2,103	0.76	1.78
Adult	2,627	2.58	6.78	—	—	—	4.75
						Total	15.18

[a] After Newbury and Bartholomew (1976). The number of naupliar stages was calculated indirectly assuming a constant mortality rate, obtained from copepodite stages I and II. N, B, and W are number of individuals, biomass, and mean weight of one individual, respectively.

[b] This stage does not feed and is eliminated from the production calculation.

specimens in field samples into three developmental groups: nauplii, copepodites (stages I–V), and adults. The numbers of these three groups were converted to biomass, integrating the apportioned numbers of the six nauplius and five copepodite stages of known body weight.[8] The necessary assumptions in these calculations are constant mortality and equal duration of stages of nauplii and copepodites. For example, let N_n, m_n, and t_n be the number of nauplii counted in field samples (without separation into stages), instantaneous mortality rate obtained from the number of nauplii and copepodites (duration of naupliar and copepodite stages are estimated from the experimental results mentioned below), and duration of each stage, respectively. The relationship between the number of naupliar stages I (N_{n1}), II (N_{n2}), III (N_{n3}), . . . , VI (N_{n6}) is then

$$N_{n2} = N_{n1}e^{-m_n t_n}$$

$$N_{n3} = N_{n2}e^{-m_n t_n}$$

$$\vdots$$

$$N_{n6} = N_{n5}e^{-m_n t_n}$$

These equations were combined and solved for N_{n1}:

$$N_n = N_{n1} + N_{n1}e^{-m_n t_n} + N_{n1}e^{-2m_n t_n} + \cdots + N_{n1}e^{-5m_n t_n}$$
$$= N_{n1}(1 + e^{-m_n t_n} + e^{-2m_n t_n} + \cdots + e^{-5m_n t_n}) \qquad (9.48)$$

Then

$$N_{n1} = N_n/(1 + e^{-m_n t_n} + e^{-2m_n t_n} + \cdots + e^{-5m_n t_n})$$
$$N_{n2} = N_{n1}e^{-m_n t_n}$$

and so on for the following stages. Simultaneous to field sampling, Heinle (1966) observed the growth rate of A. tonsa from egg to adult in seawater containing natural food assemblages. With the results from field sampling and growth rate experiments, he calculated instantaneous mortality rates between nauplii and copepodites and between copepodites and adults using the exponential equation:

$$N_i = N_{i-1}e^{-mt}$$

Then the instantaneous mortality rate (m) was converted to the finite mortality rate (M):

$$M = 1 - e^{-m} \qquad (9.49)$$

Turnover time (T) was then calculated by

$$T = \frac{1}{M} \qquad (9.50)$$

Production (P) of A. tonsa (nauplii to adults) is given by the following equation:

$$P = \frac{B_n}{T_n} + \frac{B_c}{T_c} + \frac{B_a}{T_a} \tag{9.51}$$

where B_n, B_c, and B_a are biomasses of nauplii, copepodites, and adults and T_n, T_c, and T_a are turnover times (T_a was assumed to be equal to T_c). As Heinle (1966) noted, the computation of T is unnecessary and P can be obtained more directly by multiplying biomass (B) by mortality rate (M). Similar estimates of production for *Acartia tonsa* and *Eurytemora affinis* from mortality rates were made by Allan et al. (1976).

9.5.2 *Laboratory-oriented Budgetary Method*

As the growth of zooplankton is primarily dependent on the assimilation and incorporation of organic matter in food, growth (production: P) can be calculated indirectly from a knowledge of feeding (F), assimilation efficiency (A), and metabolic expenditure (R). We assume here that soluble organic excretion by zooplankton is negligible. The balanced equation among these parameters is

$$\frac{FA}{100} = R + P \tag{9.52}$$

Then

$$P = \frac{FA}{100} - R \tag{9.53}$$

Note: Assimilation efficiency is a percentage (see Chapter 7, Section 7.6.1). Calories, carbon, or nitrogen are commonly used as the basis for each term.

Conover (1956) calculated the daily production of *Acartia clausi* and *A. tonsa* in Long Island Sound as follows. The average biomass of two *Acartia* was estimated as 20 g wet wt·m^{-2} from a field survey over a two-year period. Laboratory experiments showed a filtration rate for the *Acartia* of 85 ml·mg wet wt^{-1}·day^{-1} at 10.6°C, the annual mean water temperature. Hence, $20 \times 1000 \times 85 = 1.7 \times 10^6$ ml·m^{-2}·day^{-1} = 1700 liters of water was filtered by the *Acartia* under each square meter of water per day. Since the average phytoplankton content of the water was about 5 µg chl a·liter^{-1} or 333 µg C·liter^{-1}, the daily intake of the *Acartia* was $333 \times 1700 \times 1/1000 = 0.57$ g C·m^{-2}. The respiration rate of the *Acartia* measured in laboratory experiments was 0.028 ml O$_2$·mg wet wt^{-1}·day^{-1}, resulting in 560 ml O$_2$·m^{-2}·day^{-1} for the biomass of *Acartia* (20 g wet wt·m^{-2}). A respiration rate of 560 ml O$_2$·m^{-2}·day^{-1} equals a loss of 290 mg C·m^{-2}·day^{-1}. The difference between carbon incorporated and respired is $570 - 290 = 280$ (mg C·m^{-2}·day^{-1}). If we assume an assimilation efficiency of 80%, the difference is $570 \times 0.8 - 290 = 166$ $(mg\ C \cdot m^{-2} \cdot day^{-1})$. This, following Equation (9.53), is production. Neither soluble organic excretion nor molting loss are considered in this calculation.

Lasker (1966) performed extensive laboratory experiments on the feeding, assimilation, respiration, molting, and growth of *Euphausia pacifica*. His results indicate that 5% of an individual's body carbon is used daily for growth, respiration, and molting. As the biomass of *E. pacifica* in the North Pacific was estimated as $0.11\ g\ C \cdot m^{-2}$, the assimilated carbon required daily to maintain this biomass was calculated to be $0.11 \times 0.05 \times 1000 = 5.5\ mg\ C \cdot m^{-2}$. About 9% of the assimilated carbon is used for growth by this animal (excluding eggs and molts). Therefore, the production of *E. pacifica* is $5.5 \times 0.09 = 0.5\ mg\ C \cdot m^{-2} \cdot day^{-1}$.

McAllister (1969) estimated the annual production of mixed species of herbivorous zooplankton at Station P in the northern North Pacific. Data used in his calculation were from a year-round field survey of the standing stocks of phytoplankton and zooplankton and primary productivity and an assumed respiration rate (7% body $wt \cdot day^{-1}$) and assimilation efficiency (65%). Feeding of herbivorous zooplankton was not measured but equated with primary productivity, assuming that herbivorous zooplankton ingest all the primary production. As a result, McAllister obtained a figure of $13\ g\ C \cdot m^{-3} \cdot yr^{-1}$ as an estimate of herbivorous zooplankton production there.

The investigators mentioned above made simplifying assumptions about zooplankton populations or assemblages, that is, the average biomass of a single species or assemblage was integrated in time and/or space. However, significant effects of body size on metabolic (Ikeda, 1974) and feeding (Ikeda, 1977c; Conover and Huntley, 1980) rates have been shown. Thus, a knowledge of the size composition of a given zooplankton assemblage may enable a more accurate estimate of the metabolic or feeding rate of the entire assemblage to be made.

Ikeda and Motoda (1978a) developed a method to calculate the feeding and production of zooplankton. This method uses both laboratory data on the respiration rate as a function of body size and temperature and field data on the size distribution of zooplankton in the Kuroshio region. The biomass of zooplankton collected with a NORPAC net with a 350-μm mesh aperture at each station was divided by the number of individuals in the sample. Frequency distributions of the average body weight of zooplankton at 311 stations during the warm season and 117 stations during the cold season were obtained (Fig. 9.14). These skewed frequency curves were converted to normal distributions by logarithmic transformation. Size distribution analyses of the normalized curves are shown in Table 9.5. The respiration rate for the median body weight of each size class at the given temperature was derived from the equation shown in Figure 8.9 and multiplied by the frequency of each size class. The respiration rates thus obtained were integrated over the six size classes to

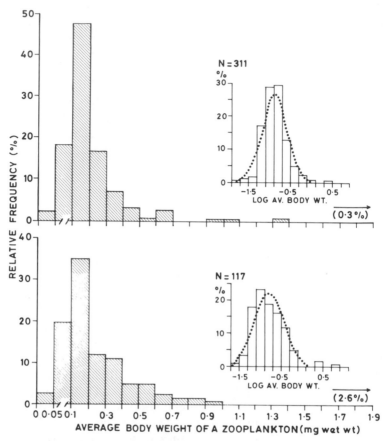

Figure 9.14. Relative frequency of the average size of zooplankton (biomass/no. of individuals, for each sampling) in warm (June–October, upper figure) and cold (December–April, lower figure) seasons in the Kuroshio and adjacent seas. A normalized frequency distribution fitted by logarithmic transformation of body weights is superimposed on the right side of each figure. N is number of sampling stations. (After Ikeda and Motoda, 1978a.)

provide a respiration rate representative of the zooplankton community sampled by the NORPAC net. Feeding and production were computed from this integrated respiration rate using the following sets of balanced equations:

$$\frac{FA}{100} = R + P$$

$$K_1 = \frac{P}{F} \times 100\%$$

and

$$K_2 = \frac{100P}{FA} \times 100\%$$

Table 9.5. Analysis of body size distribution of zooplankton in the Kuroshio and adjacent seas from a normalized catch distribution curve[a]

Class interval		Median body size		Median body size equivalent (mg wet wt · animal⁻¹)		Theoretical frequency (%)
Number	SD	W_i	SD	Warm	Cold	f
1	−3 to −2	W_1	−2.5	0.030	0.020	2.15
2	−2 to −1	W_2	−1.5	0.058	0.051	13.59
3	−1 to 0	W_3	−0.5	0.113	0.120	34.13
4	0 to 1	W_4	0.5	0.218	0.282	34.13
5	1 to 2	W_5	1.5	0.422	0.662	13.59
6	2 to 3	W_6	2.5	0.815	1.553	2.15
				$\sum W_i = 19.63$	26.79	$\sum f = 99.74$

[a] Warm season (June–October), $\mu = -0.8033$, SD $= 0.2856$; cold season (December–April), $\mu = -0.7350$, SD $= 0.3705$. The interval of $\mu \pm 3$ SD of the normal curve was equally divided by the SD class intervals (1–6), and median value in each class interval was taken as the representative body size ($W_1 - W_6$) for that class interval (after Ikeda and Motoda, 1978a).

where K_1 and K_2 are the gross and net growth efficiencies, respectively (see Chapter 7, Section 7.6.2). Rearranging these equations and solving for F and P, we obtain

$$F = \frac{100R}{A(100 - K_2)/100} = \frac{100R}{A - K_1} \qquad (9.54)$$

and

$$P = \frac{K_2R}{100 - K_2} = \frac{K_1R}{A - K_1} \qquad (9.55)$$

Ikeda and Motoda (1978a) assumed K_1 and A were 30% and 70% from the results of laboratory experiments by various workers (see Table 7.4). Thus, Equations (9.54) and (9.55) were simplified to

$$F = 2.5R \qquad (9.56)$$

and

$$P = 0.75R \qquad (9.57)$$

zooplankton biomass was divided into carnivores (29%) and herbivores (71%) and the production rate of the herbivores was corrected for natural physiological mortality. These results suggested herbivores ingested 36–214 mg $C\cdot m^{-2}\cdot day^{-1}$ and produced 10–60 mg $C\cdot m^{-2}\cdot day^{-1}$ in the Kuroshio region (Fig. 9.15). The feeding rate is equivalent to 18–72% of the primary productivity in the area.

Based on the same idea as Ikeda and Motoda (1978a), Conover and Huntley (1980) recently proposed an equation to calculate the production of particle-feeding zooplankton. They also included a term for feeding as a function of food concentration. Their equation therefore includes only one assumed term, A:

$$P = \sum_{i=1}^{j} N_i \left[\frac{A}{100} \left(\sum_{s=1}^{h} f \log\left(\frac{\overline{V}_s}{V_0}\right) e^{0.17T} \overline{W}_i^{0.82} C_s \right) - k\overline{W}_i^{\,r} \right] \qquad (9.58)$$

where N_i is the number of animals in weight class i; j is the number of weight classes of animals in the assemblage; $\overline{W}_i$ is the geometric mean weight of an animal in the weight class i (mg dry wt); A is the assimilation efficiency (assumed here to be 70%); f is the feeding constant (4.85); $\overline{V}_s$ is the geometric mean volume of a particle in size class s; V_0 is the volume of the smallest filterable particle (approximately 25 μm^3); h is the number of size classes of particles which are fed upon; k is the respiration constant ($k = 0.349e^{0.056T}$, where T is temperature in Centigrade); r is the respiration exponent ($r = 0.92e^{-0.016T}$, where T is temperature in Centigrade); and C_s is the concentration of particulate carbon in the size category s (μg $C\cdot ml^{-1}$). Results from the application of this equation have not yet appeared.

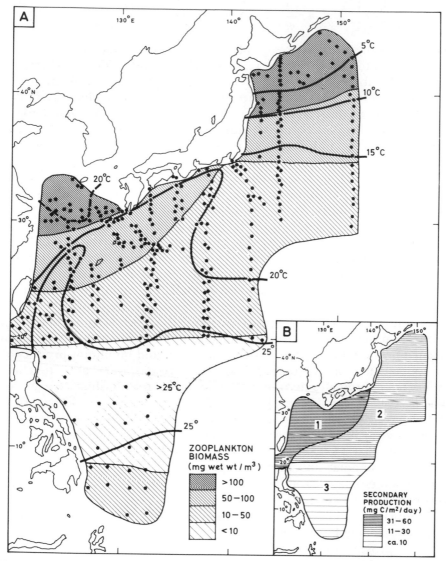

Figure 9.15. (A) Sampling stations, zooplankton biomass, and isotherms (100-m depth, continuous lines; 50-m depth, broken line) during warm season (June–October) in the Kuroshio and adjacent seas. (B) Distribution of estimated secondary production in the same area. (After Ikeda and Motoda, 1978a.)

9.5.3 *Other Methods*

From the analysis of the relationship between phytoplankton biomass (Bp) and zooplankton biomass (Bz, collected with a 300-μm mesh net) on Fladen Ground in the northern North Sea, Steele (1958) derived the following empirical equation:

$$\frac{dBz}{dt} = 0.1Bp - 0.01Bz^2 \tag{9.59}$$

The first term, $0.1Bp$, is the zooplankton production rate and is directly proportional to phytoplankton biomass. The second term, $-0.01Bz^2$, is a negative feedback term (mortality) on the zooplankton production rate and is proportional to the square of the zooplankton biomass. Mullin (1969) explained the second term as likely competition among zooplankton for a limited amount of phytoplankton food. Razouls (1975) used this empirical equation in his study of copepod production in the neritic zone of the Gulf of Lion.

Cushing (1971) made a rough estimate of zooplankton production in an upwelling region from the average zooplankton biomass multiplied by the estimated number of generations during the upwelling season. Compared with other systems, upwelling regions are considered to contain zooplankton, which have greater concentrations of phytoplankton food and less mortality due to predation. The generation length of zooplankton was derived from the growth data of *Calanus finmarchicus* (Marshall and Orr, 1955b) and *Pseudocalanus minutus* (McLaren, 1965) at various temperatures.

In contrast to Steele (1958) and Cushing (1971), who estimated realistic production values for zooplankton, Ikeda and Motoda (1978b) used similar calculations to estimate the possible maximum and minimum production of zooplankton in the Bering Sea. Their results were used to check independent estimates derived from respiration relationships (see above). The potential maximum production (27 g $C \cdot m^{-2} \cdot yr^{-1}$) was derived from the observed primary production by assuming a 100% assimilation efficiency and a 30% net growth efficiency, while the potential minimum production (3.3 g $C \cdot m^{-2} \cdot yr^{-1}$) was estimated from the peak zooplankton biomass in summer and neglected all mortality prior to that time. The production estimated from respiration (13.3 g $C \cdot m^{-2} \cdot yr^{-1}$) fell favorably between these two extreme estimates.

Production estimates in which zooplankton of various species with different feeding habits are treated as a single entity are regarded only as crude approximations. Despite their inherent inaccuracy and uncertainty, calculations based on mixed zooplankton are simple and often provide a clear picture of potential production on a global scale, which is impossible from single species studies unless all components are considered and integrated. The size integration approaches of Ikeda and Motoda (1978a) and Conover and Huntley (1980) mentioned above may be acceptable compromises between the crude "total-zooplankton" approach and the laborious "species-by-species" approach.

The ratio of production to average biomass over the duration of a cohort is the *P/B ratio* and is commonly used by Russian workers to calculate the production of another cohort of the same species from the knowledge of its average biomass (Winberg, 1971). An important feature of the P/B ratio is its constancy: 20 for multivoltine zooplankton (about 4–5 generations·yr^{-1}), 5 for univoltine zoobenthos, 10 for multivoltine zoobenthos (2–3 generations·yr^{-1}), 1.2 for hemivoltine zoobenthos (2–3 year life span), and 0.6 for warm water fishes (≤10 years life span) (Waters, 1977). Although there is an apparent need to examine its constancy for various zooplankton species, the P/B ratio may be a useful method to roughly estimate cohort production.

We now have the theories and appropriate mathematics to calculate zooplankton production, but accurate estimates of the parameters and variables in each mathematical equation are difficult to obtain from field studies and/or laboratory experiments. As stated by Mullin (1969, p. 310): "The single most important problem in research on the production of zooplankton is how adequately to take and evaluate samples from the same discrete population of zooplanktonic organisms on successive occasions. The mechanics of sampling are coming under control, so that the samples are valid estimates in a statistical sense, but sequential sampling of a pelagic population is still extremely difficult because of the movement of water and the patchiness of the populations within it. Even when a particular parcel of water can be tagged and followed, there is no guarantee that the planktonic organisms originally living therein have been followed equally well." Another problem is the difficulty of extrapolating the results of laboratory experiments (feeding, respiration, excretion, growth, etc.) to field populations, as mentioned earlier. With no exception, all calculations of zooplankton production made thus far have been based on several assumptions; thus, the accuracy of the results depends on how well the assumptions are justified. However, justification is difficult in most cases due to our lack of knowledge. Much effort is needed in the future to obtain accurate values of parameters and variables from field studies and laboratory experiments.

For methods to calculate the production of aquatic animals in general, we advise the readers to consult the books of Winberg (1971) and Edmondson and Winberg (1971). Reviews by Mullin (1969) and Tranter (1976) of marine zooplankton production discuss the problems in detail.

Notes

[1] Organic matter leaking from fecal pellets and diffusing out from the body surface (e.g., amino acids) are included. Direct measurements of these losses have not been made for the zooplankton, but a substantial amount (10–30% of assimilated organic matter) has been reported for benthic animals (Kofoed, 1975).

[2] The terms *direct development* (*type*) and *nondirect development* (*type*) used here are analogous to "holometabola" and "hemimetabola," respectively, used to describe the development of insects (Abercrombie et al., 1951).

[3] Thomson (1947) classified developmental stages of Chaetognatha into five "age classes" on the basis of the status of maturation of the ovary. Kotori (1976) divided development into eight stages based on external morphological characters.

[4] Crustaceans are of the nondirect development type and develop in a stepwise manner by molting. However, larvae of Cladocera and Mysidacea after their release from the female body have forms similar to the adults. Many other crustaceans also lose much of their morphological distinctiveness after they reach the juvenile stage. Therefore, age class analysis based on morphological characters is difficult to apply for some nondirect development types. In this case, growth of larvae or juveniles can be estimated from the increase in body size as is done for animals belonging to the direct development type.

[5] Winberg (1971, p. 115) notes that the growth curve of copepods is more likely to be parabolic than S shaped, particularly if the production of eggs is included in the curve.

[6] Among the many laboratory conditions known to affect to the growth of zooplankton, food (quantity and quality) and temperature appear to be the most important factors (see Paffenhöfer and Harris, 1979). When food is optimal, growth of animals is primarily controlled by temperature, and this relationship is expressed by Bělehrádek's (1935) function:

$$D = a(T - \alpha)^{-b} \tag{9.22}$$

where D is development time (days), T is temperature (°C), a is a constant, b is a common constant for animal groups adapted to a certain temperature range, and α is a constant called the *biological zero*.

[7] Certain situations can lead to conceptual difficulties with this equation and other methods. Some female invertebrates lose a considerable fraction of their body weight as eggs at the time of spawning. If W_{i-1} was measured before and W_i after spawning, the investigator might erroneously conclude that production had been negative during the interval $i - 1$ to i.

[8] The reason for this assumption is the difficulty of separating and counting the six naupliar stages in the samples. Mullin and Brooks (1970a) also grouped the nauplier stages of *Calanus helgolandicus* in their production study. Recently, Durbin and Durbin (1981) counted separate naupliar stages of *Acartia hudsonica* and *A. tonsa* in Narragansett Bay, Rhode Island, to compute production.

[9] Shushkina (1968) used the equation containing K_2 to estimate the production of *Haloptilus longicornis* in the Fiji Sea. The values of K_1 and K_2 by Shushkina (1968) and Ikeda and Motoda (1978a), respectively, are from laboratory experiments with abundant food and, therefore, do not consider the possible effects of food limitation. We consider that the use of set values for A and K_1 is more advantageous than the use of a single value of K_2, as the former allows the calculations of feeding and production simultaneously.

CHAPTER 10 _____

Distribution
and Community
Structure

Unlike the previous chapters, which concern methodology, this chapter includes conceptual information. We introduce methods to describe and analyze certain aspects of distributional pattern and community (or assemblage) structure.[1]

Popularization of computers has made the application of sophisticated analytical procedures easy today, and it is sometimes felt that valuable information might be extracted from poor data by using sophisticated mathematical methods. This is wrong because the information content of data cannot be increased above its original level by any means.

As described in Chapter 2, temporal and spatial variation of plankton samples are always considerable. The range of variation between the samples taken at the same station with few hour intervals is often large because of various factors, such as a tidal advection and patchy distribution of plankton. Furthermore, there is the possibility that organisms of the different communities that inhabit different waters are mixed during sampling, and thus the sample consists of an "artificial community." Sophisticated mathematical procedures are very useful when dealing with ecological situations where a pattern is not clear. However, the greatest understanding is achieved only after the analysis of data obtained in a careful and detailed sampling program.

10.1   Distributional Patterns in Relation to Environmental Conditions

The preference for and tolerance of a certain environmental factor differs between each zooplankton species and each stage in its life cycle. Thus, organisms are classified by the extent to which the species may be widely (eury-) or narrowly (steno-) tolerant of factors such as temperature and salinity. The occurrence of species, their distributional patterns, and variations in plankton communities and assemblages are influenced to some extent by environmental factors that can be measured. But organisms are also affected by complicated interactions between the known environmental factors and others that are unmeasureable or unsuspected. Furthermore, a species may reflect in its distribution the climatic, geographical, and biological conditions that it experienced in the past more than those of its present environment. There is the danger of misinterpretation when the relationship between the distribution and environment of the zooplankton is discussed only on the basis of observations of its present condition.

For marine zoogeography the reader is referred to the excellent texts by Ekman (1953), Van der Spoel and Pierrot-Bults (1979), and Nishimura (1981). The primary biogeographical classification of a zooplankton species is based on its regional distribution range and is characterized in terms of both its near-shore versus offshore occurrence and its latitudinal distribution. A species may thus be described as being neritic, coastal, or oceanic and also as being arctic, antarctic, subarctic, subantarctic, temperate, subtropical, or equatorial.

Many investigators have concluded that in the open ocean there are characteristic communities in each of the different water masses (i.e., McGowan, 1974; Reid et al., 1978b). In the North Pacific Ocean, for example, species are classified as subarctic species, transitional water species, central water species, equatorial species, and eastern Pacific tropical species, as shown in Figure 10.1.

The gross patterns of geostrophic oceanic circulation form the basis for these different biological provinces. The principal cyclonic gyres occur at subarctic and subantarctic latitudes and extend toward the equator along the eastern boundaries. The surface waters of these cyclonic gyres are cold, high in nutrients, and undergo large seasonal changes. A relatively small number of species are indigenous to these gyres, but the total zooplankton biomass is relatively large. On the other hand anticyclonic gyres occur in the subtropical zones. Their surface waters are warm, low in nutrients, and have less seasonal variations than occurs at higher latitudes. The number of species is great but the community biomass is small. The equatorial zone is the warmest of the zones at the surface, but the water beneath the upper layer is cooler than in the anticyclonic gyres. It also contains a large number of species and a large biomass, but there is substantial east–west variation, with some species confined to eastern longitudes. Recently, detailed morphological comparisons of epipelagic species previously considered to be ubiquitously distributed have demonstrated genetic divergence between different oceans or both sides of an ocean. Between the north and south latitudes of 20° to 30°, the continents are a barrier to the spread of species, especially for zooplankton which

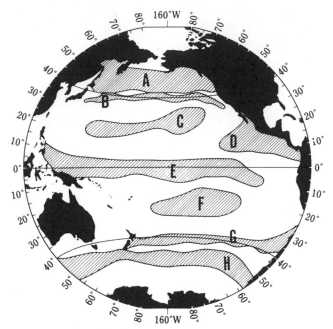

Figure 10.1. The distributional patterns of the basic biotic provinces of the oceanic Pacific. (A) Subarctic fauna; (B) North Pacific transitional zone fauna; (C) North Pacific central fauna; (D) Eastern Pacific tropical fauna; (E) Equatorial fauna; (F) South Pacific central fauna; (G) South Pacific transitional zone fauna; (H) Subantarctic fauna. (After McGowan, 1974.)

reproduce in the surface layers of tropical and subtropical regions, and so few species occur in more than one ocean (Fleminger and Hulsemann, 1973). In coastal waters, where the local geographical environment greatly affects the isolation of a community, species composition varies significantly from place to place and the differentiation of species is even more evident than in the open sea (Omori, 1977).

Figure 10.2 shows the regional distribution range of copepods, tunicates, and mollusks, sampled by Hardy continuous plankton recorder, in the Northeast Atlantic Ocean. It is clear that a particular species may be associated with a particular area, and hence the distribution of certain plankton species whose habitat preferences are known may provide insight into the distributioins of water masses and their changes. We call such organisms *indicator species*.

The conventional method of investigating environmental conditions with indicator species is to use either presence–absence data for one or more indicator species within a sample or the ratio of the abundance of one or more indicator species to the abundance of all zooplankton in the assemblage. In the former case species with very weak tolerances to

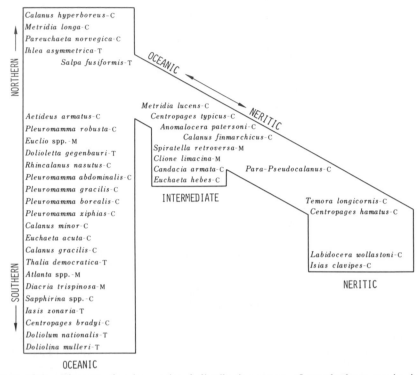

Figure 10.2. Diagram showing regional distribution range of zooplankton species in the eastern North Atlantic. Letters following the specific names indicate systematic groups: C, Copepoda; M, Mollusca; T, Tunicata. (Modified after Colebrook et al., 1961.)

changes of one or more environmental factors are used. In the latter case quantitative relationships between the factors and the assemblage's variability are used; for example, some species are widely distributed and in low abundance under normal conditions but are abundant under specific environmental conditions. These approaches require an investigator to have good insight into the regional and seasonal occurrence and abundance of the zooplankton, especially of the species in question, as well as a knowledge of any relevant autoecological studies.

There have also been attempts to use distributional data for particular species to determine the movement and velocity of specific water masses. Omori (1967) found that the distribution of *Calanus cristatus,* a typical subarctic Pacific copepod species, extends from the Oyashio region to the southern coast of Honshu, Japan. He deduced from geographical variations in its depth distribution, population density, degree of maturity, weight, and chemical composition, that individuals produced north of 40° N descend and drift to the south with the movement of the cold water mass. From this, he calculated that the average velocity of the water mass is 5.2 cm·sec^{-1}. The analysis of problems involving the dispersion and transportation of a zooplankton species is possible only when the life history is thoroughly known and considerable autoecological information is available.

Plankton species will continue to be useful for general monitoring of certain aspects of the environment, such as hydrographic events, eutrophication, pollution, warming trends, and long-term changes symptomatic of environmental disturbances. Organisms respond to a vast array of environmental conditions, more than can be analyzed on a routine bases. Marked abnormalities of species composition may indicate the need for more intensive physical and chemical analyses. However, analysis of water masses by the use of indicator species is indirect when compared to the use of physical and chemical data, and it is also time consuming. Further, it must always be remembered that organisms have the ability to adapt to changing environments and there is always a considerable time lag in the response of biological populations when compared to that of physicochemical variables. Even if a particular species disappears from certain waters, it is often difficult to determine whether this indicates the present actual environmental changes in the sampling area. Some species used as indicators today were originally selected without statistical analysis of records of their occurrence, and their biological characteristics should now be verified for their continued use as indicators.

10.2 Associations

The concept of an association of species is based on the similar responses of individual species to variations in natural properties of the environment. The recurrence of an association is assumed to indicate predictable interactions among species and between species and their environment. By investigating the similarities of species associations between samples from different locations and times, it may be possible to deduce characteristics of the physicochemical and/or biological environments of the waters investigated.

Generally, grouping involves two steps; the first step is to measure the degree of similarity (association) between the samples (or species) to be grouped and the second step is to determine the groups of similar samples (or species) by applying a clustering analysis using a correlation matrix. Many procedures have been developed to measure similarity (association); some of these methods of analysis are also applicable to numerical taxonomy. We deal with several methods that are frequently used.

10.2.1 Correlation Matrix

Correlation of occurrence between species is generally expressed by a numerical index for each pair. Consider a data set consisting of counts of

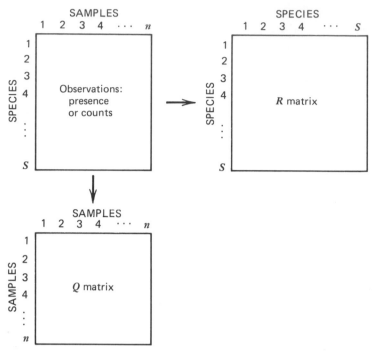

Figure 10.3. Derivation of Q and R matrices for Q (similarity between samples) and R (association between species) analyses.

s species from n different sampling stations. These data can be represented as an $n \times s$ matrix in which each element is a species count for a sample. There are two main methods of proceeding: by forming either an $s \times s$ matrix of correlations between species or an $n \times n$ matrix of correlations between samples (Fig. 10.3). The first type of analysis is called R analysis and is commonly used to reveal species groups within particular habitats, whereas the second type, or Q analysis, is instructive in helping to group samples with similar communities and indicate common ecological conditions between samples from the relative distribution of species.

Figure 10.4 contains a correlation matrix of an R analysis based on the distribution of 22 species of epipelagic zooplankton around the United Kingdom (Colebrook, 1964). Colebrook obtained correlation indices for each species pair (see original article for details of the grouping procedure) and classified the species by association into five groups: group A, northern oceanic species; group B, southern oceanic species; group C, southern intermediate species; group D, neritic species; and group E, northern intermediate species.

Species may be ordered within a matrix by trial and error so that as many as possible of the highly positive correlations appear near the

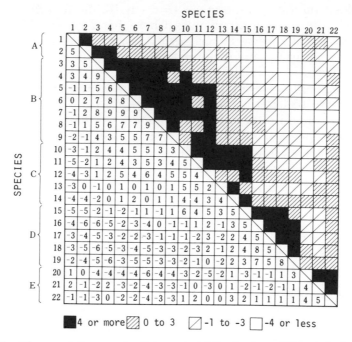

Figure 10.4. The correlation matrix of the regional distributions of 22 zooplankton species occurring in the waters around the U.K. The correlation indices are given to one decimal place and multiplied by 10. (Modified after Colebrook, 1964.) Species are: (1) *Pareuchaeta norvegica*; (2) *Acartia* spp. (Atl.); (3) *Pleuromamma robusta*; (4) *Rhincalanus nasutus*; (5) *Calanus gracilis*; (6) *Calanus minor*; (7) *Pleuromamma gracilis*; (8) *P. borealis*; (9) *Centropages bradyi*; (10) *C. typicus*; (11) *Candacia armata*; (12) *Corycaeus* spp.; (13) *Euchaeta hebes*; (14) *Calanus helgolandicus*; (15) *Paracalanus* spp. and *Pseudocalanus* spp; (16) *Acartia* spp. (N. Sea); (17) *Labidocera wollastoni*; (18) *Centropages hamatus*; (19) *Temora longicornis*; (20) *Calanus finmarchicus*; (21) *Clione limacina*; (22) *Spiratella retroversa*.

principal diagonal. An interactive procedure and other approaches to produce results equivalent to maximizing the diagonal of the correlation matrix have also been proposed (Beum and Brundage, 1950; Margalef and González, 1969).

10.2.2 *Similarity Indices*

There are two main approaches to assess association: first, by indices based on presence–absence or binary data and, second, by indices based on quantitative (count) data. Here we explain the measurement of association using the first approach.

Similarity may be based on the comparison of the number of joint occurrences of species with their total number of occurrences. When two species *A* and *B* are compared on the basis of their presence or absence in samples from different locations (R analysis), the following values are computed: the number of samples where both species occur (a); the number of samples where either species A or B occurs without the other (b and c, respectively); and the number of double absences (d). If N is the total number of samples, then $N = a + b + c + d$.

When similarity of two samples A and B is measured (Q analysis), the total number of species in data set N is $a + b + c + d$; a is the number of species common to both samples, b and c are the number of species that occur in one sample but not the other, and d is number of species present in the data set but not present in these two samples.

The simplest expression of the similarity between samples (or species) is Jaccard's (1902) coefficient of similarity (C):

$$C = \frac{a}{a + b + c} \tag{10.1}$$

Other simple similarity coefficients are Sørensen's (1948) quotient of similarity (Q) and Mountford's (1962) index of similarity (I):

$$Q = \frac{2a}{2a + b + c} \times 100 \tag{10.2}$$

$$I = \frac{2a}{2bc - (b + c)a} \tag{10.3}$$

If $a = 0$, that is, when there are no common samples (or species), all of the above coefficients are zero.

A more sophisticated coefficient has been proposed by Fager (1957) and Fager and McGowan (1963) in a study of zooplankton species associations in the North Pacific. It is called "recurrent group analysis" and is a measure of how often species occur together. The similarity index of affinity (S) between two species A and B in a series of samples is computed as follows:

$$S = \frac{a}{\sqrt{(a + b)(a + c)}} - \frac{1}{2\sqrt{a + c}} \qquad c \geq b \tag{10.4}$$

Species designations must be assigned so that $c \geq b$. Fager and McGowan (1963) used 0.5 as a clustering bound and considered pairs of species which had a value of S greater than or equal to this predetermined value to show positive affinity for each other. Species are then arranged into the largest possible groups within which all species pairs show positive affinity. Species that are not assigned to a group but show positive affinities with species within one or more groups are considered associates of those groups. The preparation of the incidence matrix and grouping procedure have been outlined by Fager (1957) and Fager and McGowan (1963). Their method has been made operational in a computer program, but any agglomerative clustering procedure, for example, complete-linkage clustering, could be used, refining the groups of associated species by hand. The result indicates which species are common elements of each other's environment but not which species respond similarly to that environment. This "recurrent group analysis" has been used often by American plankton researchers. In addition to Fager and McGowan (1963), Venrick (1971) on diatom species, and Rottman (1978), Brinton (1979), and McGowan and Walker (1979) on zooplankton are useful references.

Using the index of affinity based upon presence–absence data, rather than quantitative estimates, the effect of sampling errors may be greatly reduced. However, with this correlation method, a very rare species will often not be shown to associate with a more common species with which it always co-occurs because the number of joint occurrences between the two species may be an insignificant portion of the total number of the occurrences of the more common species. For this reason it is recommended to use only fairly abundant and frequently occurring species in this correlation analysis and to eliminate rare species. One must note, however, that this procedure cannot avoid the loss of information inherent in minor species counts which are disregarded but likely to reflect real differences in distributional patterns. On the other hand, presence–absence similarity measures in Q analysis tend to overemphasize rare relative to dominant species, for similarity is calculated from the total number and number of jointly occurring species, assuming all species are equivalent. In zooplankton samples at higher latitudes, most individuals often belong to a few or even a single species. If two samples have many rare species in common but also contain a small number of dominant species that are not shared, the similarity between the two samples will be estimated to be high.

Taking these factors into account, Morisita (1959) compared the similarity between two quantitative data sets using Simpson's (1949) diversity index (λ; see Section 10.3.1) and a related similarity index (C_λ);

$$\lambda_1 = \frac{\sum\limits_{i=1}^{S} n_{1,i}(n_{1,i} - 1)}{N_1(N_1 - 1)} \quad \text{and} \quad \lambda_2 = \frac{\sum\limits_{i=1}^{S} n_{2,i}(n_{2,i} - 1)}{N_2(N_2 - 1)}$$

$$C_\lambda = \frac{2 \sum\limits_{i=1}^{S} n_{1,i} n_{2,i}}{(\lambda_1 + \lambda_2) N_1 N_2} \qquad 0 \le C_\lambda \le 1 \tag{10.5}$$

where N_1 and N_2 are the total number of individuals in samples A and B, $n_{1,i}$ and $n_{2,i}$ are the number of individuals of species i in the two samples, respectively, and S is the total number of species to be compared.

Table 10.1.　Example of species composition

Species	Sample A abundance (n_1)	Sample A % abundance $[(n_1/N_1)100]$	Sample B abundance (n_2)	Sample B % abundance $[(n_2/N_2)100]$
a	26	13	46	23
b	28	14	38	19
c	52	26	38	19
d	60	30	44	22
e	28	14	12	6
f	6	3	22	11
Total	$N_1 = 200$		$N_2 = 200$	

Kimoto (1967) recommends the C_π index as a revised form of the equation above:

$$\sum \pi_1{}^2 = \frac{\sum_{i=1}^{S}(n_{1,i})^2}{N_1{}^2} \quad \text{and} \quad \sum \pi_2{}^2 = \frac{\sum_{i=1}^{S}(n_{2,i})^2}{N_2{}^2} \tag{10.6}$$

$$C_\pi = \frac{2\sum_{i=1}^{S}n_{1,i}n_{2,i}}{(\sum \pi_1{}^2 + \sum \pi_2{}^2)N_1 N_2} \qquad 0 \le C_\pi \le 1 \tag{10.7}$$

If the species in the two samples are identical, C_π becomes 1, and if the two samples do not contain any common species, it becomes 0.

As an example, assuming we obtained the data shown in Table 10.1 for six species of zooplankton in two samples, C_λ and C_π are calculated as follows:

$$\lambda_1 = 0.2107$$

$$\lambda_2 = 0.1851$$

$$C_\lambda = \frac{2 \times (26 \times 46 + 28 \times 38 + 52 \times 38 + 60 \times 44 + 28 \times 12 + 6 \times 22)}{(0.2107 + 0.1851) \times 200 \times 200} = 0.9277$$

$$\sum \pi_1{}^2 = 0.2140$$

$$\sum \pi_2{}^2 = 0.1892$$

$$C_\pi = \frac{2 \times (26 \times 46 + 28 \times 38 + 52 \times 38 + 60 \times 44 + 28 \times 12 + 6 \times 22)}{(0.2140 + 0.1892) \times 200 \times 200} = 0.9107$$

The percent similarity index (Whittaker, 1952) is also an easy and frequently used method of analysis. Two samples are compared with respect to the percent composition of their component species. Within each sample species abundances are converted to percent of the total abundance of all species, that is, $(n/N) \times 100$. The percentage index (P) is calculated as follows:

$$P = 100 - 0.5 \sum_{i=1}^{S} | P_{a,i} - P_{b,i} |$$

$$= \sum_{i=1}^{S} \min(P_{a,i}, P_{b,i}) \tag{10.8}$$

where $P_{a,i}$ and $P_{b,i}$ are the percent abundances of species i in samples A and B, respectively, and S is the total number of species to be compared. If we again take the data in Table 10.1 as an example, P is $13 + 14 + 19 + 22 + 6 + 3 = 77\%$. When the similarity is high, P approaches 100%. A defect of this method of calculation may be that in some sense rare species contribute little to the index as the measure often tends to be biased towards the more abundant, common species.[2]

10.2.3 *Use of Similarity Indices for Numerical Taxonomy*

In general, the morphological heterogeneity of a taxon, for example family or genus, increases in relation to the number of species included. In order to clarify relationships among species and, if possible, divide taxonomic groups into subgroups on the basis of morphology, a similarity analysis may be used. Such subgrouping is useful not only in taxonomy but also in the interpretation of ecological relationships between species.

Some similarity indices are used for numerical taxonomy (see Sneath and Sokal, 1973, for the principles of numerical taxonomy). A matrix is set up consisting of a number of (ideally more than 20) morphological characters and a number of operational taxonomic units composed of several species. All characters are weighted equally to eliminate any subjective bias other than that involved in their selection and they are compared one by one for each combination of two taxonomic units.

We show a case of the morphological similarity between two species expressed by a nonparametric index of similarity (Sheals, 1964; Matthews, 1972). Morphological characters (e.g., the mean body length) for which there is a range of values on a linear scale are treated quantitatively:

$$S_i = 1 - \frac{| X_{i,j} - X_{i,h} |}{r} \tag{10.9}$$

where S_i is the coefficient of similarity, $X_{i,j}$ and $X_{i,h}$ are the ranked or absolute measurements of the ith character for the species j and h, and r

(>0) is the difference in the observed range of the character over all species under consideration. Purely qualitative, unranked characters (e.g., the shape of distal margin of the last thoracic segment; round or pointed) are scored on a match or mismatch basis, so that when characters match, $S_i = 1$, and when they do not, $S_i = 0$. The similarity of the two species is then calculated as

$$S_{j,h} = \frac{S_1 + S_2 + \cdots + S_n}{n} \tag{10.10}$$

where $S_1, S_2, \ldots, S_n$ are the scores (coefficient of similarity) for each of the characters used in the comparison, and n is the total number of characters considered when there are no missing values.

For example, when two species (j and h) have the characters listed below, $S_{j,h}$ is determined as follows.

Character: 1. Mean body length (mm); 2. Shape of distal margin of the last thoracic segment; 3. Rostral filament; 4. Number of spines on the genital segment; 5. Number of exopodal segments of the first leg.

Character	1	2	3	4	5
Species j	6.6	round	not exist	4	2
Species h	10.0	round	exist	5	3
Range of the character over all species	6.0–11.5	round or pointed	exist or not exist	2–6	1–3
Difference (r)	5.5	1	1	4	2
Score	0.38	1	0	0.75	0.50

$$S_{j,h} = \frac{0.38 + 1 + 0 + 0.75 + 0.50}{5}$$

$$= 0.53$$

The matrix of similarity indices is then examined for taxonomic groups based upon S values by the use of one of the clustering methods described in the following section.

10.2.4 Cluster Analysis

When the number of samples or species to be compared is numerous, linkage clustering within a similarity matrix is often a convenient method of illustrating relationships. The result of clustering analysis is generally a dendrogram. There are several strategies for determining clusters and

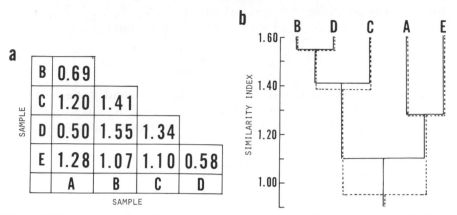

Figure 10.5. An example to illustrate a clustering analysis with five samples. (a) Similarity matrix. (b) Clustering with simple-linkage method (solid line) and Mountford's method (dotted line).

each of these grouping methods has characteristics and limitations, so when interpreting dendrograms it is necessary to be aware of the limitations of the methods used (see Sneath and Sokal, 1973). We explain the simple-linkage clustering, Mountford's (1962) method which is a type of average-linkage clustering, and complete-linkage clustering.

In simple-linkage the sample (or species) pair with the highest similarity value is chosen and then the group is progressively enlarged by adding the sample or species with the next highest similarity to any of the clustered group. For example, in the similarity matrix shown in Figure 10.5, the highest value is 1.55 for B and D, and thus B and D are linked on the 1.55 level. The second highest similarity is 1.41, between B and C, so C is linked to BD on the 1.41 level, forming a BCD group. The third highest link, at 1.34, is between C and D, but these are already in the same group. Next a new group is formed at the 1.28 level by A and E. Finally, at the similarity value of 1.10 for C and E, the BCD group is linked to the AE group, thus completing the clustering (Fig. 10.5b). This method is very simple and is the easiest to translate into computer algorithms. Clustering is, however, influenced greatly by the similarity value for a single pair among the members of each group.

In Mountford's method the pair with the highest value in the matrix is first chosen, and the similarity between this combined pair and the remaining unit is then calculated to form a new matrix using the mean similarity between each unclustered unit and all of the clustered group. The sample or species pair with the highest value in the new matrix is then picked out, and the process is repeated until the clustering is complete.

The similarity coefficient (I) between group A (A_1, A_2, ..., A_m) and B (B_1, B_2, ..., B_n) is calculated anew using the following equation:

$$I(A_1, \ldots, A_m : B_1, \ldots, B_n) = \frac{1}{mn} \sum_{i=1}^{m} \sum_{j=1}^{n} I(A_i B_j) \qquad (10.11)$$

where m and n are the number of individuals in A and B, respectively. For example, consider the data in Figure 10.5. First, a matrix is formed by calculating the similarities between BD (the highest value) and A, C, and E:

$$I(BD:A) = \frac{I(BA) + I(DA)}{2} = \frac{0.69 + 0.50}{2} = 0.60$$

$$I(BD:C) = \frac{I(BC) + I(DC)}{2} = \frac{1.41 + 1.34}{2} = 1.38$$

$$I(BD:E) = \frac{I(BE) + I(DE)}{2} = \frac{1.07 + 0.58}{2} = 0.83$$

BD	0.60		
C	1.20	1.38	
E	1.28	0.83	1.10
	A	BD	C

Since the similarity between BD and C is the greatest of this matrix, $(BD)C$ is formed and the similarity coefficients among $(BD)C$ and A and E are calculated. By proceeding in this manner to link the highest levels in each matrix, that is, AE at 1.28 and then $(AE)(BD)C$ at 0.94, a dendrogram is created (Fig. 10.5b).

Complete-linkage clustering is the direct antithesis of the simple-linkage technique. The initial linkage is between mutually most similar pairs, which are the same as by the previous methods (BD and AE, respectively). Then a sample (or species) that is candidate for admission to an extant cluster has similarity to that cluster equal to its similarity to the least similar member within the cluster (CD at the 1.34 level). When two clusters join, their similarity is that existing between the least similar pair of members, one in each cluster (DE at the 0.58 level).

10.2.5 Principal Component and Factor Analyses

Principal component and factor analyses are multivariate techniques which generate a sequence of variates known as components or factors in a correlation matrix. These analytical methods have been successfully used in terrestrial and marine ecology since Goodall's (1954) initial study of vegetative populations. In zooplankton investigations these analyses have been used by Colebrook (1964), Ibanez and Seguin (1972), and Angel and Fasham (1973) to extract information on the spatial patterns of species from complex correlation matrices. Colebrook (1977) also used principal component analysis to determine major taxonomic groups of zooplankton which undergo yearly fluctuations in biomass in the California Current.

The theory and analytical (mathematical) procedures of the analyses are complicated and detailed explanation is beyond the scope of this book. We discuss only concepts of the principal component and factor analyses and suggest that the reader consult additional references (e.g., Lawley and Maxwell, 1963; Harman, 1967) as necessary.

The basic difference between principal component and factor analyses is that in the former it is assumed that all the variance is common and its orthogonal components can be extracted, whereas in the latter the variance common to all variables is separated from the specific and error (residual) variances. In principal component analysis the components are row vectors corresponding to rectangles. The first latent vector generates a component (component I) which explains the largest possible variance in the data. The second vector generates a component (component II) which has the largest possible variance in the residual following the removal of the variability associated with the first component, and so on. Each vector is orthogonal and the components derived from respective vectors are not correlated. If the original data are coherent to any extent, the first few components account for a large proportion of the variability of the original data array.

As a simple example of principal component analysis, we show the results of Reid et al. (1978a) who investigated phytoplankton assemblages off southern California. They investigated spatial relationships among species based upon their numerical abundances in samples taken from the chlorophyll maximum layer and the surface and from offshore and inshore stations. Table 10.2 shows the results of an analysis of the 14 most important species using log-transformed, numerical data.[3] We see that half a dozen or so species (dinoflagellates and the coccolithophorid *Calciosolenia murrayi*) are positively correlated among themselves and contribute most to component I while a different half dozen (diatoms and the ciliate *Mesodinium rubrum*) contribute most to component II. Together, components I and II account for 72% of the variance in the system. Components III and greater appear to be mostly noise and the consequence of nonlinearity. In Figure 10.6, 58 samples from Santa Monica Bay and the lower (southern) coastal region are plotted in the space defined by components I and II. There are three clusters of samples: the uppermost cluster consists of samples from Santa Monica Bay; the lower left cluster consists of all the remaining surface samples and three samples from the chlorophyll maximum layer in the farthest offshore stations; whereas the lower right cluster consists of all the remaining samples from the chlorophyll maximum layer.

Table 10.2. **Principal component analysis of the 14 most important species in the surface and chlorophyll maximum layer off southern California**[a]

Species	Component I	Component II	Component III
Prorocentrum sp.	0.597	−0.077	0.068
Gymnodinium sp.	0.416	−0.063	0.064
Ceratium kofoidii	0.324	0.158	0.165
Gymnodinium splendens	0.318	0.096	0.326
Eucampia zodiacus	0.283	0.123	−0.757
Cochlodinium catenatum	0.266	0.016	0.046
Calciosolenia murrayi	0.228	−0.104	0.106
Eutreptiella gymnastica	0.123	0.285	0.177
Rhizosolenia fragilissima	0.111	0.365	−0.315
Hemiaulus sinensis	0.000	0.434	−0.076
Mesodinium rubrum	−0.003	0.382	−0.049
Leptocylindrus danicus	−0.027	0.203	−0.025
Skeletonema costatum	−0.099	0.497	0.363
Thalassiothrix frauenfeldii	−0.151	0.304	−0.010
% of variance	41	31	9

[a] After Reid et al. (1978a).

Evidently, the surface and chlorophyll maximum layer samples separate along the axis of component I. The group of dinoflagellates contributing to component I is more strongly represented in the chlorophyll maximum layer than in the surface waters. The three anomalous samples from the chlorophyll maximum layer which clustered with the surface samples indicate that the floristic distinctness of the chlorophyll maximum layer is

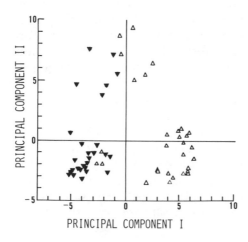

Figure 10.6. Principal component II vs. principal component I. △, Samples from the chlorophyll maximum layer; ▼, surface samples. (After Reid et al., 1978a.)

a near-shore phenomenon and the composition of surface and deep assemblages in terms of the abundances of important species becomes similar offshore. The Santa Monica Bay samples separate from the others along the axis of component II and are dominated by a group of diatoms and a ciliate. For the Santa Monica Bay samples, surface and chlorophyll maximum layer samples continued to separate along the component I axis but less so than for the southern stations. This indicates that characteristics of a chlorophyll maximum layer assemblage in Santa Monica Bay were not as strongly developed as in other waters.

10.3 Diversity

The number of species and their relative abundances are the basis of such descriptions as "simple," "complex," or "dominated by one or a few species," each of which refers to the general term *diversity*. Thus, diversity indicates the degree of complexity of community structure. It is a function of two elements, one being the number of species (richness) and the other the "equitability" (evenness) with which individuals are distributed among species. Diversity is a concise expression of how individuals in a community are distributed within subsets of species. Diversity decreases when a community becomes dominated by one or a few species, when individuals of rare species are replaced by individuals of a more common species, or when one or a few species rapidly reproduce. Diversity is often related to certain environmental characteristics of water masses and the degree of complexity of the flow of energy within the community. The measurement of the temporal variation of diversity provides useful information on the succession of community structure. It may also be used as an index for assessing the degree of environmental pollution. When polluted and unpolluted waters are compared, the diversity is generally lowest in polluted waters.

Species diversity is generally greater in the waters of lower latitudes than in higher latitudes. It is often higher in marine environments than in freshwater and is lowest in brackish water regions. Diversity is also reported to be higher in the deep sea than in shallow waters, although the deep-sea diversity then declines in the great depths. These correlations have not yet been clearly explained, but higher diversity tends to be seen in areas of greater environmental stability (or predictability), especially if associated with higher levels of productivity. References are, for example, Connell (1978), Dumbar (1960), Hessler and Sanders (1967), Menge and Sutherland (1976), Paine (1966), and Patten (1962).

As in the analysis of similarity, it is often practically impossible to measure the diversity of an entire community with consideration of all species. Therefore, the analysis is made only on taxonomic or ecological groups of organisms thought to be important, and conclusions are drawn about the entire community (or assemblage) on the basis of these results.

Diversity should be investigated with more than one index, so as to clarify function of the above-mentioned two elements, that is, species richness and evenness.

10.3.1 *Indices of the Number of Species*

Margalef (1951) created a diversity index (D) for a relatively small community, considering the number of species in a collection of replicate samples approximately proportions to the logarithm of the number of individuals:

$$D = \frac{S - 1}{\ln N} \qquad (10.12)$$

where S is the number of species and N is the total number of individuals in the sample. When the community is composed of a single species, $D = 0$. This is the simplest method of expression but is not often used because it is unsuitable for comparing samples of greatly different sample size (number of individuals). This index is also especially sensitive and variable when the sample size is small and should not be used when N is smaller than 100.

Simpson's index (1949) does not require the assumption of a regular distribution of individuals within species. It expresses the probability that two specimens taken at random belong to the same species. Such a probability (λ) is

$$\lambda = \sum_{i=1}^{S} \frac{n_i(n_i - 1)}{N(N - 1)} \qquad (10.13)$$

where S is the total number of species, N is the total number of individuals, and n_i is the number of individuals of the species i. For a large number of species and individuals, setting $P_i = n_i/N$, the following approximation is adequate:

$$\lambda \simeq \sum_{i=1}^{S} P_i^2 \qquad (10.14)$$

The numerical value of this expression is low for a high diversity and its maximum value is 1 for a community of one species, so it is more intuitive to use $1 - \lambda$ as a measure of diversity.

10.3.2 *Indices of Information Content*

The mathematical theory of information provides us with an index that is also useful in other fields of science as a measure of the possibilities of choice. Margalef (1957, 1958) expressed diversity as the information content per individual (H); the units of H are "bits":[4]

$$H = \frac{1}{N} \log_2 \frac{N!}{n_1! n_2! \cdots n_s!}$$

$$= \frac{3.321928}{N} \left(\log_{10} N! - \sum_{i=1}^{S} \log_{10} n_i! \right) \tag{10.15}$$

where N is the total number of individuals and $n_1, n_2, \ldots, n_s$ are the respective numbers of individuals of each species. The term H indicates the uncertainty involved in predicting whether an individual chosen at random belongs to a specific species (average information content), and thus the prediction becomes less certain as H increases.

When N and n_i are sufficiently large, the information content per individual can be approximated from the following equation adopted by Shannon and Weaver (1949):

$$H' = - \sum_{i=1}^{S} P_i \log_2 P_i \tag{10.16}$$

where S is total number of species and P_i is the proportion of the number of individuals of species i to the total number of individuals, that is, $P_i = n_i/N$; then one might as well cast the equation for H' directly in terms of the observed values of n_i:

$$H' \approx \frac{3.321928}{N} \left(N \log_{10} N - \sum_{i=1}^{S} n_i \log_{10} n_i \right) \tag{10.17}$$

The functions $\log_{10} n!$ and $n \log_{10} n$ for all integers from $n = 1$ to $n = 1050$ have been provided by Lloyd et al. (1968) and are given in the Appendix (Table A.1). The maximum value of H' is indicated as H'_{max}.

$$H'_{max} = - \sum_{i=1}^{S} \frac{1}{S} \ln \frac{1}{S}$$

$$= \ln S \tag{10.18}$$

where H'_{max} increases with increasing S.

Sheldon (1969) proposed e^H (and $e^{H'}$) as a diversity index in which the base of natural logarithms is used; the units of e^H are "nits." In this index the upper limit is always the total number of species:

$$e^{H_{max}} = e^{\ln S} = S$$

As an example, consider a sample containing 106 individuals belonging to five species of zooplankton, as follows:

Species	Individuals	$\log_{10} n_i!$	$n_i \log_{10} n_i$
n_1	40	47.9116	64.0824
n_2	30	32.4237	44.3134
n_3	22	21.0508	29.5333
n_4	10	6.5598	10.0000
n_5	4	1.3802	2.4082
		Σ 109.3261	150.3773
N	106	170.0593	214.6824

H and H' are obtained as follows:

$$H = \frac{3.321928}{106} (170.0593 - 109.3261) \simeq 1.90344 \text{ bits}$$

$$H' = \frac{3.321928}{106} (214.6824 - 150.3773) \simeq 2.01525 \text{ bits}$$

and

$$e^H = e^{1.90334} = 6.709 \text{ bits}$$

When expressed as nits[4]

$$H = 1.90344 \log_e 2 = 1.31972$$

$$e^H = e^{1.31972} = 3.741 \text{ nits}$$

To investigate the structure of zooplankton assemblages in the northern Indian Ocean, Timonin (1971) measured diversities in terms of species and trophic levels (feeding types) of zooplankton (mainly copepods, euphausiids, and chaetognaths) collected from the research vessel *Vitiaz*. At each sampling station he calculated the information content per unit biomass for all feeding type groups, using the following equation:

$$H_{tr} = - \sum_{i=1}^{S} \frac{b_i}{B} \log_2 \frac{b_i}{B} \tag{10.19}$$

where H_{tr} is the trophic diversity index, B is the total biomass of the sample, b_i is the biomass of feeding type group i, and S is total number of feeding types. In this way two types of assemblages were identified, together with their relationships with the physical characteristics of the water (Fig. 10.7). The first type was characterized by the largest biomass and lowest diversities, both in terms of the number of species and trophic levels, and the assemblage was dominated by a large quantity of herbivorous (filter-feeding) zooplankton. This assemblage occurred in a strong divergence (upwelling) area. The second assemblage type was characterized by small biomass and high diversity in terms of the number of species and

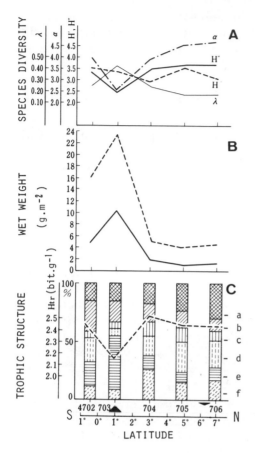

Figure 10.7. Structural differences of zooplankton assemblages in the northern Indian Ocean. (A) species diversity: λ, Simpson's index; α, Fisher's index; H' and H'', information indices, bit·ind^{-1} and bit·g^{-1} respectively. (B) Total zooplankton biomass (dotted line) and herbivore biomass (solid line). (C) Trophic groups and diversity of trophic structure (H_{tr}): a, seizing and swallowing carnivores; b, seizing and masticating carnivores; c, piercing and sucking carnivores; d, omnivores; e, coarse-filter feeders (herbivores); f, fine-filter feeders (herbivores); ▲, intensive divergence; ▼, weak convergence. (After Timonin, 1971.)

trophic levels and was found in vertically stratified waters or areas of weak convergence.

10.3.3 *Indices of the Abundance of Species*

The diversity indices proposed so far tend to increase with an increase in the number of species and, hence, the number of samples. The reason is that many indices do not distinguish between two elements of diversity, related to the species richness and evenness, and measure them together. We often see variations in diversity which are difficult to interpret for it is not clear which function of the two elements is varying. The measurement of equitability may be expressed as the difference between the actual diversity and the hypothetical diversity resulting from all species being either equally frequent or distributed according to a given hypothesis. The indices ε (Lloyd and Ghelardi, 1964), J' (Pielou, 1966), and E (Sheldon, 1969) are often used for this measurement.

The term ϵ is derived from the Shannon-Weaver function, using the models of MacArthur (1957, 1960) concerning the numbers of species and individuals. The hypothetical diversity, $M(S)$, in MacArthur's model is expressed as

$$M(S) = - \sum_{r=1}^{S} \pi_r \log_2 \pi_r \qquad (10.20)$$

where π_r is the number of individuals of the rth species, beginning with the species having the fewest individuals. The value of $M(S)$ is derived from the total number of species, S, using a table provided by Lloyd and Ghelardi (1964) (Appendix, Table A.2). The actual diversity, $M(S')$, is obtained as follows:

$$M(S') = - \sum_{i=1}^{S} P_i \log_2 P_i \qquad (10.21)$$

where P_i is the proportion of the number of individuals of species i to the total number of individuals. Lloyd and Ghelardi (1964) considered that the ratio of S', the hypothetical number of species according to MacArthur's model, to S is more appropriate than the ratio of $M(S')$ to $M(S)$ as a measure of diversity. Equitability ϵ is thus expressed as

$$\epsilon = \frac{S'}{S} \qquad (10.22)$$

The term S' can be derived from $M(S')$ using the Appendix Table A.2. For example, if there is a community in which $S = 44$ and $M(S') = 4.16$ bits, the hypothetical community value (S') is 26 species (Table A.2). Therefore, $\epsilon = {}^{26}\!/_{44} = 0.59$.

Pielou's J' (evenness component diversity) is expressed as

$$J' = \frac{H'}{H'_{\max}} = \frac{H'}{\log_2 S} \qquad (10.23)$$

where S is the number of species. The units of H' are bits. When the value of J' is great, the difference in the number of individuals among species is small. Using the example in Section 10.3.2 ($S = 5, H' = 2.01525$), J' is obtained as follows:

$$J' = \frac{2.01525}{3.3219 \log_{10} 5} \simeq 0.8651$$

On the other hand Sheldon's E is expressed as

$$E = \frac{e^H}{S} \qquad (10.24)$$

Sheldon (1969) compared these three measures of equitability from the

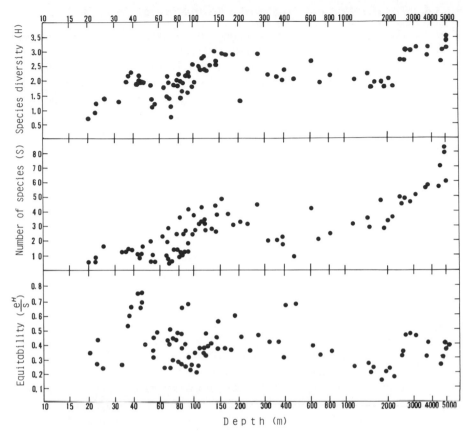

Figure 10.8. Species diversity, number of species, and equitability of Foraminifera in relation to depth in the western North Atlantic Ocean. (Modified after Buzas and Gibson, 1969.)

view point of the dependence on the species count and concluded that, although each of the measures is dependent on the species count to some degree, J' is the most stable and best suited for general use. The indices ϵ and E are particularly dependent on S when S is smaller than 10, and this dependence increases as equitability deviates from unity.

As an example of actual data, Figure 10.8 shows the variation in species diversity of benthic Foraminifera in the western North Atlantic according to depth (Buzas and Gibson, 1969). The diversity in terms of information content was largest at depths greater than 2500 m. High values were also shown by samples from depths of 35–45 m and 100–200 m. The peaks at 100–200 m and >2500 m depths were due to an increase in the number of species while the high diversity at depths of 35–45 m was due to high equitability.

10.3.4 *Rarefaction Method and Assumption of Sample Size*

The diversity indices discussed in preceding sections are based upon the total number of the species or individuals in the sample and is, therefore, a function of sample size. Sanders (1968) compared the diversity of organisms in benthic samples by use of the "shape" of the species abundance curve. The rarefaction method is demonstrated by considering a sample of 1000 individuals comprising 40 species (Table 10.3). First, the sample size is hypothetically reduced to 25 individuals without changing the percentage composition of the respective species, so that one individual now corresponds to 4% of the total sample. In the original sample seven species each comprised 4% or more and jointly made up 76% of the total number of individuals. Thus, each of these seven dominant species is included in the rarefied sample of 25 individuals. The remaining 24% is comprised of 33 less abundant species. Because each of these rare species comprises less than 4% of the number of individuals in the original sample, not all of them can be represented in the rarefied sample. In the original sample the number of species which may occur in the remaining 24% will be 24%/4%, or six, and thus the number of species present in the rarefied sample of 25 individuals is 7 + 6 = 13. Next consider a hypothetical sample consisting of 100 individuals, each individual corresponding to 1% of the sample; there are 15 species which individually comprise more than 1% of the fauna and cumulatively account for 92.1% of the sample. The residual 7.9% of the rarefied sample is made up of 7.9/1.0 = 7.9 species, thus there are 15 + 7.9 = 22.9 species in the rarefied sample of 100 individuals. Using this technique we can compute and graph the number of species corresponding to samples of a given number of individuals (Fig. 10.9).

The rarefaction method can be used to compare the species richness curve of the different collections by scaling down sample sizes to similar numbers (presumably that in the smallest collection). This scaling is necessary because large collections tend to contain more species than small ones even if they are drawn from the same assemblage or community. It was pointed out, however, that Sanders' scaling procedure consistently overestimates the number of species expected in a sample of n individuals drawn randomly from a larger collection (e.g., Hurlbert, 1971).

The true value of the expected number of species $E(S_n)$ in a sample of n individuals selected at random from a collection containing N individuals and S species is determined by a hypergeometric distribution[5]:

$$E(S_n) = S - \binom{N}{n}^{-1} \sum_{i=1}^{S} \binom{N - N_i}{n} \qquad (10.25)$$

where N_i is the number of individuals of species i in the original sample. If n is much smaller than N, drawing n individuals randomly can be approximated by sampling with replacement. That is, if successive draw-

Table 10.3 Hypothetical sample with 1000 individuals and 40 species to illustrate the rarefaction method[a]

Rank of species by abundance	Number of individuals	Percent of sample	Cumulative percent of sample
1	365	36.5	36.5
2	112	11.2	47.7
3	81	8.1	55.8
4	61	6.1	61.9
5	55	5.5	67.4
6	46	4.6	72.0
7	40	4.0	76.0
8	38	3.8	79.8
9	29	2.9	82.7
10	23	2.3	85.0
11	21	2.1	87.1
12	15	1.5	88.6
13	13	1.3	89.9
14	12	1.2	91.1
15	10	1.0	92.1
16–27	8–3	0.8–0.3	92.9–98.3
28–33	2 each	0.2 each	99.3
34–40	1 each	0.1 each	100.0
Total	1000		

[a] Modified after Sanders (1968).

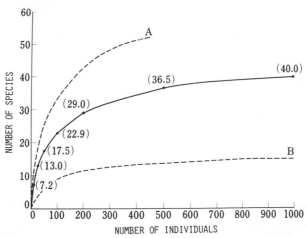

Figure 10.9. Plot of the number of species for different sample sizes using the rarefaction method of Sanders. Solid curve is estimate from Table 10.3; A is an example of a higher diversity sample; B is an example of a lower diversity sample.

ings of individuals for the smaller sample can be assumed to not affect the distribution of individuals with species in the original, larger sample, then a multinomial distribution applies instead of a hypergeometric one:

$$E(S_n) = S - \sum_{i=1}^{S} \left(1 - \frac{N_i}{N}\right)^n \tag{10.26}$$

$$\text{var}(S_n) = \sum_{i=1}^{S} \left(1 - \frac{N_i}{N}\right)^n \left[1 - \left(1 - \frac{N_i}{N}\right)^n\right]$$

$$+ 2 \sum_{j=2}^{S} \left[\left(1 - \frac{N_i}{N} - \frac{N_j}{N}\right)^n - \left(1 - \frac{N_i}{N}\right)^n \left(1 - \frac{N_j}{N}\right)^n\right] \tag{10.27}$$

where var (S_n) is the variance of the expected number of species, N_j is the number of individuals of species j in the original sample, and $j > i$ (Heck et al., 1975).

As suggested in Figure 10.9 species richness curves rise rapidly at first and then flatten out, so the results of species richness comparisons tend to stabilize at sufficiently large sizes. Therefore, if a single measure of species richness is desired, as when richness is being examined for correlation with other factors, one might compare $E(S_n)$ values calculated for some high, standardized value of n.

The rarefaction diversity measure can also be used to determine the sample size (e.g., the number of samples or individuals to be collected). The most commonly used method to determine sample size is to plot the number of species obtained per unit sample as a function of the number of samples or the volume of water filtered; an asymptote indicates a sufficient sample size. In order to design a sampling scheme, Heck et al. (1975) suggest to first take a large sample in the area of interest to estimate the total number of species and distribution of individuals among species. If this original sample is sufficiently large so that nearly all the species present both temporally and spatially have been enumerated, then by using the above equations, one can estimate a sample size which allows, on the average, a desired portion of the species to be taken. In some situations one might be satisfied to collect 70–80% of the total number of species known to occur in a given area as long as the most common species were obtained. If the analysis showed that the greatest, sustained, and affordable sampling effort would only sample an unacceptably small portion of the total species pool, one would change the sampling strategy.

Notes

[1] Although the terms *community* and *assemblage* are used somewhat loosely in the book, we consider that a community refers to all organisms in an ecosystem, plants and animals inclusive, while an assemblage consists of a group of taxonomically and/or behaviorally related organisms (e.g., copepod assemblage or zooplankton assemblage).

[2] It depends on investigator's perspective. A 5% difference between rare species counts the same as a 5% difference between common species. To that extent they are weighed equally.

[3] Depending on data, correlation coefficients may be calculated using log-transformed or ranked sample count. Distributions of sample counts (numbers of individuals or biomass) for zooplankton species are often overdispersed so that their variances exceed their means. Cassie (1962) has shown that an approximate normalizing transformation for these distributions is given by the following equation:

$$Y = \log_{10}(1 + X) \tag{10.28}$$

where X is the raw sample count. Thus, conversion to logarithms is a common *ad hoc* expedient for reducing the relative influence of larger values in a set of data. Among a suite of opportunistic organisms living in a fluctuating environment, log abundances will be more nearly linearly related to the various causes of commonness and rareness than will the untransformed abundances. Conversion of data from counts to rank order of abundances or groups of abundance classes also reduces the influence of apparent variations in abundance due to inefficient sampling and the patchy distribution of the zooplankton. Another advantage of methods which employ ranks is that it is unnecessary to assume that the species abundances are normally distributed. The reader should refer to textbooks of statistics for further methods of data standardization.

[4] Information is commonly expressed in "bits" when using $\log_2$, "dits" when using $\log_{10}$, and "nits" when using $\log_e$:

$\log_e 10 = 2.3026$
$\log_e 2 = 0.6931$
$\log_2 e - 1.4427$
$\log_{10} e = 0.4343$
$e = 2.7183$
$\log_2 a = 3.3219 \log_{10} a = 1.4427 \log_e a$
$\log_e a = 2.3026 \log_{10} a = 0.6931 \log_2 a$

[5] $\binom{N}{n}$ and $\binom{N - N_i}{n}$ symbolize two-dimensional column vectors.

References

Abercrombie, M., Hickman, C. J., and Johnson, M. L. *A Dictionary of Biology*, 6th ed. Hunt Barnard, New York, 1951.

Adams, H. R., Flerchinger, A. P., and Steedman, H. F. Ctenophora fixation and preservation. In H. F. Steedman (ed.), *Zooplankton Fixation and Preservation*. Monographs on Oceanographic Methodology 4. UNESCO, Paris, 1976, pp. 270, 271.

Adams, J. A., and Steele, J. H. Shipboard experiments on the feeding of *Calanus finmarchicus* (Gunnerus). In H. Barnes (ed.), *Some Contemporary Studies in Marine Science*. Allen and Unwin, London, 1966, pp. 19–45.

Ahlstrom, E. H., and Thrailkill, T. R. Plankton volume loss with time of preservation. *Rep. Calif. Coop. Oceanic Fish. Invest.* **9,** 57–73 (1963).

Ahmed, S. I., Kenner, R. A., and King, F. D. Preservation of enzyme activity in marine plankton by low temperature freezing. *Mar. Chem.* **4,** 133–139 (1976).

Alcaraz, M., Paffenhöfer, G.–A., and Strickler, J. R. Catching the algae: A first account of visual observations on filter-feeding calanoids. In W. C. Kerfoot (ed.), *Evolution and Ecology of Zooplankton Communities*. The University Press of New England, Hanover, NH, 1980.

Allan, J. D. Life history patterns in zooplankton. *Am. Natur.* **110,** 165–180 (1976).

Allan, J. D., Kinsey, T. G., and James, M. C. Abundances and production of copepods in the Rhode River subestuary of Chesapeake Bay. *Chesapeake Sci.* **17,** 86–92 (1976).

Allan, J. D., Richman, S., Heinle, D. R., and Huff, R. Grazing in juvenile stages of some estuarine calanoid copepods. *Mar. Biol.* **43,** 317–331 (1977).

Alldredge, A. L. Abandoned larvacean houses: A unique food source in the pelagic environment. *Science* **177,** 885–887 (1972).

Alldredge, A. L. The impact of appendicularian grazing on natural food concentrations *in situ*. *Limnol. Oceanogr.* **26,** 247–257 (1981).

Alldredge, A. L., and King, J. M. Distribution, abundance, and substrate perferences of demersal reef zooplankton at Lizard Island lagoon, Great Barrier Reef. *Mar. Biol.* **41,** 317–333 (1977).

Allen, E. J., and Nelson, E. W. On the artificial culture of marine plankton organisms. *J. Mar. Biol. Assoc. U.K.* **8,** 421–474 (1910).

Allen, K. R. The computation of production in fish populations. *New Zealand Sci. Rev.* **8,** 89 (1950).

Alvarez, V., and Matthews, J. B. L. Experimental studies on the deep-water pelagic community of Korsfjorden, western Norway. Feeding and assimilation by *Chiridius armatus* (Crustacea, Copepoda). *Sarsia* **58,** 67–78 (1975).

Ambler, J. W., and Frost, B. W. The feeding behavior of a predatory planktonic copepod, *Tortanus discaudatus*. *Limnol. Oceanogr.* **19,** 446–451 (1974).

Angel, M. V. Observations on the behaviour of *Conchoecia spinirostris*. *J. Mar. Biol. Assoc. U.K.* **50,** 731–736 (1970).

Angel, M. V., and Fasham, M. J. R. SOND cruise 1965: Factor and cluster analyses of the plankton results. A general summary. *J. Mar. Biol. Assoc. U.K.* **53,** 185–231 (1973).

Anraku, M. Influence of the Cape Cod Canal on the hydrography and on the copepods in Buzzards Bay and Cape Cod Bay, Massachusetts. II. Respiration and feeding. *Limnol. Oceanogr.* **9,** 195–206 (1964a).

Anraku, M. Some technical problems encountered in quantitative studies of grazing and predation by marine planktonic copepods. *J. Oceanogr. Soc. Japan* **20,** 221–231 (1964b).

Anraku, M. Continuous culture systems of zooplankton. *Bull. Plankton Soc. Japan* **20,** 12–18 (1973) (in Japanese).

Anraku, M. (ed.). Mass-culture of zooplankton as food for fish larvae. *Suisan Zoyoshoku Sosho* **28,** 1979 (in Japanese).

Anraku, M., and Kozasa, E. The effect of heated effluents on the production of marine plankton (Takashima Nuclear Power Station -1). *Bull. Plankton Soc. Japan* **25,** 93–110 (1978) (in Japanese).

Anraku, M., and Omori, M. Preliminary survey of the relationship between the feeding habit and the structure of the mouthparts of marine copepods. *Limnol. Oceanogr.* **8,** 116–126 (1963).

Ansell, A. D. The adenosine triphosphate content of some marine bivalve molluscs. *J. Exp. Mar. Biol. Ecol.* **28,** 269–283 (1977).

Aron, W. The use of a large capacity portable pump for plankton sampling, with notes on plankton patchiness. *J. Mar. Res.* **16,** 158–173 (1958).

Aron, W. Some aspects of sampling the macroplankton. *Rapp. R.-v. Reun. Cons. Perm. Int. Explor. Mer* **153,** 29–38 (1962).

Atkinson, D. E. Adenine nucleotides as stoichiometric coupling agents in metabolism and as regulatory modifiers: The adenylate energy charge. In H. J. Vogel (ed.), *Metabolic Regulation,* Vol. 5 of *Metabolic Pathways.* Academic Press, New York 1971, pp. 1–21.

Azam, F., and Chisholm, S. W. Silicic acid uptake and incorporation by natural marine phytoplankton populations. *Limnol. Oceanogr.* **21,** 427–435 (1976).

Baker, A de C., Clarke, M. R., and Harris, M. J. The N.I.O. combination net (RMT 1 + 8) and further developments of rectangular midwater trawls. *J. Mar. Biol. Assoc. U.K.* **53,** 167–184 (1973).

Balch, N. ATP content of *Calanus finmarchicus*. *Limnol. Oceanogr.* **17,** 906–908 (1972).

Båmstedt, U. Biochemical studies on the deep-water pelagic community of Korsfjorden, western Norway. Methodology and sampling design. *Sarsia* **56,** 71–86 (1974).

Båmstedt, U. Studies on the deep-water pelagic community of Korsfjorden, western Norway. Changes in the size and biochemical composition of *Meganyctiphanes norvegica* (Euphausiacea) in relation to its life cycle. *Sarsia* **61,** 15–30 (1976).

Båmstedt, U. Studies on the deep-water pelagic community of Korsfjorden, western Norway. Seasonal variation in weight and biochemical composition of *Chiridius armatus* (Copepoda), *Boreomysis arctica* (Mysidacea), and *Eukrohnia hamata* (Chaetognatha) in relation to their biology. *Sarsia* **63,** 145–154 (1978).

Båmstedt, U. ETS activity as an estimator of respiratory rate of zooplankton populations. The significance of variations in environmental factors. *J. Exp. Mar. Biol. Ecol.* **42,** 267–283 (1980).

Båmstedt, U., and Matthews, J. B. L. Studies on the deep-water pelagic community of Korsfjorden, western Norway. The weight and biochemical composition of *Euchaeta norvegica* Boeck in relation to its life cycle. In H. Barnes (ed.), *Proceeding of the 9th European Marine Biology Symposium.* Aberdeen University Press, 1975, pp. 311–327.

Båmstedt, U. and Skjoldal, H. R. RNA concentration of zooplankton: Relationship with size and growth. *Limnol. Oceanogr.* **25,** 304–316 (1980).

Barica, J. Reliability of an ammonia probe for electrometric determination of total ammonia nitrogen in fish tanks. *J. Fish. Res. Bd. Canada* **30,** 1389–1392 (1973).

Beers, J. R. Ammonia and inorganic phosphorus excretion by the planktonic chaetognath, *Sagitta hispida* Conant. *J. Cons. Perm. Int. Explor. Mer* **29,** 123–129 (1964).

Beers, J. R. Studies on the chemical composition of the major zooplankton groups in the Sargasso Sea off Bermuda. *Limnol. Oceanogr.* **11,** 520–528 (1966).

Beers, J. R., and Stewart, G. L. Micro-zooplankton and its abundance relative to the larger zooplankton and other seston components. *Mar. Biol.* **4,** 182–189 (1969).

Beers, J. R., and Stewart, G. L. The ecology of the plankton off La Jolla, California in the period April through September, 1967. Part VI. Numerical abundance and estimated biomass of microzooplankton. *Bull. Scripps Inst. Oceanogr.* **17,** 67–87 (1970).

Beers, J. R., Stewart, G. L., and Strickland, J. D. H. A pumping system for sampling small plankton. *J. Fish. Res. Bd. Canada* **24,** 1811–1818 (1967).

Beklemishev, C. W. Superfluous feeding of marine herbivorous zooplankton. *Rapp. P.-v. Reun. Cons. Perm. Int. Explor. Mer* **153,** 108–113 (1962).

Belcher, R. *Instrumental Organic Elemental Analysis.* Academic Press, New York, 1977.

Bělehrádek, J. *Temperature and Living Matter.* Protoplasma Monograph 8. Borntraeger, Berlin, 1935.

Bell, G. R. A guide to the properties, characteristics, and uses of some general anaesthetics for fish. *Bull. Fish. Res. Bd. Canada.* **148,** 1–4 (1964).

Bercaw, J. S. A folding midwater trawl depressor. *Limnol. Oceanogr.* **11,** 633–635 (1966).

Berkes, F. Production of the euphausiid crustacean *Thysanoëssa raschii* in the Gulf of St. Lawrence. *J. Fish. Res. Bd. Canada* **34,** 443–446 (1977).

Bernard, M., Möller, F., Nassogne, A., and Zettera, A. Influence of pore size of plankton nets and towing speed on the sampling performance of two high-speed samplers (Delfino I and II) and its consequences for the assessment of plankton populations. *Mar. Biol.* **20,** 109–136 (1973).

Berner, A. Feeding and respiration in the copepod *Temora longicornis* (Muller). *J. Mar. Biol. Assoc. U.K.* **42,** 625–640 (1962).

Bernhard, M. Chemical contamination of culture media: Assessment, avoidance and control. In O. Kinne (ed.), *Marine Ecology 3. Cultivation.* John Wiley & Sons, New York, 1977, pp. 1459–1499.

Bertalanffy, L. von, A quantitative theory of organic growth. *Human Biol.* **10,** 181–213 (1938).

Beum, C. P., and Brundage, E. G. A method for analyzing the sociomatrix. *Sociometry* **13,** 141–145 (1950).

Bieri, R. A method for the microscopic examination and manipulation of plankton on board ship. *J. Cons. Perm. Int. Explor. Mer* **22,** 38–41 (1956).

Biggs, D. C. Respiration and ammonium excretion by open ocean gelatinous zooplankton. *Limnol. Oceanogr.* **22,** 108–117 (1977).

Bogorov, B. G. On the standardization of marine plankton investigations. *Int. Rev. Ges. Hydrobiol. Hydrogr.* **44,** 621–642 (1959).

Bower, C. E., and Bidwell, J. P. Ionization of ammonia in seawater: Effects of temperature, pH, and salinity. *J. Fish. Res. Bd. Canada* **35,** 1012–1016 (1978).

Bowman, T. E., and Gruner, H.-E. The families and genera of Hyperiidea (Crustacea: Amphipoda). *Smithsonian Contrib. Zool.* **146,** 1–64 (1973).

Boyd, C. M. Selection of particle sizes by filter-feeding copepods: A plea for reason. *Limnol. Oceanogr.* **21,** 175–180 (1976).

Boyd, C. M., Smith, S. L., and Cowles, T. J. Grazing patterns of copepods in the upwelling system off Peru. *Limnol. Oceanogr.* **25,** 583–596 (1980).

Bridger, J. P. On efficiency tests made with a modified Gulf III high-speed tow net. *J. Cons. Perm. Int. Explor. Mer* **23,** 357–365 (1958).

Brinton, E. Parameters relating to the distributions of planktonic organisms, especially euphausiids in the eastern tropical Pacific. *Progress Oceanogr.* **8,** 125–189 (1979).

Brown, D. M., and Cheng, L. New net for sampling the ocean surface. *Mar. Ecol. Prog. Ser.* **5,** 225–227 (1981).

Buck, J. D., and Cleverdon, R. C. The spread plates as a method for the enumeration of marine bacteria. *Limnol. Oceanogr.* **5,** 78–80 (1960).

Bulnheim, H.-P. Vergleichende Undersuchungen zur Atmungsphysiologie euryhaliner Gammariden unter besonderer Berücksichtigung der Salzgehaltsanpassung. *Helgoländer Wiss. Meeresunters.* **23,** 485–534 (1972).

Butler, E. I., Corner, E. D. S., and Marshall, S. M. On the nutrition and metabolism of zooplankton. VI. Feeding efficiency of *Calanus* in terms of nitrogen and phosphorus. *J. Mar. Biol. Assoc. U.K.* **49,** 977–1001 (1969).

Butler, E. I., Corner, E. D. S., and Marshall, S. M. On the nutrition and metabolism of zooplankton. VII. Seasonal survey of nitrogen and phosphorus excretion by *Calanus* in the Clyde Sea-area. *J. Mar. Biol. Assoc. U.K.* **50,** 525–560 (1970).

Buzas, M. A., and Gibson, T. G. Species diversity: Benthic Foraminifera in western North Atlantic. *Science* **193,** 72–75 (1969).

Caperon, J., Schell, D., Hirota, J., and Laws, E. Ammonium excretion rates in Kaneohe Bay, Hawaii, measured by a ^{15}N isotope dilution technique. *Mar. Biol.* **54,** 33–40 (1979).

Carpenter, E. J., Peck, B. B., and Anderson, S. J. Survival of copepods passing through a nuclear power station on northeastern Long Island Sound, USA. *Mar. Biol.* **24,** 49–55 (1974).

Cassie, R. M. Some uses of probability paper in the analysis of size frequency distributions. *Aust. J. Mar. Freshwat. Res.* **5,** 513–522 (1954).

Cassie, R. M. Frequency distribution in the ecology of plankton and other organisms. *J. Animal Ecol.* **31,** 65–82 (1962).

Cassie, R. M. Multivariate analysis in the interpretation of numerical plankton data. *New Zealand J. Sci.* **6,** 36–59 (1963).

Cassie, R. M. Sampling and statistics. In W. T. Edmondson, and G. G. Winberg (eds.), *A Manual on Methods for the Assessment of Secondary Productivity in Fresh Waters.* IBP Handbook No. 17. Blackwell Scientific Publications, London, 1971, pp. 174–209.

Caswell, H. On instantaneous and finite birth rates. *Limnol. Oceanogr.* **17,** 787–791 (1972).

Champalbert, G., and Kerambrun, P. Influence du mode de conservation sur la composition chimique élémentaire de *Pontella mediterranea* (Copepoda: Pontellidae). *Mar. Biol.* **51,** 357–360 (1979).

Chapman, A. G., Fall, L., and Atkinson, D. E. Adenylate energy charge in *Escherichia coli* during growth and starvation. *J. Bact.* **108,** 1072–1086 (1971).

Checkley, D. M. Jr. The egg production of a marine planktonic copepod in relation to its food supply: Laboratory studies. *Limnol. Oceanogr.* **25,** 430–446 (1980).

Checkley, D. M. Jr. Selective feeding by Atlantic herring (*Clupea harengus*) larvae on zooplankton in natural assemblages. *Mar. Ecol. Prog. Ser.* **9,** 245–253 (1982).

Chesson, J. Measuring preference in selective predation. *Ecology* **59,** 211–215 (1978).

Childress, J. J. Respiratory adaptations to the oxygen minimum layer in the bathypelagic mysid *Gnathophausia ingens. Biol. Bull.* **141,** 109–121 (1971).

Childress, J. J. The respiratory rates of midwater crustaceans as a function of depth of occurrence and relation to the oxygen minimum layer off southern California. *Comp. Biochem. Physiol.* **50A,** 787–799 (1975).

Childress, J. J., Barnes, A. T., Quetin, L. B., and Robison, B. H. Thermally protecting codends for the recovery of living deep-sea animals. *Deep-Sea Res.* **25,** 419–422 (1978).

Christensen, J. P., and Packard, T. T. Respiratory electron transport activities in phytoplankton and bacteria: Comparison of methods. *Limnol. Oceanogr.* **24,** 576–583 (1979).

Chuang, S. H. *C. S. K. Zooplankton Data Report,* No. 4. Singapore University Press, Singapore, 1974.

Clarke, G. L., and Bonnet, D. D. The influence of temperature on the survival, growth and respiration of *Calanus finmarchicus. Biol. Bull.* **76,** 371–383 (1939).

Clarke, G. L., and Bumpus, D. F. The plankton sampler—an instrument for quantitative plankton investigations. *Spec. Publs. Am. Soc. Limnol. Oceanogr.* **5,** 1–8 (1950).

Clarke, M. R. A new midwater trawl for sampling discrete depth horizons. *J. Mar. Biol. Assoc. U.K.* **49,** 945–960 (1969).

Clarke, W. D. The Jet Net, a new high-speed plankton sampler. *J. Mar. Res.* **22,** 284–287 (1964).

Clutter, R. I., and Anraku, M. Avoidance of samplers. In D. J. Tranter and J. H. Fraser (eds.), *Zooplankton Sampling.* Monographs on Oceanographic Methodology 2. UNESCO, Paris, 1968, pp. 57–76.

Clutter, R. I., and Theilacker, G. H. Ecological efficiency of a pelagic mysid shrimp; estimates from growth, energy budget, and mortality studies. *Fishery Bull. U.S.* **69,** 93–115 (1971).

Cochran, W. G. *Sampling Techniques*, 2nd ed. John Wiley & Sons, New York, 1963.

Colebrook, J. M. Continuous plankton records: A principal component analysis of the geographical distribution of zooplankton. *Bull. Mar. Ecol.* **6**, 78–100 (1964).

Colebrook, J. M. Annual fluctuations in biomass of taxonomic groups of zooplankton in the California current, 1955–59. *Fishery Bull. U.S.* **75**, 357–368 (1977).

Colebrook, J. M., Glover, R. S., and Robinson, G. A. Continuous plankton records, 1948–1956: Contributions towards a plankton atlas of the north-eastern Atlantic and the North Sea. General introduction. *Bull. Mar. Ecol.* **5**, 67–80 (1961).

Comita, G. W. A study of calanoid copepod population in an arctic lake. *Ecology* **37**, 576–591 (1956).

Connell, J. H. Diversity in tropical rain forests and coral reefs. *Science* **199**, 1302–1310 (1978).

Conover, R. J. Oceanography of Long Island Sound, 1952–1954. VI. Biology of *Acartia clausi* and *A. tonsa*. *Bull. Bingham Oceanogr. Coll.* **15**, 156–233 (1956).

Conover, R. J. Regional and seasonal variation in the respiratory rate of marine copepods. *Limnol. Oceanogr.* **4**, 259–268 (1959).

Conover, R. J. Metabolism and growth in *Calanus hyperboreus* in relation to its life cycle. *Rapp. P.-v. Reun. Cons. Perm. Int. Explor. Mer* **153**, 190–197 (1962).

Conover, R. J. Food relations and nutrition of zooplankton. *Proc. Symp. Exp. Mar. Ecol., Univ. RI, Occ. Publ.* **2**, 81–91 (1964).

Conover, R. J. Assimilation of organic matter by zooplankton. *Limnol. Oceanogr.* **11**, 338–345 (1966a).

Conover, R. J. Factors affecting the assimilation of organic matter by zooplankton and the question of superfluous feeding. *Limnol. Oceanogr.* **11**, 346–354 (1966b).

Conover, R. J. Transformation of organic matter. In O. Kinne (ed.), *Marine Ecology 4. Dynamics*. John Wiley & Sons, New York, 1978, pp. 221–499.

Conover, R. J., and Corner, E. D. S. Respiration and nitrogen excretion by some marine zooplankton in relation to their life cycles. *J. Mar. Biol. Assoc. U.K.* **48**, 49–75 (1968).

Conover, R. J., and Francis, V. The use of radioactive isotopes to measure the transfer of materials in aquatic food chains. *Mar. Biol.* **18**, 272–283 (1973).

Conover, R. J., and Huntley, M. E. General rules of grazing in pelagic ecosystems. In P. Falkowski (ed.), *Primary Productivity in the Sea*. Plenum Press, New York, 1980, pp. 461–485.

Conover, R. J., and Lalli, C. M. Feeding and growth in *Clione limacina* (Phipps), a pteropod mollusc. *J. Exp. Mar. Biol. Ecol.* **9**, 279–302 (1972).

Conover, R. J., and Lalli, C. M. Feeding and growth in *Clione limacina* (Phipps), a pteropod mollusc. II. Assimilation, metabolism, and growth efficiency. *J. Exp. Mar. Biol. Ecol.* **16**, 131–154 (1974).

Cooper, W. E. Dynamics and production of a natural population of a fresh-water amphipod, *Hyalella azteca*. *Ecol. Monogr.* **35**, 377–394 (1965).

Corkett, C. J. Techniques for breeding and rearing marine calanoid copepods. *Helgoländer Wiss. Meeresunters.* **20**, 318–324 (1970).

Corliss, J. O. *The Ciliated Protozoa: Characterization, Classification and Guide to the Literature*, 2nd ed. Pergamon Press, New York, 1979.

Corner, E. D. S. On the nutrition and metabolism of zooplankton. I. Preliminary observations on the feeding of the marine copepod, *Calanus helgolandicus* (Claus). *J. Mar. Biol. Assoc. U.K.* **41,** 5–16 (1961).

Corner, E. D. S., and Cowey, C. B. Some nitrogenous constituents of the plankton. In H. Barnes (ed.), *Oceanogr. Mar. Biol. Ann. Rev.*, Vol. 2. Allen and Unwin, London, 1964, pp. 147–167.

Corner, E. D. S., Cowey, C. B., and Marshall, S. M. On the nutrition and metabolism of zooplankton. III. Nitrogen excretion by *Calanus*. *J. Mar. Biol. Assoc. U.K.* **45,** 429–442 (1965).

Corner, E. D. S., Cowey, C. B., and Marshall, S. M. On the nutrition and metabolism of zooplankton. V. Feeding efficiency of *Calanus finmarchicus*. *J. Mar. Biol. Assoc. U.K.* **47,** 259–270 (1967).

Corner, E. D. S., and Davies, A. C. Plankton as a factor in the nitrogen and phosphorus cycles in the sea. *Adv. Mar. Biol.* **9,** 101–204 (1971).

Corner, E. D. S., Head, R. N., and Kilvington, C. C. On the nutrition and metabolism of zooplankton. VIII. The grazing of *Biddulphia* cells by *Calanus helgolandicus*. *J. Mar. Biol. Assoc. U.K.* **52,** 847–861 (1972).

Corner, E. D. S., Head, R. N., Kilvington, C. C., and Pennycuick, L. On the nutrition and metabolism of zooplankton. X. Quantitative aspects of *Calanus helgolandicus* feeding as a carnivore. *J. Mar. Biol. Assoc. U.K.* **56,** 345–358 (1976).

Corner, E. D. S., and Newell, B. S. On the nutrition and metabolism of zooplankton. IV. The forms of nitrogen excreted by *Calanus*. *J. Mar. Biol. Assoc. U.K.* **47,** 113–120 (1967).

Cosper, T. C., and Reeve, M. R. Digestive efficiency of the chaetognath *Sagitta hispida* Conant. *J. Exp. Mar. Biol. Ecol.* **17,** 33–38 (1975).

Costlow, J. D. Jr., Bookhout, C. G., and Monroe, R. The effect of salinity and temperature on larval development of *Sesarma cinereum* (Bosc) reared in the laboratory. *Biol. Bull.* **118,** 183–202 (1960).

Cowles, T. J. The feeding response of copepods from the Peru upwelling system: Food size selection. *J. Mar. Res.* **37,** 601–622 (1979).

Crawshay, L. R. Notes on experiments in the keeping of plankton animals under artificial conditions. *J. Mar. Biol. Assoc. U.K.* **10,** 555–576 (1915).

Crippen, R. W., and Perrier, J. L. The use of neutral red and evans blue for live-dead determinations of marine plankton. *Stain Tech.* **49,** 97–104 (1974).

Cummins, K. W., and Wuycheck, J. C. Caloric equivalents for investigations in ecological energetics. *Mitt. Internat. Verein. Limnol.* **18,** 1–158 (1971).

Curl, H. Jr. Standing crops of carbon, nitrogen, and phosphorus and transfer between trophic levels, in continental shelf waters south of New York. *Rapp. P.-v. Reun. Cons. Perm. Int. Explor. Mer* **153,** 183–189 (1962).

Currie, R. I. The Indian Ocean Standard Net. *Deep-Sea Res.* **10,** 27–32 (1963).

Cushing, D. H. The effect of grazing in reducing the primary production: A review. *Rapp. P.-v. Reun. Cons. Perm. Int. Explor. Mer* **144,** 149–154 (1958).

Cushing, D. H. Patchiness. *Rapp. P.-v. Reun. Cons. Perm. Int. Explor. Mer* **153,** 152–164 (1962).

Cushing, D. H. Upwelling and the production of fish. *Adv. Mar. Biol.* **9,** 255–334 (1971).

Cushing, D. H., Humphrey, G. F., Banse, K., and Laevastu, T. Report of the committee on terms and equivalents. *Rapp. P.-v. Reun. Cons. Perm. Int. Explor. Mer* **144,** 15, 16 (1958).

Dagg, M. J. Loss of prey body contents during feeding by an aquatic predator. *Ecology* **55**, 903–906 (1974).

Dagg, M. J., and Grill, D. W. Natural feeding rates of *Centropages typicus* females in the New York Bight. *Limnol. Oceanogr.* **25**, 597–609 (1980).

Dagg, M. J., and Littlepage, J. L. Relationships between growth rate and RNA, DNA, protein and dry weight in *Artemia salina* and *Euchaeta elongata. Mar. Biol.* **17**, 162–170 (1972).

Dan, K., and Okazaki, K. Cyto-embryological studies of sea urchins. III. Role of the secondary mesenchyme cells in the formation of the primitive gut in sea urchin larvae. *Biol. Bull.* **110**, 29–42 (1956).

Darbyshire, J. F. The estimation of soil protozoan populations, In R. G. Board and D. W. Lovelock (eds.), *Sampling-Microbiological Monitoring of Environments.* Academic Press, New York, 1973.

Daro, M. H. A simplified ^{14}C method for grazing measurements on natural planktonic populations. *Helgolander Wiss. Meeresunters.* **31**, 241–248 (1978).

David, P. M. A neuston net. A device for sampling the surface fauna of the ocean. *J. Mar. Biol. Assoc. U.K.* **45**, 313–320 (1965).

Davies, I. E., and Barham, E. G. The Tucker opening-closing micronekton net and its performance in a study of the deep scattering layer. *Mar. Biol.* **2**, 127–131 (1969).

Deason, E. E. Potential effect of phytoplankton colony breakage on the calculation of zooplankton filtration rates. *Mar. Biol.* **57**, 279–286 (1980).

Degobbis, D. On the storage of seawater samples for ammonia determination. *Limnol. Oceanogr.* **18**, 146–150 (1973).

Dixon, M. *Manometric Method.* 3rd ed. Cambridge University Press, Cambridge, MA, 1934.

Donaghay, P. L., and Small, L. F. Food selection capabilities of the estuarine copepod *Acartia clausi. Mar. Biol.* **52**, 137–146 (1979).

Dressel, D. M., Heinle, D. R., and Grote, M. C. Vital staining to sort dead and live copepods. *Chesapeake Sci.* **13**, 156–159 (1972).

Dumbar, M. J. The evolution of stability in marine environments. Natural selection at the level of the ecosystem. *Am. Natur.* **94**, 129–136 (1960).

Durbin, A. G., and Durbin, E. G. Standing stock and estimated production rates of phytoplankton and zooplankton in Narragansett Bay, Rhode Island. *Estuaries* **4**, 24–41 (1981).

Dussart, B. M. Les différentes catégories de plancton. *Hydrobiologia* **26**, 72–74 (1965).

Duval, W. S., and Geen, G. H. Diel feeding and respiration rhythms in zooplankton. *Limnol. Oceanogr.* **21**, 823–829 (1976).

Edmondson, W. T. Reproductive rate of planktonic rotifers in natural populations. *Mem. Ist. Ital. Idrobiol.* **12**, 21–77 (1960).

Edmondson, W. T. A graphical model for evaluating the use of the egg ratio for measuring birth and death rates. *Oecologia* **1**, 1–37 (1968).

Edmondson, W. T. Instar analysis, In W. T. Edmondson, and G. G. Winberg (eds.), *A Manual on Methods for the Assessment of Secondary Productivity in Fresh Waters.* IBP Handbook No. 17. Blackwell Scientific Publications, Oxford, 1971, pp. 149–157.

Edmondson, W. T., Comita, G. W., and Anderson, G. C. Reproductive rate of copepods in nature and its relation to phytoplankton populations. *Ecology* **43**, 625–634 (1962).

Edmondson, W. T., and Winberg, G. G. (eds.). *A Manual on Methods for the Assessment of Secondary Productivity in Fresh Waters*. IBP Handbook No. 17. Blackwell Scientific Publications, Oxford, 1971.

Ekman, S. *Zoogeography of the Sea*. Sidgwick and Jackson, London, 1953.

Ellertsen, B. A new apparatus for sampling surface fauna. *Sarsia* **63**, 113, 114 (1977).

Epp, R. W., and Lewis, W. M. Jr. The nature and ecological significance of metabolic changes during the life history of copepods. *Ecology* **61**, 259–264 (1980).

Eppley, R. W. Temperature and phytoplankton growth in the sea. *Fishery Bull. U.S.* **70**, 1063–1085 (1972).

Fager, E. W. Determination and analysis of recurrent groups. *Ecology* **38**, 586–595 (1957).

Fager, E. W. Estimation of mortality coefficients from field samples of zooplankton. *Limnol. Oceanogr.* **18**, 297–301 (1973).

Fager, E. W., Flechsig, A. O., Ford, R. F., Clutter, R. I., and Chelardi, R. J. Equipment for use in ecological studies using SCUBA. *Limnol. Oceanogr.* **11**, 503–509 (1966).

Fager, E. W., and McGowan, J. A. Zooplankton species groups in the North Pacific. *Science* **140**(3566), 453–460 (1963).

Fauchald, K., and Jumars, P. A. The diet of worms: A study of polychaete feeding guilds. *Oceanogr. Mar. Biol. Ann. Rev.* **17**, 193–284 (1979).

Ferguson, C. F., and Raymont, J. K. B. Biochemical studies on marine zooplankton. XII. Further investigations on *Euphausia superba* Dana. *J. Mar. Biol. Assoc. U.K.* **54**, 719–725 (1974).

Fernandez, F. Efecto de la intensidad de luz natural en la actividad metabólica y en la alimentación de varias especies de copépods planctónicos. *Invest. Pesquera* **41**, 575–602 (1977).

Fleming, J. M., and Coughlan, J. Preservation of vitally stained zooplankton for live/dead sorting. *Estuaries* **1**, 135–137 (1978).

Fleminger, A., and Hulsemann, K. Relationship of Indian Ocean epipelagic calanoids to the world oceans. In B. Zeitzschel (ed.), *The Biology of the Indian Ocean*. Springer-Verlag, Berlin, 1973, pp. 339–347.

Flood, P. R. Filter characteristics of appendicularian food catching nets. *Experientia* **34**, 173–175 (1978).

Flüchter, J. Fish: Conservation and culture in the laboratory, In C. Schlieper (ed.), *Research Methods in Marine Biology*. Sidgwick and Jackson, London, 1972, pp. 236–247.

Fogg, G. E. *Algal Cultures and Phytoplankton Ecology*. University of Wisconsin Press, Madison, WI, 1965.

Föyn, B. Lebenszyklus, Cytologie und sexualität der Chlorophycee *Cladophora suhriana* Kützing. *Arch. Protistenkd.* **83**, 1–56 (1934).

Fraser, J. H. Standardization of zooplankton sampling methods at sea. In D. J. Tranter and J. H. Fraser (eds.), *Zooplankton Sampling*. Monographs on Oceanographic Methodology 2. UNESCO, Paris, 1968, pp. 145–169.

Friedman, M. M., and Strickler, J. R. Chemoreceptors and feeding in calanoid copepods (Arthropoda: Crustacea). *Proc. Natl. Acad. Sci. USA* **72**, 4185–4188 (1975).

Frost, B. W. Effects of size and concentration of food particles on the feeding behavior of the marine planktonic copepod *Calanus pacificus. Limnol. Oceanogr.* **17,** 805–815 (1972).

Frost, B. W. Feeding processes at lower trophic levels in pelagic communities. In C. B. Miller (ed.), *The Biology of the Oceanic Pacific.* Oregon State University Press, Corvallis, OR, 1974, pp. 59–77.

Frost, B. W. Feeding behavior of *Calanus pacificus* in mixtures of food particles. *Limnol. Oceanogr.* **22,** 472–491 (1977).

Fudge, H. Biochemical analysis of preserved zooplankton. *Nature, Lond.* **219,** 380, 381 (1968).

Fuller, J. L. Feeding rate of *Calanus finmarchicus* in relation to environmental conditions. *Biol. Bull.* **72,** 233–246 (1937).

Fuller, J. L., and Clarke, G. L. Further experiments on the feeding of *Calanus finmarchicus. Biol. Bull.* **70,** 308–320 (1936).

Furukawa, I., and Hidaka, K. Technical problems encountered in the mass culture of the rotifer using marine yeast as food organisms. *Bull. Plankton Soc. Japan* **20,** 61–71 (1973) (in Japanese).

Gagnon, M., and Lacroix, G. Zooplankton sample variability in a tidal estuary: An interpretative model. *Limnol. Oceanogr.* **26,** 401–413 (1981).

Gaudy, R. Feeding four species of pelagic copepods under experimental conditions. *Mar. Biol.* **25,** 125–141 (1974).

Gauld, D. T. A peritrophic membrane in calanoid copepods. *Nature, Lond.* **179,** 325, 326 (1957).

Gauld, D. T., and Raymont, J. E. G. The respiration of some planktonic copepods. II. The effect of temperature. *J. Mar. Biol. Assoc. U.K.* **31,** 447–460 (1953).

Gehringer, J. W. An all-metal plankton sampler (model Gulf III). *Spec. Scient. Rep. U.S. Fish Wildl. Serv. Fisheries* **88,** 7–12 (1952).

Gerber, R. P., and Gerber, M. B. Ingestion of natural particulate organic matter and subsequent assimilation, respiration and growth by tropical lagoon zooplankton. *Mar. Biol.* **52,** 33–43 (1979).

Gillespie, D. M., and Benke, A. C. Methods of calculating cohort production from field data—some relationships. *Limnol. Oceanogr.* **24,** 171–176 (1979).

Gilmartin, M. Changes in inorganic phosphate concentration occurring during seawater sample storage. *Limnol. Oceanogr.* **12,** 325–328 (1967).

Glover, R. S. The continuous plankton recorder. *Rapp. R.-v Reun. Cons. Perm. Int. Explor. Mer* **153,** 8–15 (1962).

Gold, K. Some observations on the biology of *Tintinnopsis* sp. *J. Protozool.* **15,** 193, 194 (1968).

Gold, K. The preservation of tintinnids. *J. Protozool.* **16,** 126–128 (1969a).

Gold, K. Tintinnida: Feeding experiments and lorica development. *J. Protozool.* **16,** 507–509 (1969b).

Gold, K. Growth characteristics of the mass-reared tintinnid *Tintinnopsis beroidea. Mar. Biol.* **8,** 105–108 (1971).

Gold, K. Methods for preserving Tintinnida. In H. F. Steedmann (ed.), *Zooplankton Fixation and Preservation.* Monographs on Oceanographic Methodology 4. UNESCO, Paris, 1976, pp. 236–239.

Goodall, D. W. Objective methods for the classification of vegetation. III. An essay in the use of factor analysis. *Aust. J. Botany* **2,** 304–324 (1954).

Gophen, M., and Harris, R. P. Visual predation by a marine cyclopoid copepod, *Corycacus anglicus. J. Mar. Biol. Assoc. U.K.* **61,** 391–399 (1981).

Gnaiger, E. Calculation of energetic and biochemical equivalents of respiratory oxygen consumption. In E. Gnaiger and H. Forstner (eds.), *Polarographic Oxygen Sensors.* Springer-Verlag, Berlin, 1983, pp. 337–345.

Greenhalgh, G. N., and Evans, L. V. Electron microscopy. In C. Booth (ed.), *Methods in Microbiology,* Vol. 4. Academic Press, New York, 1971, pp. 518–565.

Greve, W. The "Planktonkreisel," a new device for culturing zooplankton. *Mar. Biol.* **1,** 201–203 (1968).

Greve, W. Cultivation experiments on North Sea ctenophores. *Helgoländer Wiss. Meeresunters.* **20,** 304–317 (1970).

Grice, G. D., and Hülsemann, K. New species of bottom-living calanoid copepods collected in deepwater by the DSRV Alvin. *Bull. Mus. Comp. Zool. Harvard Univ.* **139,** 185–227 (1970).

Grice, G. D., and Marcus, N. H. Dormant eggs of marine copepods. *Oceanogr. Mar. Biol. Ann. Rev.* **19,** 125–140 (1981).

Griffiths, F. B., and Caperon, J. Description and use of an improved method for determining estuarine zooplankton grazing rates on phytoplankton. *Mar. Biol.* **54,** 301–309 (1979).

Guillard, R. R. L. Culture of phytoplankton for feeding marine invertebrates. In W. L. Smith and M. H. Chanley (eds.), *Culture of Marine Invertebrate Animals.* Plenum Press, New York, 1975, pp. 29–60.

Guillard, R. R. L., and Ryther, J. H. Studies of marine planktonic diatoms. 1, *Cyclotella nana* Hustedt and *Detonulla confervacea* (Cleve) Gran. *Can J. Microbiol.* **8,** 229–239 (1962).

Gulland, J. A. Manual of methods for fish stock assessment. Part 1. Fish population analysis. *FAO Manuals in Fisheries Science* **4,** 1969.

Hairston, N. G. Jr. On the diel variation of copepod pigmentation. *Limnol. Oceanogr.* **25,** 742–747 (1980).

Halcrow, K. Acclimation to temperature in the marine copepod, *Calanus finmarchicus* (Gunner.). *Limnol. Oceanogr.* **8,** 1–8 (1963).

Hall, D. J. An experimental approach to the dynamics of a natural population of *Daphnia galeata mendotae. Ecology* **45,** 94–112 (1964).

Hallegraeff, G. M. Seasonal study of phytoplankton pigments and species at a coastal station off Sydney: Importance of diatoms and the nanoplankton. *Mar. Biol.* **61,** 107–118 (1981).

Hamburger, K. A gradient diver for measurement of respiration in individual organisms from the micro- and meiofauna. *Mar. Biol.* **61,** 179–183 (1981).

Hamilton, A. L. On estimating annual production. *Limnol. Oceanogr.* **14,** 771–782 (1969).

Hamilton, R. D., and Preslan, J. E. Cultural characteristics of a pelagic marine hymenostome ciliate, *Uronema* sp. *J. Exp. Biol. Ecol.* **4,** 90–99 (1969).

Hamner, W. M. Blue-water plankton. *Natl. Geogr. Mag.* **146,** 530–545 (1974).

Hamner, W. M. Underwater observations of blue-water plankton. Logistics, techniques, and safety procedures for divers at sea. *Limnol. Oceanogr.* **20,** 1045–1051 (1975).

Hamner, W. M. Observations at sea of live, tropical zooplankton, In *Proceedings of the Symposium on Warm Water Zooplankton.* NIO, Goa, 1977, pp. 284–296.

Hamner, W. M., and Carleton, J. Copepod swarms: Attributes and role in coral reef ecosystems. *Limnol. Oceanogr.* **24,** 1–14 (1979).

Hamner, W. M., Madin, L. P., Alldredge, A. L., Gilmer, R. W., and Hamner, P. P. Underwater observations of gelatinous zooplankton. Sampling problems, feeding biology, and behavior. *Limnol. Oceanogr.* **20,** 907–917 (1975).

Hanna, B. A., and Lilly, D. M. Axenic culture of *Uronema marinum. Am. Zool.* **10,** 539, 540 (1970).

Hanna, B. A., and Lilly, D. M. Growth of *Uronema marinum* in chemically defined medium. *Mar. Biol.* **26,** 153–160 (1974).

Haq, S. M. Nutritional physiology of *Metridia lucens* and *M. longa* from the Gulf of Maine. *Limnol. Oceanogr.* **12,** 40–51 (1967).

Harbison, G. R., and Gilmer, R. W. The feeding rates of the pelagic tunicate *Pegea confederata* and two other salps. *Limnol. Oceanogr.* **21,** 517–528 (1976).

Harbison, G. R., Madin, L. P., and Swanberg, N. R. On the natural history and distribution of oceanic ctenophores. *Deep-Sea Res.* **25,** 233–256 (1978).

Harbison, G. R., and McAlister, V. L. Fact and artifact in copepod feeding experiments. *Limnol. Oceanogr.* **25,** 971–981 (1980).

Harding, J. P. The use of probability paper for the graphical analysis of polymodal frequency distributions. *J. Mar. Biol. Assoc. U.K.* **28,** 141–153 (1949).

Hargis, J. R. Comparison of techniques for the measurement of zooplankton filtration rates. *Limnol. Oceanogr.* **22,** 942–945 (1977).

Hargrave, B. T., and Geen, G. H. Phosphorus excretion by zooplankton. *Limnol. Oceanogr.* **13,** 332–342 (1968).

Hargrave, B. T., and Geen, G. H. Effects of copepod grazing on two natural phytoplankton populations. *J. Fish. Res. Bd. Canada* **27,** 1395–1403 (1970).

Harman, H. H. *Modern Factor Analysis.* University of Chicago Press, Chicago, 1967.

Harris, E. The nitrogen cycle in Long Island Sound. *Bull. Bingham Oceanogr. Coll.* **17,** 31–65 (1959).

Harris, R. P., and Paffenhöfer, G.-A. Feeding, growth and reproduction of the marine planktonic copepod *Temora longicornis* Müller. *J. Mar. Biol. Assoc. U.K.* **56,** 675–690 (1976a).

Harris, R. P., and Paffenhöfer, G.-A. The effect of food concentration on cumulative ingestion and growth efficiency of two small marine planktonic copepods. *J. Mar. Biol. Assoc. U.K.* **56,** 875–888 (1976b).

Harvey, H. W. Note on selective feeding by *Calanus. J. Mar. Biol. Assoc. U.K.* **22,** 97–100 (1937).

Hasle, G. R. The inverted-microscope method. In A. Sournia (ed.), *Phytoplankton Manual.* Monographs on Oceanographic Methodology 6. UNESCO, Paris, 1978, pp. 88–96.

Haury, L. R. Sampling bias of a Longhurst-Hardy plankton recorder. *Limnol. Oceanogr.* **18,** 500–506 (1973).

Heck, K. L. Jr., Belle, G. A., and Simberloff, D. Explicit calculation of the rarefaction diversity measurement and the determination of sufficient sample size. *Ecology* **56,** 1459–1461 (1975).

Heinbokel, J. F. Studies on the functional role of tininnids in the Southern California Bight. II. Grazing rates of field populations. *Mar. Biol.* **47,** 191–197 (1978).

Heinle, D. R. Production of a calanoid copepod, *Acartia tonsa*, in the Patuxent River estuary. *Chesapeake Sci.* **7,** 59–74 (1966).

Heinle, D. R. Temperature and zooplankton. *Chesapeake Sci.* **10,** 186–209 (1969).

Heinle, D. R. Effects of passage through power plant cooling systems on estuarine copepods. *Environ. Pollut.* **11,** 39–58 (1976).

Heinrich, A. K. On the production of copepods in the Bering Sea. *Int. Revue Ges. Hydrobiol.* **47,** 465–469 (1962).

Hempel, G., and Weikert, H. The neuston of the subtropical and boreal North-eastern Atlantic Ocean. A review. *Mar. Biol.* **13,** 70–88 (1972).

Herman, A. W., and Dauphinee, T. M. Continuous and rapid profiling of zooplankton with an electronic counter mounted on a "Batfish" vehicle. *Deep-Sea Res.* **27A,** 79–96 (1980).

Herman, A. W., and Mitchell, M. R. Counting and identifying copepod species with an *in situ* electronic zooplankton counter. *Deep-Sea Res.* **28A,** 739–755 (1981).

Heron, A. C. Population ecology of a colonizing species: The pelagic tunicate *Thalia democratica.* I. Individual growth rate and generation time. *Oecologia* **10,** 269–293 (1972).

Hessler, R. R., and Sanders, H. L. Faunal diversity in the deep-sea. *Deep-Sea Res.* **14,** 65–78 (1967).

Hesthagen, I. H. On the near-bottom plankton and benthic invertebrate fauna of the Josephine Seamount and the Great Meteor Seamount. *"Meteor" -Forschungsergeb. D* **8,** 61–70 (1970).

Hirata, H. An attempt to apply an experimental microcosm for the mass culture of marine rotifer, *Brachionus plicatilis* Müller. *Mem. Fac. Fish. Kagoshima Univ.* **23,** 163–172 (1974).

Hirayama, K., and Ogawa, S. Fundamental studies on physiology of rotifer for its mass culture I. Filter feeding of rotifer. *Bull. Japanese Soc. Sci. Fish.* **38,** 1207–1214 (1972).

Hirota, J. Laboratory culture and metabolism of the planktonic ctenophore, *Pleurobrachia bachei* A. Agassiz. In A. Y. Takenouchi et al. (eds.), *Biological Oceanography of the Northern North Pacific Ocean.* Idemitsu-Shoten, Tokyo, 1972, pp. 465–484.

Hirota, J. Quantitative natural history of *Pleurobrachia bachei* in La Jolla bight. *Fishery Bull. U.S.* **72,** 295–335 (1974).

Hobbie, J. E., Holm-Hansen, O., Packard, T. T., Pomeroy, L. R., Sheldon, R. W., Thomas, J. P., and Wiebe, W. J. A study of the distribution and activity of microorganisms in ocean water. *Limnol. Oceanogr.* **17,** 544–555 (1972).

Hobson, E. S., and Chess, J. R. Zooplankters that emerge from the lagoon floor at night at Kure and Midway atolls, Hawaii. *Fishery Bull. U.S.* **77,** 275–280 (1979).

Holeton, G. F. Metabolic cold adaptation of polar fish: Fact or artifact? *Physiol. Zool.* **47,** 137–152 (1974).

Holme, N. A. Methods of sampling the benthos. *Adv. Mar. Biol.* **2,** 171–260 (1964).

Holmes, R. W. The preparation of marine phytoplankton and enumeration on molecular filters. *Spec. Scient. Rep. U.S. Fish. Wildl. Serv. Fisheries* **433,** 1–6 (1962).

Holm-Hansen, O. Determination of microbial biomass in ocean profiles. *Limnol. Oceanogr.* **14,** 740–747 (1969).

Holm-Hansen, O., and Booth, C. R. The measurement of adenosine triphosphate in the ocean and its ecological significance. *Limnol. Oceanogr.* **11,** 510–519 (1966).

Hopkins, T. L. Carbon and nitrogen content of fresh and preserved *Nematoscelis difficilis*, a euphausiid crustacean. *J. Cons. Perm. Int. Explor. Mer* **31**, 300–304 (1968).

Hopkins, T. L., Baird, R. C., and Milliken, D. M. A messenger-operated closing trawl. *Limnol. Oceanogr.* **18**, 488–490 (1973).

Hopwood, D. Some aspects of fixation with glutaraldehyde. *J. Anat.* **101**, 83–92 (1967).

Humes, A. G., and Gooding, R. U. A method for studying the external anatomy of copepods. *Crustaceana* **6**, 238–240 (1964).

Hündgen, M. Der Einfluss verschiedener Aldehyde auf die Strukturerhaltung gezüchteter Zellen und auf die Darstellbarkeit von vier Phosphatasen. *Histochemie* **15**, 46–61 (1968).

Huntley, M. Nonselective, nonsaturated feeding by three calanoid copepod species in the Labrador Sea. *Limnol. Oceanogr.* **26**, 831–842 (1981).

Hurlbert, S. H. The non-concept of species diversity: A critique and alternative parameters. *Ecology* **52**, 577–586 (1971).

Ibanez, F. Contribution a l'analyse mathematique des evenements en ecologie planctonique. Optimisations methodologiques, etude expérimentale en continu á petite échelle de l'hétérogéneité du plancton côtier. *Bull. Inst. Océanogr. Monaco* **72**, 1–96 (1976).

Ibanez, M. F., and Seguin, G. Etude du cycle anuel du zooplankton d'Abidjan. Comparison de plusieurs méthodes d'analyse multivariable: Composantes principales, correspondances, coordonnées principales. *Invest. Pesquera* **36**, 81–108 (1972).

Icanberry, J. W., and Richardson, R. W. Quantitative sampling of live zooplankton with a filter-pump system. *Limnol. Oceanogr.* **18**, 333–335 (1973).

Ikeda, T. Nutritional ecology of marine zooplankton. *Mem. Fac. Fish. Hokkaido Univ.* **22**, 1–97 (1974).

Ikeda, T. The effect of laboratory conditions on the extrapolation of experimental measurements to the ecology of marine zooplankton. I. Effect of feeding condition on the respiration rate. *Bull. Plankton Soc. Japan* **23**, 51–60 (1976).

Ikeda, T. The effect of laboratory conditions on the extrapolation of experimental measurements to the ecology of marine zooplankton. II. Effect of oxygen saturation on the respiration rate. *Bull. Plankton Soc. Japan* **24**, 19–28 (1977a).

Ikeda, T. The effect of laboratory conditions on the extrapolation of experimental measurements to the ecology of marine zooplankton. IV. Changes in respiration and excretion rates of boreal zooplankton species maintained under fed and starved conditions. *Mar. Biol.* **41**, 241–252 (1977b).

Ikeda, T. Feeding rates of planktonic copepods from a tropical sea. *J. Exp. Mar. Biol. Ecol.* **29**, 263–277 (1977c).

Ikeda, T., and Mitchell, A. W. Oxygen uptake, ammonia excretion and phosphate excretion by krill and other Antarctic zooplankton in relation to their body size and chemical composition. *Mar. Biol.* **71**, 283–298 (1982).

Ikeda, T., and Motoda, S. Estimated zooplankton production and their ammonia excretion in the Kuroshio and adjacent seas. *Fishery. Bull. U.S.* **76**, 357–367 (1978a).

Ikeda, T., and Motoda, S. Zooplankton production in the Bering Sea calculated from 1956-1970 Oshoro-Maru data. *Mar. Sci. Comms.* **4**, 329–346 (1978b).

Ikeda, T., and Skjoldal, H. R. The effect of laboratory conditions on the extrapolation of experimental measurements to the ecology of marine zooplankton. VI. Changes in physiological activities and biochemical components of *Acetes sibogae australis* and *Acartia australis* after capture. *Mar. Biol.* **58,** 285–293 (1980).

Ikeda, T., Hing Fay, E., Hutchinson, S. A., and Boto, G. M. Ammonia and inorganic phosphate excretion by zooplankton from inshore waters of the Great Barrier Reef, Queensland. I. Relationship between excretion rates and body size. *Aust. J. Mar. Freshwat. Res.* **33,** 55–70 (1982).

Ikematsu, W. Ecological studies on the fauna of Macrura and Mysidacea in the Ariake Sea. *Bull. Seikai Reg. Fish. Res. Lab.* **30,** 1–124 (1963) (in Japanese).

Isaacs, J. D., and Brown, D. M. Isaacs-Brown opening-closing trawl. *Scripps Inst. Oceanogr.* Ref. 66–18, 1966.

Isaacs, J. D., and Kidd, L. W. Isaacs-Kidd midwater trawl. Final report. *Scripps Inst. Oceanogr.* Ref. 53–3 (Oceanographic equipment rept. no. 1), 1953.

Isaacs, J. D., Fleminger, A., and Miller, J. K. Distributional atlas of zooplankton biomass in the California Current region: Spring and fall 1955-1959. *Calif. Coop. Oceanic Fish. Invest. Atlas* **10,** 1–252 (1969).

Itoh, K. A consideration on feeding habits of planktonic copepods in relation to the structure of their oral parts. *Bull. Plankton Soc. Japan* **17,** 1–10 (1970) (in Japanese).

Ivlev, V. S. *Experimental Ecology of the Feeding of Fishes.* Moskva, 1955 (in Russian). English translation: D. Scott, Yale University Press, New Haven, CT, 1961.

Iwasaki, H. Physiology of phytoplankton, In S. Motoda (ed.), *Marine Plankton.* Tokai University Press, Tokyo, 1975, pp. 49–115 (in Japanese).

Jaccard, P. Gesetze der Pflanzenverteilung in der alpinen Region. *Flora* **90,** 349–377 (1902).

Jannash, H. W. Studies on planktonic bacteria by means of a direct membrane filter method. *J. Gen. Microbiol.* **18,** 609–620 (1958).

Japan Fishery Agency. *Biological Sampling: On the Kaiyo Maru Type Opening/Closing Midwater Net (KOC Net). Survey Report on Marine Biota and Background in Connection with Ocean Dumping of Low-Level Radioactive Wastes (1977–1980),* 1980 (in Japanese).

Jawed, M. Body nitrogen and nitrogenous excretion in *Neomysis rayii* Murdoch and *Euphausia pacifica* Hansen. *Limnol. Oceanogr.* **14,** 748–754 (1969).

Jawed, M. Effects of environmental factors and body size on rates of oxygen consumption in *Arachaeomysis grebnitzkii* and *Neomysis awatschensis* (Crustacea: Mysidae). *Mar. Biol.* **21,** 173–179 (1973).

Johnson, S. B., and Attramadal, Y. G. A simple protective cod-end for recovering live specimens. *J. Exp. Mar. Biol. Ecol.* **61,** 169–174 (1982).

Judkins, D. C., and Fleminger, A. Comparison of foregut contents of *Sergestes similis* obtained from net collections and albacore stomachs. *Fishery Bull. U.S.* **70,** 217–223 (1972).

Karnovsky, M. J. A formaldehyde-glutaraldehyde fixative of high osmolality for use in electron microscopy. *Abstr. 5th Ann. Meeting Am. Soc. Cell Biol.* **137** (1965).

Kasahara, S., Uye, S., and Onbé, T. Calanoid copepod eggs in sea-bottom muds. *Mar. Biol.* **26,** 167–171 (1974).

Katona, S. K. Growth characteristics of the copepods *Eurytemora affinis* and *E. herdmani* in laboratory cultures. *Helgoländer Wiss. Meeresunters.* **20,** 373–384 (1970).

Kawamura, A., and Hamaoka, S. Feeding habits of the gonostomatid fish, *Vinciguerria nimbaria,* collected from the stomach of Bryde's whales in the southwestern North Pacific. *Bull. Plankton Soc. Japan* **28**, 141–151 (1981).

Kersting, K. A nitrogen correction for calorific values. *Limnol. Oceanogr.* **17**, 643–644 (1972).

Kils, U. Performance of antarctic krill *Euphausia superba* at different levels of oxygen saturation. *Meeresforsh.* **27**, 35–48 (1979).

Kimoto, S. Some quantitative analysis on the Chrysomelid fauna of the Ryukyu Archipelago. *Esakia* **6**, 27–54 (1967).

King, F. D., and Packard, T. T. Respiration and the activity of the respiratory electron transport system in marine zooplankton. *Limnol. Oceanogr.* **20**, 849–854 (1975a).

King, F. D., and Packard, T. T. The effect of hydrostatic pressure on respiratory electron transport system activity in marine zooplankton. *Deep-Sea Res.* **22**, 99–105 (1975b).

King, J. E., and Hida, T. S. Variations in zooplankton abundance in Hawaiian waters, 1950-52. *Spec. Scient. Rep. U.S. Fish. Wildl. Serv. Fisheries* **118**, 1–66 (1954).

Kinne, O. Cultivation, Part 2. In O. Kinne, (ed.), *Marine Ecology, III.* John Wiley & Sons, New York, 1977.

Klekowski, R. Z. Cartesian diver micro-respirometry. In W. T. Edmondson, and G. G. Winberg (eds.), *A Manual on Methods for the Assessment of Secondary Productivity in Fresh Waters.* IBP Handbook No. 17. Blackwell Scientific Publications, Oxford, 1971a, pp. 290–295.

Klekowski, R. Z. Cartesian diver microrespirometry for aquatic animals. *Pol. Arch. Hydrobiol.* **18**, 93–114 (1971b).

Klekowski, R. Z. Microrespirometer for shipboard measurements of metabolic rate of microzooplankton. *Pol. Arch. Hydrobiol.* **24** (Suppl.), 455–465 (1977).

Koehl, M. A. R., and Strickler, J. R. Copepod feeding currents: Food capture at low Reynolds number. *Limnol. Oceanogr.* **26**, 1062–1073 (1981).

Kofoed, L. H. The feeding biology of *Hydrobia ventrosa* (Monotogu). II. Allocation of the components of the carbon-budget and the significance of the secretion of dissolved organic material. *J. Exp. Mar. Biol. Ecol.* **19**, 243–256 (1975).

Kokubo, S. *Marine and Freshwater Plankton Experimental Methods* (revised ed.). Koseisha Koseikaku, Tokyo, 1963 (in Japanese).

Kotori, M. The biology of Chaetognatha in the Bering Sea and the northern North Pacific Ocean, with emphasis of *Sagitta elegans*. *Mem. Fac. Fish. Hokkaido Univ.* **23**, 95–183 (1976).

Kremer, P. Respiration and excretion by the ctenophore *Mnemiopsis leidyi.* *Mar. Biol.* **44**, 43–50 (1977).

Kutty, M. N. Ammonia quotient in sockeye salmon (*Oncorhynchus nerka*). *J. Fish. Res. Bd. Canada* **35**, 1003–1005 (1978).

Lalli, C. M. Structure and function of the buccal apparatus of *Clione limacina* (Phipps) with a review of feeding in gymnosomatous pteropods. *J. Exp. Mar. Biol. Ecol.* **4**, 102–118 (1970).

Lam, R. K., and Frost, B. W. Model of copepod filtering response to changes in size and concentration of food. *Limnol. Oceanogr.* **21**, 490–500 (1976).

Lampert, W. Release of dissolved organic carbon by grazing zooplankton. *Limnol. Oceanogr.* **23**, 831–834 (1978).

Landry, M. R. Predatory feeding behavior of a marine copepod, *Labidocera trispinosa*. *Limnol. Oceanogr.* **23,** 1103–1113 (1978).

Landry, M. R. The development of marine calanoid copepods with comment on the isochronal rule. *Limnol. Oceanogr.* **28,** 614–624 (1983).

Lasker, R. Utilization of organic carbon by a marine crustacean: Analysis with carbon-14. *Science* **131,** 1098–1100 (1960).

Lasker, R. Feeding, growth, respiration and carbon utilization of a euphausiid crustacean. *J. Fish. Res. Bd. Canada* **23,** 1291–1317 (1966).

Laval, P. Hyperiid amphipods as crustacean parasitoids associated with gelatinous plankton. *Oceanogr. Mar. Biol. Ann. Rev.* **18,** 11–56 (1980).

Lawley, M. A., and Maxwell, A. E. *Factor Analysis as a Statistical Method.* Butterworth, Kent, U.K., 1963. Japanese translation: M. Okamoto, Nikkagiren Shuppan, Tokyo, 1970.

Lawton, J. H., and Richards, J. Comparability of Cartesian diver, Gilson, Warburg and Winkler methods of measuring the respiratory rates of aquatic invertebrates in ecological studies. *Oecologia* **4,** 319–324 (1970).

LeBlond, P. H., and Parsons, T. R. A simplified expression for calculating cohort production. *Limnol. Oceanogr.* **22,** 156, 157 (1977).

Le Borgne, R. P. Etude de la respiration et de l'excretion d'azote et de phosphore des populations zooplanctoniques de l'upwelling mauritanien (mars-avril, 1972). *Mar. Biol.* **19,** 249–257 (1973).

Lebour, M. V. The food of plankton organisms. *J. Mar. Biol. Assoc. U.K.* **12,** 644–677 (1922).

Lebour, M. V. The food of plankton organisms II. *J. Mar. Biol. Assoc. U.K.* **13,** 70–92 (1923).

Lee, J. J., Tietjen, J. H., and Mastropaolo, C. A. Axenic culture of the marine hymenostome ciliate *Uronema marinum* in chemically defined medium. *J. Protozool.* **18** (Suppl.), 10 (1971).

Lehman, J. T. The filter-feeder as an optimal forager, and the predicted shapes of feeding curves. *Limnol. Oceanogr.* **21,** 501–516 (1976).

Lehman, J. T. Release and cycling of nutrients between planktonic algae and herbivores. *Limnol. Oceanogr.* **25,** 620–632 (1980).

Lenz, J. A new type of plankton pump on the vacuum principle. *Deep-Sea Res.* **19,** 453–459 (1972).

Liddicoat, M. L., Tibbitts, S., and Butler, E. I. The determination of ammonia in seawater. *Limnol. Oceanogr.* **20,** 131–133 (1975).

Lloyd, M., and Ghelardi, R. J. A table for calculating the 'equitability' component of species diversity. *J. Anim. Ecol.* **33,** 217–226 (1964).

Lloyd, M., Zar, J. H., and Karr, J. R. On the calculation of information-theoretical measures of diversity. *Am. Midland Nat.* **79,** 257–272 (1968).

Lohmann, H. Über das Nannoplankton und die Zentrifugierung. *Int. Revue Ges. Hydrobiol. Hydrogr.* **4,** 1–38 (1911).

Longhurst, A. R., Reith, A. D., Bower, R. E., and Seibert, D. L. R. A new system for the collection of multiple serial plankton samples. *Deep-Sea Res.* **13,** 213–222 (1966).

Love, R. M. *The Chemical Biology of Fishes.* Academic Press, New York, 1970.

Lovegrove, T. The effect of various factors on dry weight values. *Rapp. P.-v. Reun. Cons. Perm. Int. Explor. Mer* **153,** 86–91 (1962).

Lovegrove, T. The determination of the dry weight of plankton and the effect of various factors on the values obtained. In H. Barnes (ed.), *Some Contemporary Studies in Marine Science*. Allen and Unwin, London, 1966, pp. 429–467.

Løvtrup, S., and Larsson, S. Electromagnetic recording diver balance. *Nature (London)* **208**, 1116, 1117 (1965).

Lowry, P. H., Rosenbrough, N. J., Farr, A. L., and Randall, R. J. Protein measurement with a Folin phenol reagent. *J. Biol. Chem.* **193**, 265–275 (1951).

Lund, J. W. G., Kipling, C., and LeGran, E. P. The inverted microscope method of estimating algal numbers and statistical basis of examination by counting. *Hydrobiologia* **11**, 143–170 (1958).

Lundin, A., and Thore, A. Comparison of methods for extraction of bacterial adenine nucleotides determined by firefly assay. *Appl. Microbiol.* **30**, 713–721 (1975).

Lyman, J., and Fleming, R. H. Composition of sea water. *J. Mar. Res.* **3**, 134–146 (1940).

McAlice, B. J. Phytoplankton sampling with the Sedgwick-Rafter cell. *Limnol. Oceanogr.* **16**, 19–28 (1971).

McAllister, C. D. Aspects of estimating zooplankton production from phytoplankton production *J. Fish. Res. Bd. Canada* **26**, 199–220 (1969).

McAlister, C. D. Zooplankton ratios, phytoplankton mortality and the estimation of marine production. In J. H. Steele (ed.), *Marine Food Chains*. Oliver & Boyd, Edinburgh, U.K. 1970, pp. 419–457.

MacArthur, R. H. On the relative abundance of bird species. *Proc. Natl. Acad. Sci. USA* **43**, 293–295 (1957).

MacArthur, R. H. On the relative abundance of species. *Am. Natur.* **94**, 25–36 (1960).

McGowan, J. A. The nature of oceanic ecosystems. In C. B. Miller (ed.), *The Biology of the Oceanic Pacific*. Oregon State University Press, Corvallis, OR, (1974), pp. 9–28.

McGowan, J. A., and Brown, D. M. A new opening-closing paired zooplankton net. *Univ. Calif. Scripps Inst. Oceanogr.* Ref. 66–23 (1966).

McGowan, J. A., and Walker, P. W. Structure in the copepod community of the North Pacific Central Gyre. *Ecol. Monogr.* **49**, 195–226 (1979).

McLachlan, J. Some considerations of the growth of marine algae in artificial media. *Can. J. Microbiol.* **10**, 769–782 (1964).

McLaren, I. A. Some relationships between temperature and egg size, body size, development rate, and fecundity of the copepod *Pseudocalanus*. *Limnol. Oceanogr.* **10**, 528–538 (1965).

McLaren, I. A. Population and production ecology of zooplankton in Ogac Lake, a landlocked fiord on Baffin Island. *J. Fish. Res. Bd. Canada* **26**, 1485–1559 (1969).

McLaren, I. A. Generation lengths of some temperate marine copepods: Estimation, prediction, and implications. *J. Fish. Res. Bd. Canada* **35**, 1330–1342 (1978).

Mackas, D., and Bohrer, R. Fluorescence analysis of zooplankton gut contents and an investigation of diel feeding patterns. *J. Exp. Mar. Biol. Ecol.* **25**, 77–85 (1976).

Madin, L. P. Field observations of the feeding behavior of salps (Tunicata: Thaliacea). *Mar. Biol.* **25**, 143–148 (1974).

Mahnken, C. V. W., and Jossi, J. W. Flume experiments on the hydrodynamics of plankton nets. *J. Cons. Perm. Int. Explor. Mer* **31,** 38–45 (1967).

Margalef, D. R. Diversidad de especies en les communidades naturales. *Publnes Inst. Biol. Apl., Barcelona* **9,** 5–27 (1951).

Margalef, D. R. La teoria de la informacion en ecologia. *Mems. R. Acad. Cienc. Artes Barcelona* **23,** 373–449 (1957).

Margalef, R. Temporal succession and spatial heterogeneity in phytoplankton. In A. Buzzati-Traverso (ed.), *Perspectives in Marine Biology.* University of California Press, Berkeley, 1958, pp. 323–349.

Margalef, R. Sampling design: Some examples. In A. Sournia (ed.), *Phytoplankton Manual.* Monographs on Oceanographic Methodology 6. UNESCO, Paris, 1978, pp. 17–31.

Margalef, R., and González, B. F. Grupos de especies asociadas en el fitoplancton del mar Caribe (NE de Venezuela). *Invest. Pesq.* **33,** 287–312 (1969).

Marshall, S. M. Respiration and feeding in copepods. *Adv. Mar. Biol.* **11,** 57–120 (1973).

Marshall, S. M., and Orr, A. P. On the biology of *Calanus finmarchicus.* VIII. Food uptake, assimilation and excretion in adult and stage V *Calanus. J. Mar. Biol. Assoc. U.K.* **34,** 495–529 (1955a).

Marshall, S. M., and Orr, A. P. *The Biology of a Marine Copepod.* Oliver & Boyd, Edinburgh, U.K. 1955b.

Marshall, S. M., and Orr, A. P. Experimental feeding of the copepod *Calanus finmarchicus* (Gunner) on phytoplankton cultures labelled with radioactive carbon (^{14}C). *Deep-Sea Res.* **3,** Suppl. Papers in marine biology and oceanography, 110–114 (1955c).

Marshall, S. M., and Orr, A. P. Some uses of antibiotics in physiological experiments in sea water. *J. Mar. Res.* **17,** 341–346 (1958).

Marshall, S. M., and Orr, A. P. Food and feeding in copepods. *Rap. P.-v. Reun. Cons. Perm. Int. Explor. Mer* **153,** 92–98 (1962).

Marshall, S. M., Nicholls, A. G., and Orr, A. P. On the biology of *Calanus finmarchicus.* VI. Oxygen consumption in relation to environmental conditions. *J. Mar. Biol. Assoc. U.K.* **20,** 1–28 (1935).

Martin, J. H. Phytoplankton-zooplankton relationships in Narragansett Bay. III. Seasonal changes in zooplankton excretion rates in relation to phytoplankton abundance. *Limnol. Oceanogr.* **13,** 63–71 (1968).

Matsudaira, C. Culturing of a Copepoda, *Sinocalanus tenellus. Inform. Bull. Planktol. Japan* **5,** 1–6 (1957) (in Japanese).

Matsuo, Y., Nemoto, T., and Marumo, R. A convertible neuston net for zooplankton. *Bull. Plankton Soc. Japan* **23,** 26–30 (1976).

Matthews, J. B. L. The genus *Euaugaptilus* (Crustacea, Copepoda). New descriptions and a review of the genus in relation to *Augaptilus, Haloptilus* and *Pseudaugaptilus. Bull. Br. Mus. Nat. Hist.* (Zool.) **24,** 1–71 (1972).

Mayzaud, P. Respiration and nitrogen excretion of zooplankton. II. Studies of the metabolic characteristics of starved animals. *Mar. Biol.* **21,** 19–28 (1973).

Mayzaud, P. Respiration and nitrogen excretion of zooplankton. IV. The influence of starvation on the metabolism and the biochemical composition of some species. *Mar. Biol.* **37,** 47–58 (1976).

Menge, B. A., and Sutherland, J. P. Species diversity gradients: synthesis of the roles of predation, competition and temporal heterogeneity. *Am. Natur.* **110,** 351–369 (1976).

Menzie, C. A. A note on the Hynes method of estimating secondary production. *Limnol. Oceanogr.* **25,** 770–773 (1980).

Merks, A. G. A. Determination of ammonia in sea water with an ion-selective electrode. *Neth. J. Sea Res.* **9,** 371–375 (1975).

Miller, C. B., Johnson, J. K., and Heinle, D. R. Growth rules in the marine copepod genus *Acartia*. *Limnol. Oceanogr.* **22,** 326–335 (1977).

Miller, C. B., and Judkins, D. C. Design of pumping systems for sampling zooplankton, with descriptions of two high-capacity samplers for coastal studies. *Biol. Oceanogr.* **1,** 29–56 (1981).

Miller, D. A modification of the small Hardy plankton sampler for simultaneous high-speed plankton hauls. *Bull. Mar. Ecol.* **5,** 165–172 (1961).

Mitamura, O., and Saijo, Y. Urea supply from decomposition and excretion of zooplankton. *J. Oceanogr. Soc. Japan* **36,** 121–125 (1980).

Morisita, M. Measuring of interspecific association and similarity between communities. *Mem. Fac. Sci. Kyushu Univ., Ser. E., Biol.* **3,** 65–80 (1959).

Morisita, M. Methods of production measurement. In *Methods of Biological Productivity Measurements in Freshwaters, JIBP*. Kodansha, Tokyo, 1969, pp. 357–367 (in Japanese).

Motoda, S. North Pacific standard plankton net. *Inform. Bull. Planktol. Japan* **4,** 13–15 (1957) (in Japanese).

Motoda, S. Devices of simple plankton apparatus. *Mem. Fac. Fish. Hokkaido Univ.* **7,** 73–94 (1959).

Motoda, S. Devices of simple plankton apparatus, V. *Bull. Fac. Fish. Hokkaido Univ.* **22,** 101–106 (1971).

Motoda, S. Plankton sampling. In R. Marumo (ed.), *Marine Plankton*. University of Tokyo Press, Tokyo, 1974, pp. 191–225 (in Japanese).

Motoda, S. Preliminary processing of plankton samples. *Bull. Plankton Soc. Japan* **21,** 115–134 (1975) (in Japanese).

Mountford, M. D. An index of similarity and its application to classificatory problems. In P. W. Murphy (ed.), *Progress in Soil Science*. Butterworths, Kent, 1962, pp. 43–50.

Muller, W. A., and Lee, J. J. Apparent indispensability of bacteria in foraminiferan nutrition. *J. Protozool.* **16,** 471–495 (1969).

Mullin, M. M. Some factors affecting the feeding of marine copepods of the genus *Calanus*. *Limnol. Oceanogr.* **8,** 239–250 (1963).

Mullin, M. M. Selective feeding by calanoid copepods from the Indian Ocean. In H. Barnes (ed.), *Some contemporary Studies in Marine Science*. Allen and Unwin, London, 1966, pp. 545–554.

Mullin, M. M. Production of zooplankton in the ocean. The present status and problems. *Oceanogr. Mar. Biol. Ann. Rev.* **7,** 293–314 (1969).

Mullin, M. M. Differential predation by the carnivorous marine copepod, *Tortanus discaudatus*. *Limnol. Oceanogr.* **24,** 774–777 (1979).

Mullin, M. M., and Brooks, E. R. Laboratory culture, growth rate, and feeding behavior of a planktonic marine copepod. *Limnol. Oceanogr.* **12,** 657–666 (1967).

Mullin, M. M., and Brooks, E. R. The ecology of the plankton off La Jolla, California in the period April through September, 1967. Part VII. Production of the planktonic copepod, *Calanus helgolandicus*. *Bull. Scripps Inst. Oceanogr.* **17,** 89–103 (1970a).

Mullin, M. M., and Brooks, E. R. The effect of concentration of food on body weight, cumulative ingestion, and rate of growth of the marine copepod *Calanus helgolandicus*. *Limnol. Oceanogr.* **15**, 748–755 (1970b).

Mullin, M. M., and Brooks, E. R. Some consequences of distributional heterogeneity of phytoplankton and zooplankton. *Limnol. Oceanogr.* **21**, 784–791 (1976).

Mullin, M. M., and Evans, P. M. The use of a deep tank in plankton ecology. 2. Efficiency of a plankton food chain. *Limnol. Oceanogr.* **19**, 902–911 (1974).

Mullin, M. M. Sloan, P. R., and Eppley, R. W. Relationship between carbon content, cell volume, and area in phytoplankton. *Limnol. Oceanogr.* **11**, 307–311 (1966).

Mullin, M. M., Perry, M. J., Renger, E. H., and Evans, P. M. Nutrient regeneration by oceanic zooplankton: A comparison of methods. *Mar. Sci. Commun.* **1**, 1–13 (1975a).

Mullin, M. M., Stewart, E. F., and Fuglister, F. J. Ingestion by planktonic grazers as a function of concentration of food. *Limnol. Oceanogr.* **20**, 259–262 (1975b).

Munro, H. N., and Fleck, A. The determination of nucleic acids. In D. Glick (ed.), *Methods of Biochemical Analysis*, Vol. 14. Wiley-Interscience, New York, 1966a, pp. 113–176.

Munro, H. N., and Fleck, A. Recent developments in the measurement of nucleic acids in biological materials. *Analyst* **91**, 78–88 (1966b).

Nagasawa, S., and Marumo, R. Feeding of a pelagic chaetognath *Sagitta nagae* Alvarino, in Suruga Bay, Japan. *J. Oceanogr. Soc. Japan* **28**, 181–186 (1973).

Nakai, Z. Apparatus for collecting macroplankton in the spawning surveys of Iwashi (sardine, anchovy, and round herring and others). *Bull. Tokai Reg. Fish. Res. Lab.* **9**, 221–237 (1962).

Nakai, Z., and Honjo, K. Comparative studies on measurements of the weight and the volume of plankton samples. A preliminary account. *Proc. 9th Indo-Pacif. Fish. Coun.*, Sec. 2, 9–16 (1961).

Nassogne, A. Influence of food organisms on the development and culture of pelagic copepods. *Helgoländer Wiss. Meeresunters.* **20**, 333–345 (1970).

Needham, J. G., Galtsoff, P. S., Lutz, F. E., and Welch, P. S. (eds.). *Culture Methods for Invertebrate Animals*. Dover Publications, New York, 1959.

Nemoto, T. Chlorophyll pigments in the stomach of euphausiids. *J. Oceanogr. Soc. Japan* **24**, 253–260 (1968).

Nemoto, T., and Saijo, Y. Trace of chlorophyll pigments in stomachs of deep sea zooplankton. *J. Oceanogr. Soc. Japan* **24**, 311–313 (1968).

Neunes, G. W., and Pongolini, G. F. Breeding a pelagic copepod, *Euterpina acutifrons* (Dana), in the laboratory. *Nature, Lond.* **208**, 571–573 (1965).

Newbury, T. K. Consumption and growth rates of chaetognaths and copepods in subtropical oceanic waters. *Pacif. Sci.* **32**, 61–77 (1978).

Newbury, T. K., and Bartholomew, E. F. Secondary production of microcopepods in the southern, eutrophic basin of Kaneohe Bay, Oahu, Hawaiian Islands. *Pacif. Sci.* **30**, 373–384 (1976).

Nishimura, S. *Chikyu no umi to seibutsu* [Introduction to the marine biogeography]. Kaimeisha, Tokyo, 1981 (in Japanese).

Nival, P., and Nival, S. Particle retention efficiencies of an herbivorous copepod, *Acartia clausi* (adult and copepodite stages): Effects on grazing. *Limnol. Oceanogr.* **21**, 24–38 (1976).

Nival, P., Malara, G., Charra, R., Palazzoli, I., and Nival, S. Etude de la respiration et de l'exrétion de quelques copépodes planktonique (Crustacea) dans la zone de remontée d'eau profonde des cotes Marocaines. *J. Exp. Mar. Biol. Ecol.* **15**, 231–260 (1974).

Niven, D. F., Collins, P. A., and Knowles, C. J. Adenylate energy charge during batch culture of *Beneckea natriegens*. *J. Gen. Microbiol.* **98**, 95–108 (1977).

Northby, J. A. A comment on rate measurements in open systems. *Limnol. Oceanogr.* **21**, 180–182 (1976).

O'Conners, H. B., Small, L. F., and Donaghay, P. L. Particle-size modification by two size classes of the estuarine copepod *Acartia clausi*. *Limnol. Oceanogr.* **21**, 300–308 (1976).

Omori, M. A 160 cm opening-closing plankton net—I. Description of the gear. *J. Oceanogr. Soc. Japan* **21**, 212–220 (1965).

Omori, M. *Calanus cristatus* and submergence of the Oyashio water. *Deep-Sea Res.* **14**, 525–532 (1967).

Omori, M. Weight and chemical composition of some important zooplankton in the North Pacific Ocean. *Mar. Biol.* **3**, 4–10 (1969a).

Omori, M. A bottom-net to collect zooplankton living close to the sea-floor. *J. Oceanogr. Soc. Japan* **25**, 291–294 (1969b).

Omori, M. Variations of length, weight, respiratory rate and chemical composition of *Calanus cristatus* in relation to its food and feeding. In J. H. Steele (ed.), *Marine Food Chains*. Oliver & Boyd, Edinburgh, U.K., 1970, pp. 113–126.

Omori, M. Preliminary rearing experiments on the larvae of *Sergestes lucens* (Penaeidea, Natantia, Decapoda). *Mar. Biol.* **9**, 228–234 (1971).

Omori, M. The biology of pelagic shrimps in the ocean. *Adv. Mar. Biol.* **12**, 233–324 (1974).

Omori, M. Calcium carbonate platelet deposits in preserved zooplankton. In H. F. Steedman (ed.), *Zooplankton Fixation and Preservation*. Monographs on Oceanographic Methodology 4. UNESCO, Paris, 1976a, p. 219.

Omori, M. The glass shrimp, *Pasiphaea japonica* sp. nov. (Caridea, Pasiphaeidae), a sibling species of *Pasiphaea sivado*, with note on its biology and fishery in Toyama Bay. *Bull. Natn. Sci. Mus., Tokyo, Ser. A* (Zool.) **19**, 201–218 (1976b).

Omori, M. Distribution of warm water epiplanktonic shrimps of the genera *Lucifer* and *Acetes* (Macrura, Penaeidea, Sergestidae). *Proc. Symp. Warm Water Zooplankton*. NIO, Goa, 1977, pp. 1–12.

Omori, M. Some factors affecting on dry weight, organic weight and concentrations of carbon and nitrogen in freshly prepared and in preserved zooplankton. *Int. Revue Ges. Hydrobiol.* **63**, 261–269 (1978).

Omori, M. Growth, feeding, and mortality of larval and early postlarval stages of the oceanic shrimp *Sergestes similis* Hansen. *Limnol. Oceanogr.* **24**, 273–288 (1979).

Omori, M. Abundance assessment of micronektonic sergestid shrimp in the ocean. *Biol. Oceanogr.* **2**, 199–210 (1983).

Omori, M., and Hamner, W. M. Patchy distribution of zooplankton: Behavior, population assessment and sampling problems. *Mar. Biol.* **72**, 193–200 (1982).

Onbé, T. Occurrence of the resting eggs of a marine cladoceran, *Penilia avirostris* Dana, on the sea bottom. *Bull. Japanese Soc. Sci. Fish.* **38**, 305 (1972).

Onbé, T. Preliminary notes on the biology of the resting eggs of marine cladocerans. *Bull. Plankton Soc. Japan* **20**, 74–77 (1973) (in Japanese).

Onbé, T. Sugar floating method for sorting the resting eggs of marine cladocerans and copepods from sea-bottom sediment. *Bull. Japanese Soc. Sci. Fish.* **44,** 1411 (1978).

Ong, J. E. The fine structure of the mandibular sensory receptors in the brackish water calanoid copepod *Gladioferens pectinatus* (Brady). *Z. Zellforsch. Mikrosk. Anat.* **97,** 178–195 (1969).

Oppenheimer, C. H., and ZoBell, C. E. The growth and viability of sixty-three species of marine bacteria as influenced by hydrostatic pressure. *J. Mar. Res.* **11,** 10–18 (1952).

Ortner, P. B., Cummings, S. R., Aftring, R. P., and Edgerton, H. E. Silhouette photography of oceanic zooplankton. *Nature, Lond.* **277,** 50, 51 (1979).

Ortner, P. B., Hill, L. C., and Edgerton, H. E. *In-situ* silhouette photography of Gulf Stream zooplankton. *Deep-Sea Res.* **28A,** 1569–1576 (1981).

Owen, G., and Steedman, H. F. Preservation of molluscs. *Proc. Malac. Soc. Lond.* **33,** 101–103 (1958).

Owens, T. G., and King, F. D. The measurement of respiratory electron-transport-system activity in marine zooplankton. *Mar. Biol.* **30,** 27–36 (1975).

Packard, T. T. The measurement of respiratory electron transport activity in marine phytoplankton. *J. Mar. Res.* **29,** 235–244 (1971).

Packard, T. T., Devol, A. H., and King, F. D. The effect of temperature on the respiratory electron transport system in marine plankton. *Deep-Sea Res.* **22,** 237–249 (1975).

Paffenhöfer, G.-A. Cultivation of *Calanus helgolandicus* under controlled conditions. *Helgoländer Wiss. Meeresunters.* **20,** 346–359 (1970).

Paffenhöfer, G.-A. Grazing and ingestion rates of nauplii, copepodids and adults of the marine planktonic copepod *Calanus helgolandicus*. *Mar. Biol.* **11,** 286–298 (1971).

Paffenhöfer, G.-A. Feeding, growth, and food conversion efficiency of the marine planktonic copepod *Calanus helgolandicus*. *Limnol. Oceanogr.* **21,** 39–50 (1976).

Paffenhöfer, G.-A., and Harris, R. P. Laboratory culture of marine holozooplankton and its contribution to studies of marine planktonic food webs. *Adv. Mar. Biol.* **16,** 211–308 (1979).

Paffenhöfer, G.-A., Stricker, J. R., and Alcaraz, M. Suspension-feeding by herbivorous calanoid copepods: A cinematographic study. *Mar. Biol.* **67,** 193–199 (1982).

Paine, R. Ash and caloric determinations of sponge and opisthobranch tissues. *Ecology* **45,** 384–387 (1964).

Paine, R. The measurement and application of the calorie to ecological problems. *Ann. Rev. Ecol. Systematics* **2,** 145–164 (1971).

Paine, R. T. Food web complexity and species diversity. *Am. Natur.* **100,** 65–75 (1966).

Paloheimo, J. E. Calculation of instantaneous birth rate. *Limnol. Oceanogr.* **19,** 692–694 (1974).

Paloheimo, J. E., and Dickie, L. M. Food and growth of fishes. III. Relations among food, body size, and growth efficiency. *J. Fish. Res. Bd. Canada* **23,** 1209–1248 (1966).

Pantin, C. F. A. *Notes on Microscopical Technique for Zoologists.* Cambridge University Press, Cambridge, U.K., 1964.

Paranjape, M. A. Molting and respiration of euphausiids. *J. Fish. Res. Bd. Canada* **24,** 1229–1240 (1967).

Parsons, T. R., and Takahashi, M. Environmental control of phytoplankton cell size. *Limnol. Oceanogr.* **18,** 511–515 (1973).

Parsons, T. R., Stephens, K., and Strickland, J. D. H. On the chemical composition of eleven species of marine phytoplankters. *J. Fish. Res. Bd. Canada.* **18,** 1001–1016 (1961).

Patten, B. C. Species diversity in net phytoplankton of Raritan Bay. *J. Mar. Res.* **20,** 57–75 (1962).

Pearcy, W. G., and Mesecar, R. S. Scattering layers and vertical distribution of oceanic animals off Oregon. In *Proc. Int. Symp. Biological Sound Scattering in the Ocean.* U.S. Dept. Navy, MC Rept. 005, 1971, pp. 381–394.

Pearcy, W. G., Theilacker, G. H., and Lasker, R. Oxygen consumption of *Euphausia pacifica.* The lack of a diel rhythm or light-dark effect, with a comparison of experimental techniques. *Limnol. Oceanogr.* **14,** 219–223 (1969).

Pearre, S. Jr. Feeding by Chaetognatha: Energy balance and importance of various components of the diet of *Sagitta elegans. Mar. Ecol. Prog. Ser.* **5,** 45–54 (1981).

Perry, M. J. Phosphate utilization by an oceanic diatom in phosphorus-limited chemostat culture and in the oligotrophic waters of the central North Pacific. *Limnol. Oceanogr.* **21,** 88–107 (1976).

Peters, R. H., and Rigler, F. H. Phosphorus release by *Daphnia. Limnol. Oceanogr.* **18,** 821–839 (1973).

Petipa, T. S. Trophic relationships in communities and the functioning of marine ecosystems 1. In M. J. Dunbar (ed.), *Marine Production Mechanisms.* Cambridge University Press, Cambridge, U.K., 1979, pp. 233–250.

Petrusewics, K., and MacFadyen, A. (ed.). *Productivity of Terrestrial Animals— Principles and Methods.* IBP Handbook No. 13. Blackwell Scientific Publications, Oxford, 1970.

Phillipson, J. A miniature bomb calorimeter for small biological samples. *Oikos* **15,** 130–139 (1964).

Pielou, E. C. The measurement of diversity in different types of biological collections. *J. Theor. Biol.* **13,** 131–144 (1966).

Platt, T. Analysis of the importance of spatial and temporal heterogeneity in the estimation of annual production by phytoplankton in a small, enriched, marine basin. *J. Exp. Mar. Biol. Ecol.* **18,** 99–110 (1975).

Platt, T., and Irwin, B. Caloric content of phytoplankton. *Limnol. Oceanogr.* **18,** 306–310 (1973).

Platt, T., Brawn, V. M., and Irwin, B. Caloric and carbon equivalents of zooplankton biomass. *J. Fish. Res. Bd. Canada* **26,** 2345–2349 (1969).

Platt, T., Dickie, L. M., and Trites, R. W. Spatial heterogeneity of phytoplankton in a near-shore environment. *J. Fish. Res. Bd. Canada* **27,** 1453–1473 (1970).

Pomeroy, L. R., Mathews, H. M., and Min, H. S. Excretion of phosphate and soluble organic phosphorus compounds by zooplankton. *Limnol. Oceanogr.* **8,** 50–55 (1963).

Porter, J. W., Porter, K. G., and Batac-Catalan, Z. Quantitative sampling of Indo-Pacific demersal reef plankton. In *Proc. 3rd Int. Symp. Coral Reefs, 1.* Rosenstiel School Mar. Atmos. Sci., University of Miami, 1977, pp. 105–112.

Poulet, S. A. Grazing of *Pseudocalanus minutus* on naturally occurring particulate matter. *Limnol. Oceanogr.* **18,** 564–573 (1973).

Poulet, S. A. Seasonal grazing of *Pseudocalanus minutus* on particles. *Mar. Biol.* **25,** 109–123 (1974).

Poulet, S. A. Comparison between five coexisting species of marine copepods feeding on naturally occurring particulate matter. *Limnol. Oceanogr.* **23,** 1126–1143 (1978).

Poulet, S. A. and Marsot, P. Chemosensory grazing by marine calanoid copepods (Arthropoda: Crustacea). *Science* **200,** 1403–1405 (1978).

Prosser, C. L. (ed.) *Comparative Animal Physiology*, 3rd ed. W. B. Saunders, Philadelphia, 1973a.

Prosser, C. L. Oxygen: Respiration and metabolism. In C. L. Prosser (ed.), *Comparative Animal Physiology*, 3rd ed. W. B. Saunders, Philadelphia, 1973b, pp. 165–211.

Provasoli, L., McLaughlin, J. J. A., and Droop, M. R. The development of artificial media for marine algae. *Arch. Microbiol.* **25,** 392–428 (1957).

Provasoli, L., Shiraishi, K., and Lance, J. R. Nutritional idiosyncrasies of *Artemia* and *Tigriopus* in monoxenic culture. *Ann. N.Y. Acad. Sci.* **77,** 250–261 (1959).

Raymont, J. E. G. *Plankton and Productivity in the Oceans*. Pergamon Press, New York, 1963.

Raymont, J. E. G., and Conover, R. J. Further investigations on the carbohydrate content of marine zooplankton. *Limnol. Oceanogr.* **6,** 154–164 (1961).

Raymont, J. E. G., and Gauld, D. T. The respiration of some planktonic copepods. *J. Mar. Biol. Assoc. U.K.* **29,** 681–693 (1951).

Raymont, J. E. G., and Krishnaswamy, S. Carbohydrates in some marine planktonic animals. *J. Mar. Biol. Assoc. U.K.* **39,** 239–248 (1960).

Raymont, J. E. G., Austin, J., and Linford, E. Biochemical studies on marine zooplankton. I. The biochemical composition of *Neomysis integer*. *J. Cons. Perm. Int. Explor. Mer* **28,** 354–363 (1964).

Raymont, J. E. G., Austin, J., and Linford, E. The biochemical composition of certain oceanic zooplankton decapods. *Deep-Sea Res.* **14,** 113–115 (1967).

Raymont, J. E. G., Austin, J., and Linford, E. Biochemical studies on marine zooplankton. V. The composition of the major biochemical fractions in *Neomysis integer*. *J. Mar. Biol. Assoc. U.K.* **48,** 735–760 (1968).

Raymont, J. E. G., Srinivasagam, R. T., and Raymont, J. K. B. Biochemical studies on marine zooplankton. IV. Investigations on *Meganyctiphanes norvegica* (M. Sars). *Deep-Sea Res.* **16,** 141–156 (1969).

Raymont, J. E. G., Srinivasagam, R. T., and Raymont, J. K. B. Biochemical studies on marine zooplankton. VIII. Further investigations on *Meganyctiphanes norvegica* (M. Sars). *Deep-Sea Res.* **18,** 1167–1178 (1971a).

Raymont, J. E. G., Srinivasagam, R. T., and Raymont, J. K. B. Biochemical studies on marine zooplankton. IX. The biochemical composition of *Euphausia superba*. *J. Mar. Biol. Assoc. U.K.* **51,** 581–588 (1971b).

Razouls, C. Estimation de la production globale des Copépodes planctoniques dan la province néritique du Golfe du Lion (Banyuls-sur-Mer). II. Variations annuelles de la biomasse et calcul de la production. *Vie Milieu* **25**(2B), 99–122 (1975).

Razouls, S. Influence des conditions expérimentales sur le taux respiratoire des copépodes planctoniques. *J. Exp. Mar. Biol. Ecol.* **9,** 145–153 (1972).

Redfield, A. C., Ketchum, B. H., and Richards, F. A. The influence of organisms on the composition of seawater, In M. N. Hill (ed.), *The Sea*, Vol. 2. Interscience, New York, 1963, pp. 146–149.

Reeve, M. R. The biology of Chaetognatha. I. Quantitative aspects of growth and egg production in *Sagitta hispida*. In J. H. Steele (ed.), *Marine Food Chains*. Oliver & Boyd, Edinburgh, U.K., 1970, pp. 168–189.

Reeve, M. R. Large cod-end reservoirs as an aid to the live collection of delicate zooplankton. *Limnol. Oceanogr.* **26,** 577–580 (1981).

Reeve, M. R., and Baker, L. D. Production of two planktonic carnivores (chaetognath and ctenophore) in south Florida inshore waters. *Fishery Bull. U.S.* **73,** 238–248 (1975).

Reeve, M. R., and Cosper, E. Acute effects of heated effluents on the copepod *Acartia tonsa* from the sub-tropical bay and some problems of assessment. In M. Ruivo (ed.), *Marine Pollution and Sea Life*. Fishing News (Books), Surrey, U.K., 1972, pp. 250–252.

Reeve, M. R., Raymont, J. E. G., and Raymont, J. K. B. Seasonal biochemical composition and energy sources of *Sagitta hispida*. *Mar. Biol.* **6,** 357–364 (1970).

Reeve, M. R., and Walter, M. A. Conditions of culture, food size selection, and the effects of temperature and salinity on growth rate and generation time in *Sagitta hispida* Conant. *J. Exp. Mar. Biol. Ecol.* **9,** 191–200 (1972).

Reeve, M. R., and Walter, M. A. Observations on the existense of lower threshold and upper critical food concentrations for the copepod *Acartia tonsa* Dana. *J. Exp. Mar. Biol. Ecol.* **29,** 211–221 (1977).

Reeve, M. R., and Walter, M. A. Nutritional ecology of ctenophores—A review of recent research. *Adv. Mar. Biol.* **15,** 249–287 (1978).

Reid, F. M. H., Stewart, E., Eppley, R. W., and Goodman, D. Spatial distribution of phytoplankton species in chlorophyll maximum layers off southern California. *Limnol. Oceanogr.* **23,** 219–226 (1978a).

Reid, J. L., Brinton, E., Fleminger, A., Venrick, E. L., and McGowan, J. A. Ocean circulation and marine life. In H. Chanock and G. Deacon (eds.), *Advances in Oceanography*. Plenum Publishing, New York, 1978b, pp. 65–130.

Rice, A. L. and, Williamson, D. I. Methods for rearing larval decapod Crustacea. *Helgoländer Wiss. Meeresunters.* **20,** 417–434 (1970).

Richards, F. A. Anoxic basins and fjords. In J. P. Riley and G. Skirrow (eds.), *Chemical Oceanography*, Vol. 1. Academic Press, New York, 1965, pp. 611–645.

Richman, S. Calorimetry. In W. T. Edmondson and G. G. Winberg (eds.), *A Manual on Methods for the Assessment of Secondary Productivity in Fresh Waters*. IBP Handbook No. 17. Blackwell Scientific Publications, Oxford, 1971, pp. 146–149.

Richman, S., and Rogers, J. N. The feeding of *Calanus helgolandicus* on synchronously growing populations of the marine diatom *Ditylum brightwellii*. *Limnol. Oceanogr.* **14,** 701–709 (1969).

Richman, S., Heinle, D. R., and Huff, R. Grazing by adult estuarine calanoid copepods of the Chesapeake Bay. *Mar. Biol.* **42,** 69–84 (1977).

Rigler, F. H. Feeding rate. In W. T. Edmondson and G. G. Winberg (eds.), *A Manual on Methods for the Assessment of Secondary Productivity in Fresh Waters*. IBP Handbook No. 17. Blackwell Scientific Publications, Oxford, 1971, pp. 228–256.

Rigler, F. H., and Cooley, J. M. The use of field data to derive population statistics of multivoltine copepods. *Limnol. Oceanogr.* **19,** 636–655 (1974).

Roe, H. S. J., and Shale, D. M. A new multiple rectangular midwater trawl (RMT 1 + 8*M*) and some modifications to the Institute of Oceanographic Science's RMT 1 + 8. *Mar. Biol.* **50,** 283–288 (1979).

Roff, J. C. Oxygen consumption of *Limnocalanus macrurus* Sars (Calanoida, Copepoda) in relation to environmental conditions. *Can. J. Zool.* **51,** 877–885 (1973).

Roger, C. Azote et phosphore chez un crustacé macroplanctonique *Meganyctephanes norvegica* (M. Sars) (Euphausiacea): Excrétion minérale et constitution. *J. Exp. Mar. Biol. Ecol.* **33,** 57–83 (1978).

Rogers, C. G. *Textbook of Comparative Physiology.* McGraw-Hill, New York, 1927.

Roman, M. R., and Rublee, P. A. Containment effects in copepod grazing experiments: A plea to end the black box approach. *Limnol. Oceanogr.* **25,** 982–990 (1980).

Roman, M. R., and Rublee, P. A. A method to determine *in situ* zooplankton grazing rates on natural particle assemblages. *Mar. Biol.* **65,** 303–309 (1981).

Rosenberg, G. G. Filmed observations of filter feeding in the marine planktonic copepod *Acartia clausii. Limnol. Oceanogr.* **25,** 738–742 (1980).

Rothlisberg, P. C., and Pearcy, W. G. An epibenthic sampler used to study the ontogeny of vertical migration of *Pandalus jordani* (Decapoda, Caridea). *Fishery Bull. U.S.* **74,** 994–997 (1977).

Rottman, M. L. Ecology of recurrent groups of pteropods, euphausiids, and chaetognaths in the Gulf of Thailand and the South China Sea. *Mar. Biol.* **48,** 63–78 (1978).

Runge, J. A. Effects of hunger and season on the feeding behavior of *Calanus pacificus. Limnol. Oceanogr.* **25,** 134–145 (1980).

Runham, N. W., Isarankura, K., and Smith, B. J. Methods for narcotizing and anaesthetizing gastropods. *Malacologia* **2,** 231–238 (1965).

Russell, F. S. A net for catching plankton near the bottom, *J. Mar. Biol. Assoc. U.K.* **15,** 105–108 (1928).

Sameoto, D. D. Life history, ecological production, and an empirical mathematical model of the population of *Sagitta elegans* in St. Margaret's Bay, Nova Scotia. *J. Fish. Res. Bd. Canada* **28,** 971–985 (1971).

Sameoto, D. D. Tidal and diurnal effects on zooplankton sampling variability in a nearshore marine environment. *J. Fish. Res. Bd. Canada* **32,** 347–366 (1975).

Sameoto, D. D., Jaroszynski, L. O., and Franser, W. B. A multiple opening and closing plankton sampler based on the MOCNESS and N.I.O. nets. *J. Fish. Res. Bd. Canada* **34,** 1230–1235 (1977).

Sanders, H. L. Marine benthic diversity: A comparative study. *Am. Natur.* **102,** 243–282 (1968).

Sandifer, P. F., Zielinski, P. B., and Castro, W. E. Enhanced survival of larval grass shrimp in dilute solutions of the synthetic polymer, polyethylene oxide. *Fishery Bull. U.S.* **73,** 678–680 (1975).

Sato, H., and Ito, T. *Methods of Sampling, Breeding and Experiment of Invertebrates.* Hokuryukan, Tokyo, 1961 (in Japanese).

Satomi, M., and Pomeroy, L. R. Respiration and phosphorus excretion in some marine populations. *Ecology* **46,** 877–881 (1965).

Shannon, C. E., and Weaver, W. *The Mathematical Theory of Communication.* University of Illinois Press, Champaign, 1949.

Sheals, J. G. The application of computer techniques to acarine taxonomy: A preliminary examination with species of the *Hypoaspis-Androlaelaps* complex (Acarina). *Proc. Linn. Soc. Lond.* **176,** 11–21 (1964).

Sheldon, A. L. Equitability indices: Dependence on the species count. *Ecology* **50,** 466, 467 (1969).

Shuman, F. R., and Lorenzen, C. L. Quantitative degradation of chlorophyll by a marine herbivore. *Limnol. Oceanogr.* **20,** 580–586 (1975).

Shushkina, E. A. Calculation of copepod production based on metabolic features and the coefficient of the utilization of assimilated food for growth. *Oceanology* **8,** 126–138 (1968).

Siegel, S. *Nonparametric Statistics for the Behavioral Sciences.* McGraw-Hill, New York, 1956.

Simpson, E. H. Measurement of diversity. *Nature, Lond.* **163,** 688 (1949).

Skjoldal, H. R., and Båmstedt, U. Ecobiochemical studies on the deep-water pelagic community of Korsfjorden, western Norway. Adenine nucleotides in zooplankton. *Mar. Biol.* **42,** 197–211 (1977).

Smith, D. F., and Wiebe, W. J. ^{14}C-labelling of the compounds excreted by phytoplankton for employment as a realistic tracer in secondary productivity measurements. *Microb. Ecol.* **4,** 1–8 (1977).

Smith, D. F., Bulleid, N. C., Campbell, R., Rowe, F., Transter, D. J., and Tranter, H. Marine food-web analysis: An experimental study of demersal zooplankton using isotopically labelled prey species. *Mar. Biol.* **54,** 49–59 (1979).

Smith, P. E., Counts, R. C., and Clutter, R. I. Changes in filtering efficiency of plankton nets due to clogging under tow. *J. Cons. Perm. Int. Explor. Mer* **32,** 232–248 (1968).

Smithe, F. B. *Naturalist's Color Guide.* American Museum of Natural History, New York, Part 1, 1975; Part 2, 1974; Part 3, 1981.

Sneath, P. H. A., and Sokal, R. R. *Numerical Taxonomy: The Principles and Practice of Numerical Classification.* W. H. Freeman, San Francisco, 1973.

Snedecor, G. W., and Cochran, W. G. *Statistical Methods.* Iowa State University Press, Ames, 1967.

Sokal, R. R., and Rohlf, F. J. *Biometry.* W. H. Freeman, San Francisco, 1969.

Sorokin, J. I. Carbon-14 method in the study of the nutrition of aquatic animals. *Int. Rev. Ges. Hydrobiol.* **51,** 209–224 (1966).

Sorokin, J. I. The use of C^{14} in the study of the nutrition of aquatic animals. *Internat. Assoc. Theor. and Appl. Limnol. Comm.* **16,** (1968).

Sørensen, T. A method of establishing groups of equal amplitude in plant sociology based on similarity of species content. *K. Danske Vidensk. Selsk. Skr.* **5,** 1–34 (1948).

Spittler, P. Feeding experiments with tintinnids. *Oikos* **15** (Suppl.), 128–132 (1973).

Steedman, H. F. Laboratory methods in the study of marine zooplankton. *J. Cons. Perm. Int. Explor. Mer* **35,** 351–358 (1974).

Steedman, H. F. (ed.). *Zooplankton Fixation and Preservation.* Monographs on Oceanographic Methodology 4. UNESCO, Paris, 1976a.

Steedman, H. F. Narcotizing agents and methods. In H. F. Steedman (ed.), *Zooplankton Fixation and Preservation.* Monographs on Oceanographic Methodology 4. UNESCO, Paris, 1976b, pp. 87–94.

Steedman, H. F. Permanent mounting media. In H. F. Steedman (ed.), *Zooplankton Fixation and Preservation.* Monographs on Oceanographic Methodology 4. UNESCO, Paris, 1976c, pp. 189–195.

Steele, J. H. Plant production in the northern North Sea. *Scot. Home Dep. Mar. Res.* **7,** 1–36 (1958).

Stoll, N. R., Dollfus, R. Ph., Forest, J., Riley, N. D., Sabrosky, C. W., Wright, C. W., and Melville, R. V. (eds.). *International Code of Zoological Nomenclature Adopted by the XV International Congress of Zoology.* International Trust for Zoological Nomenclature, London, 1964.

Strathmann, R. R. Estimating the organic carbon content of phytoplankton from cell volume or plasma volume. *Limnol. Oceanogr.* **12,** 411–418 (1967).

Strickland, J. D. H., and Parsons, T. R. *A Practical Handbook of Seawater Analysis,* 2nd ed. *Bull. Fish. Res. Bd. Canada* **167,** (1972).

Sullivan, B. K. In situ feeding behavior of *Sagitta elegans* and *Eukrohnia hamata* (Chaetognatha) in relation to the vertical distribution and abundance of prey at Ocean Station P. *Limnol. Oceanogr.* **25,** 317–326 (1980).

Sullivan, C. W. Diatom mineralization of silicic acid. I. Si(OH)$_4$ transport characteristics in *Navicula pelliculosa. J. Phycology* **12,** 390–396 (1976).

Sutcliffe, W. H. Jr. Relationship between growth rate and ribonucleic acid concentrations in some invertebrates. *J. Fish. Res. Bd. Canada* **27,** 606–609 (1970).

Swanberg, N. The feeding behavior of *Beroe ovata. Mar. Biol.* **24,** 69–76 (1974).

Sysoeva, T. K., and Degtereva, A. A. The relation between the feeding of cod larvae and pelagic fry and the distribution and abundance of their principal food organisms. *Spec. Publs. Int. Commn. NW. Atlant. Fish.* **6,** 411–416 (1965).

Szyper, J. P. Feeding rate of the Chaetognath *Sagitta enflata* in nature. *Estuar. Coast. Mar. Sci.* **7,** 567–575 (1978).

Szyper, J. P., Hirota, J., Caperon, J., and Ziemann, D. A. Nutrient regeneration by the larger net zooplankton in the southern basin of Kaneohe Bay, Oahu, Hawaiian Island. *Pacif. Sci.* **30,** 363–372 (1976).

Taga, N., and Yasuda, K. Basic studies on the production of microbial flock and its utilization as food for marine animals. *Res. Rept. Setonaikai Fish. Farm. Assoc.* 13 (1979) (in Japanese).

Taguchi, S. Relationship between photosynthesis and cell size of marine diatoms. *J. Phycology* **12,** 185–189 (1976).

Taguchi, S., and Ishii, H. Shipboard experiments on respiration, excretion, and grazing of *Calanus cristatus* and *C. plumchrus* (Copepoda) in the northern North Pacific. In A. Y. Takenouti et al. (eds.), *Biological Oceanography of the Northern North Pacific Ocean.* Idemitsu Shoten, Tokyo, 1972, pp. 419–431.

Takahashi, M., and Ikeda, T. Excretion of ammonia and inorganic phosphorus by *Euphausia pacifica* and *Metridia pacifica* at different concentrations of phytoplankton. *J. Fish. Res. Bd. Canada* **32,** 2189–2195 (1975).

Tanaka, H., Imayama, Y., Azeta, M., and Anraku, M. A hydrodynamic study of a modified model of the Clarke Jet Net. *Mar. Biol.* **1,** 204–209 (1968).

Tangen, K. A device for safe sedimentation using the Utermöhl technique. *J. Cons. Perm. Int. Explor. Mer* **36,** 282–284 (1976).

Tate, M. W., and Clelland, R. C. *Nonparametric and Shortcut Statistics in the Social, Biological and Medical Sciences.* Interstate, Danville, Ill., 1957.

Taylor, B. E., and Slatkin, M. Estimating birth and death rates of zooplankton. *Limnol. Oceanogr.* **26,** 143–158 (1981).

Teal, J. M. Pressure effects on the respiration of vertically migrating decapod Crustacea. *Am. Zool.* **11,** 571–576 (1971).

Teal, J. M., and Carey, F. G. Respiration of a euphausiid from the oxygen minimum layer. *Limnol. Oceanogr.* **12,** 548–550 (1967).

Teal, J. M., and Halcrow, K. A technique for measurement of respiration of single copepods at sea. *J. Cons. Perm. Int. Explor. Mer* **27,** 125–128 (1962).

Theilacker, G. H., and McMaster, M. F. Mass culture of the rotifer *Brachionus plicatilis* and its evaluation as a food for larval anchovies. *Mar. Biol.* **10**, 183–188 (1971).

Thomson, J. M. The Chaetognatha of south-eastern Australia. *Rep. Coun. Scient. Ind. Res. Aust., Div. Fish.* **14**, 1–43 (1947).

Threlkeld, S. T. Estimating cladoceran birth rates: The importance of egg mortality and the egg age distribution. *Limnol. Oceanogr.* **24**, 610–612 (1979).

Throndsen, J. Bestemmelse av marine nakne flagellater (Identification of marine naked flagellates). *Blyttia* **38**, 189–207 (1980).

Timonin, A. G. The structure of plankton communities of the Indian Ocean. *Mar. Biol.* **9**, 281–289 (1971).

Toyama, K., and Miyoshi, G. Prevention from color fading of aquatic animals under preservation I. Test on preservatives to retain red color in fish and crustacean specimens. *J. Tokyo Univ. Fish.* **50**, 43–48 (1963) (in Japanese).

Traganza, E. D., and Graham, K. J. Carbon/adenosine triphosphate ratios in marine zooplankton. *Deep-Sea Res.* **24**, 1187–1193 (1977).

Tranter, D. J. Herbivore production. In D. H. Cushing and J. J. Walsh (eds.), *The Ecology of the Seas*. Blackwell Scientific Publications, Oxford, 1976, pp. 186–224.

Tranter, D. J., and Smith, P. E. Filtration performance. In D. J. Tranter and J. H. Fraser (eds.), *Zooplankton Sampling*. Monographs on Oceanographic Methodology 2. UNESCO, Paris, 1968, pp. 27–56.

Trelstad, R. L. The effect of pH on the stability of purified glutaraldehyde. *J. Histochem. Cytochem.* **17**, 756, 757 (1969).

Tucker, G. H. Relation of fishes and other organisms to the scattering of underwater sound. *J. Mar. Res.* **10**, 215–238 (1951).

Tuffrau, M. Perfectionnements et pratique de la technique d'imprégnation au protagol des infusoires ciliés. *Protistologica* **3**, 91–98 (1967).

Tyler, A. Prolongation of life-span of sea urchin spermatozoa, and improvement of the fertilization-reaction, by treatment of spermatozoa and eggs with metalchelating agents (amino acids, versene, dedtc, oxine, cupron). *Biol. Bull.* **104**, 224–239 (1953).

Uhlig, G. Die mehrgliedrige Kultur litoraler Folliculiniden. *Helgoländer Wiss. Meeresunters.* **12**, 52–60 (1965).

Uhlig, G. Bottom-living animals: Protozoa. In C. Schlieper (ed.), *Research Methods in Marine Biology*. Sidgwick and Jackson, London, 1972, pp. 129–141.

Uno, S. Turbidometric continuous culture of phytoplankton: Constructions of the apparatus and experiments on the daily periodicity in photosynthetic activity of *Phaeodactylum tricornutum* and *Skeletonema costatum*. *Bull. Plankton Soc. Japan* **18**, 14–27 (1971).

Uno, S., and Ueno, S. Diurnal periodicity of the marine diatom *Chaetoceros debilis* in turbidometric continuous culture. II. Chlorophyll *a*, carbon and adenosine triphosphate in the cell. *Bull. Plankton Soc. Japan* **28**, 103–109 (1981).

Utermöhl, H. Neue Wege in der quantitativen Erfassung des Planktons. *Verh. Int. Verein. Theor. Angew. Limnol.* **5**, 567–597 (1931).

Utermöhl, H. Zur Vervolkommung der quantitativen Phytoplankton-Methodik. *Mitt. Int. Verein. Theor. Angew. Limnol.* **9**, 1–38 (1958).

Uye, S. Population dynamics and production of *Acartia clausi* Giesbrecht (Copepoda: Calanoida) in inlet waters. *J. Exp. Mar. Biol. Ecol.* **57**, 55–83 (1982).

Uye, S., and Kasahara, S. Life history of marine planktonic copepods in neritic region with special reference to the role of resting eggs. *Bull. Plankton Soc. Japan* **25,** 109–122 (1978) (in Japanese).

Van der Spoel, S., and Pierrot-Bults, A. C. (eds.) *Zoogeography and Diversity of Plankton.* Bunge Scientific Publishers, Utrecht, 1979.

Van Dorn, W. G. Large-volume water sampler. *Trans. Am. Geophys. Union* **37,** 682–684 (1957).

Vannucci, M. Loss of organisms through the meshes. In D. J. Tranter, and J. H. Fraser, (eds.), *Zooplankton Sampling.* Monographs on Oceanographic Methodology 2. UNESCO, Paris, 1968.

Venrick, E. L. Recurrent groups of diatom species in the North Pacific. *Ecology* **52,** 614–625 (1971).

Venrick, E. L. Sampling strategies. In A. Sournia (ed.), *Phytoplankton Manual.* Monographs on Oceanographic Methodology 6. UNESCO, Paris, 1978a, pp. 7–16.

Venrick, E. L. How many cells to count? In A. Sournia (ed.), *Phytoplankton Manual.* Monographs on Oceanographic Methodology 6. UNESCO, Paris, 1978b, pp. 167–180.

Vidal, J. Physioecology of zooplankton. III. Effects of phytoplankton concentration, temperature, and body size on the metabolic rate of *Calanus pacificus. Mar. Biol.* **56,** 195–202 (1980).

Wallace, R. A., Jared, D. W., and Roesel, M. E. New source for MS-222. *Science* **158,** 1524 (1967).

Waller, R. A., and Eschmeyer, W. N. A method for preserving color in biological specimens. *BioScience* **15**(5), 361 (1965).

Waters, T. F. Secondary production in inland waters. *Adv. Ecol. Res.* **10,** 91–164 (1977).

Webb, K. L., and Johannes, R. E. Studies of the release of dissolved free amino acids by marine zooplankton. *Limnol. Oceanogr.* **12,** 376–382 (1967).

Webb, K. L., and Johannes, R. E. Do marine crustaceans release dissolved amino acids? *Comp. Biochem. Physiol.* **29,** 875–878 (1969).

Whittaker, R. H. A study of summer foliage insect communities in the Great Smoky Mountains. *Ecol. Monogr.* **22,** 1–44 (1952).

Wiebe, P. H. A field investigation of the relationship between length of tow, size of net and sampling error. *J. Cons. Perm. Int. Explor. Mer* **34,** 268–275 (1972).

Wiebe, P. H., Burt, K. H., Boyd, S. H., and Morton, A. W. A multiple opening/closing net and environmental sensing system for sampling zooplankton. *J. Mar. Res.* **34,** 313–326 (1976).

Williamson, D. I. An automatic sampler for use in surveys of plankton distribution. *Rapp. P.-v. Reun. Cons. Perm. Int. Explor. Mer* **153,** 16–18 (1962).

Winberg, G. G. *Methods for the Estimation of Production of Aquatic Animals.* Academic Press, New York, 1971.

Winberg, G. G., Patalas, K., Wright, J. C., Hillbricht-Ilkowska, A., Cooper, W. E., and Mann, K. H. Methods for calculating productivity. In W. T. Edmondson and G. G. Winberg (eds.), *A Manual on Methods for the Assessment of Secondary Productivity in Fresh Waters.* IBP Handbook No. 17. Blackwell Scientific Publications, Oxford, 1971, pp. 296–317.

Wishner, K. F. The biomass of the deep-sea benthopelagic plankton. *Deep-Sea Res.* **27A,** 203–216 (1980).

Yablonskaya, E. A. Study of the seasonal population dynamics of the plankton copepods as a method of determination of their production. *Rapp. P.-v. Reun. Cons. Perm. Int. Explor. Mer* **153,** 224–226 (1962).

Yasuda, K. and Taga, N. Culture of *Brachionus plicatilis* Müller using bacteria as food. *Bull. Japanese Soc. Sci. Fish.* **46,** 933–939 (1980) (in Japanese).

Youngbluth, M. J. Sampling demersal zooplankton: A comparison of field collections using three different emergence traps. *J. Exp. Mar. Biol. Ecol.* **61,** 111–124 (1982).

Zeiss, F. R. Jr. Effects of population densities on zooplankton respiration rates. *Limnol. Oceanogr.* **8,** 110–115 (1963).

Zeuthen, E. Body size and metabolic rate in the animal kingdom with special regard to the marine micro-fauna. *C. r. Trav. Lab. Carlsberg, Ser. Chem.* **26,** 17–161 (1947).

Zillioux, E. J. A continuous recirculating culture system for planktonic copepods. *Mar. Biol.* **4,** 215–218 (1969).

Zillioux, E. J. Ingestion and Assimilation in Laboratory Cultures of *Acartia. Rosenstiel School of Marine and Atmospheric Science, University of Miami,* Tech. Rep., 1970.

Zillioux, E. J. Diver-operated versus remote-controlled chambers used in underwater tracer experiments. *Helgoländer Wiss. Meeresunters.* **24,** 112–119 (1973).

Zimmerman, S. T. The transformation of energy by *Lucifer chacei* (Crustacea, Decapoda). *Pacif. Sci.* **27,** 247–259 (1973).

Appendix

Table 1. Functions useful for calculating species diversity[a]

n	log n!	n log n	n	log n!	n log n
1	0.0000	0.0000	14	10.9404	16.0458
2	0.3010	0.6021	15	12.1165	17.6414
3	0.7782	1.4314	16	13.3206	19.2659
4	1.3802	2.4082	17	14.5511	20.9176
5	2.0792	3.4949	18	15.8063	22.5949
6	2.8573	4.6689	19	17.0851	24.2963
7	3.7024	5.9157	20	18.3861	26.0206
8	4.6055	7.2247	21	19.7083	27.7666
9	5.5598	8.5882	22	21.0508	29.5333
10	6.5598	10.0000	23	22.4125	31.3197
11	7.6012	11.4553	24	23.7927	33.1251
12	8.6803	12.9502	25	25.1906	34.9485
13	9.7943	14.6813	26	26.6056	36.7893

[a] After Lloyd et al., 1968. Logarithms are to base 10. Table values are accurate to within ±1 in the eigth significant figure.

Table 1. (*Continued*)

n	log n!	n log n	n	log n!	n log n
27	28.0370	38.6468	84	126.5204	161.6395
28	29.4841	40.5204	85	128.4498	164.0006
29	30.9465	42.4095	86	130.3843	166.3669
30	32.4237	44.3136	87	132.3238	168.7382
31	33.9150	46.2322	88	134.2683	171.1145
32	35.4202	48.1648	89	136.2177	173.4957
33	36.9387	50.1110	90	138.1719	175.8818
34	38.4702	52.0703	91	140.1310	178.2728
35	40.0142	54.0424	92	142.0948	180.6685
36	41.5705	56.0269	93	144.0632	183.0689
37	43.1387	58.0235	94	146.0364	185.4740
38	44.7185	60.0318	95	148.0141	187.8837
39	46.3096	62.0515	96	149.9964	190.2980
40	47.9116	64.0824	97	151.9831	192.7169
41	49.5244	66.1241	98	153.9744	195.1402
42	51.1477	68.1765	99	155.9700	197.5679
43	52.7811	70.2391	100	157.9700	200.0000
44	54.4246	72.3119	101	159.9743	202.4365
45	56.0778	74.3946	102	161.9829	204.8772
46	57.7406	76.4869	103	163.9958	207.3222
47	59.4127	78.5886	104	166.0128	209.7715
48	61.0939	80.6996	105	168.0340	212.2249
49	62.7841	82.8196	106	170.0593	214.6824
50	64.4831	84.9485	107	172.0887	217.1441
51	66.1906	87.0861	108	174.1221	219.6098
52	67.9066	89.2322	109	176.1595	222.0795
53	69.6309	91.3866	110	178.2009	224.5532
54	71.3633	93.5493	111	180.2462	227.0309
55	73.1037	95.7199	112	182.2955	229.5124
56	74.8519	97.8985	113	184.3485	231.9979
57	76.6077	100.0849	114	186.4054	234.4872
58	78.3712	102.2788	115	188.4661	236.9803
59	80.1420	104.4803	116	190.5306	239.4771
60	81.9202	106.6891	117	192.5988	241.9777
61	83.7055	108.9051	118	194.6707	244.4821
62	85.4979	111.1283	119	196.7462	246.9901
63	87.2972	113.3585	120	198.8254	249.5017
64	89.1034	115.5955	121	200.9082	252.0170
65	90.9163	117.8394	122	202.9945	254.5359
66	92.7359	120.0899	123	205.0844	257.0583
67	94.5619	122.3470	124	207.1779	259.5843
68	96.3945	124.6106	125	209.2748	262.1138
69	98.2333	126.8806	126	211.3751	264.6467
70	100.0784	129.1569	127	213.4790	267.1831
71	101.9297	131.4393	128	215.5862	269.7229
72	103.7870	133.7279	129	217.6967	272.2661
73	105.6503	136.0226	130	219.8107	274.8126
74	107.5196	138.3231	131	221.9280	277.3625
75	109.3946	140.6296	132	224.0485	279.9158
76	111.2754	142.9418	133	226.1724	282.4723
77	113.1619	145.2598	134	228.2995	285.0320
78	115.0540	147.5834	135	230.4298	287.5951
79	116.9516	149.9125	136	232.5634	290.1613
80	118.8547	152.2472	137	234.7001	292.7307
81	120.7632	154.5873	138	236.8400	295.3033
82	122.6770	156.9327	139	238.9830	297.8791
83	124.5961	159.2835	140	241.1291	300.4579

Table 1. (*Continued*)

n	log n!	n log n		n	log n!	n log n
141	243.2783	303.0399		198	370.2970	454.7397
142	245.4306	305.6249		199	372.5959	457.4718
143	247.5860	308.2131		200	374.8969	460.2060
144	249.7443	310.8042		201	377.2001	462.9424
145	251.9057	313.3984		202	379.5054	465.6810
146	254.0700	315.9955		203	381.8129	468.4217
147	256.2374	318.5956		204	384.1226	471.1646
148	258.4076	321.1987		205	386.4343	473.9095
149	260.5808	323.8048		206	388.7482	476.6566
150	262.7569	326.4137		207	391.0642	479.4059
151	264.9359	329.0255		208	393.3822	482.1572
152	267.1177	331.6402		209	395.7024	484.9106
153	269.3024	334.2578		210	398.0246	487.6661
154	271.4899	336.8782		211	400.3489	490.4236
155	273.6803	339.5014		212	402.6752	493.1832
156	275.8734	342.1274		213	405.0036	495.9449
157	278.0693	344.7562		214	407.3340	498.7085
158	280.2679	347.3878		215	409.6664	501.4743
159	282.4693	350.0221		216	412.0009	504.2420
160	284.6735	352.6592		217	414.3373	507.0118
161	286.8803	355.2990		218	416.6758	509.7835
162	289.0898	357.9414		219	419.0162	512.5573
163	291.3020	360.5866		220	421.3587	515.3330
164	293.5168	363.2344		221	423.7031	518.1107
165	295.7343	365.8849		222	426.0494	520.8904
166	297.9544	368.5379		223	428.3977	523.6720
167	300.1771	371.1936		224	430.7480	526.4556
168	302.4024	373.8520		225	433.1002	529.2411
169	304.6303	376.5129		226	435.4543	532.0285
170	306.8608	379.1763		227	437.8103	534.8179
171	309.0938	381.8423		228	440.1682	537.6091
172	311.3293	384.5109		229	442.5281	540.4023
173	313.5674	387.1820		230	444.8898	543.1974
174	315.8079	389.8556		231	447.2534	545.9944
175	318.0509	392.5317		232	449.6189	548.7932
176	320.2965	395.2102		233	451.9862	551.5939
177	322.5444	397.8913		234	454.3555	554.3965
178	324.7948	400.5748		235	456.7265	557.2009
179	327.0477	403.2607		236	459.0994	560.0072
180	329.3030	405.9491		237	461.4742	562.8154
181	331.5606	408.6398		238	463.8508	565.6253
182	333.8207	411.3330		239	466.2292	568.4371
183	336.0832	414.0285		240	468.6094	571.2507
184	338.3480	416.7265		241	470.9914	574.0661
185	340.6152	419.4268		242	473.3752	576.8833
186	342.8847	422.1294		243	475.7608	579.7023
187	345.1565	424.8344		244	478.1482	582.5231
188	347.4307	427.5417		245	480.5374	585.3457
189	349.7071	430.2513		246	482.9283	588.1700
190	351.9859	432.9632		247	485.3210	590.9961
191	354.2669	435.6774		248	487.7154	593.8240
192	356.5502	438.3938		249	490.1116	596.6536
193	358.8358	441.1126		250	492.5096	599.4850
194	361.1236	443.8335		251	494.9093	602.3181
195	363.4136	446.5567		252	497.3107	605.1529
196	365.7059	449.2822		253	499.7138	607.9895
197	368.0003	452.0098		254	502.1186	610.8278

Table 1. (*Continued*)

n	log n!	n log n		n	log n!	n log n
255	504.5252	613.6677		312	644.3226	778.1762
256	506.9334	616.5094		313	646.8182	781.1054
257	509.3433	619.3528		314	649.3151	784.0359
258	511.7549	622.1979		315	651.8134	786.9678
259	514.1682	625.0446		316	654.3131	789.9011
260	516.5832	627.8931		317	656.8142	792.8358
261	518.9999	630.7432		318	659.3166	795.7718
262	521.4182	633.5949		319	661.8204	798.7092
263	523.8381	636.4484		320	664.3255	801.6480
264	526.2597	639.3034		321	666.8320	804.5881
265	528.6830	642.1602		322	669.3399	807.5296
266	531.1078	645.0185		323	671.8491	810.4724
267	533.5344	647.8785		324	674.3596	813.4166
268	535.9625	650.7401		325	676.8715	816.3621
269	538.3922	653.6034		326	679.3847	819.3089
270	540.8236	656.4682		327	681.8993	822.2571
271	543.2566	659.3347		328	684.4152	825.2066
272	545.6912	662.2027		329	686.9324	828.1575
273	548.1273	665.0724		330	689.4509	831.1096
274	550.5651	667.9437		331	691.9707	834.0631
275	553.0044	670.8165		332	694.4918	837.0178
276	555.4453	673.6909		333	697.0143	839.9739
277	557.8878	676.5669		334	699.5380	842.9313
278	560.3318	679.4445		335	702.0631	845.8900
279	562.7774	682.3236		336	704.5894	848.8500
280	565.2246	685.2042		337	707.1170	851.8113
281	567.6733	688.0865		338	709.6460	854.7738
282	570.1235	690.9702		339	712.1762	857.7377
283	572.5753	693.8556		340	714.7076	860.7028
284	575.0287	696.7424		341	717.2404	863.6692
285	577.4835	699.6308		342	719.7744	866.6369
286	579.9399	702.5207		343	722.3097	869.6059
287	582.3977	705.4121		344	724.8463	872.5761
288	584.8571	708.3050		345	727.3841	875.5476
289	587.3180	711.1995		346	729.9232	878.5203
290	589.7804	714.0954		347	732.4635	881.4943
291	592.2443	716.9929		348	735.0051	884.4696
292	594.7097	719.8918		349	737.5479	887.4461
293	597.1766	722.7922		350	740.0920	890.4238
294	599.6449	725.6941		351	742.6373	893.4028
295	602.1147	728.5975		352	745.1838	896.3830
296	604.5860	731.5023		353	747.7316	899.3645
297	607.0588	734.4087		354	750.2806	902.3472
298	609.5330	737.3164		355	752.8308	905.3311
299	612.0087	740.2257		356	755.3823	908.3162
300	614.4858	743.1364		357	757.9349	911.3026
301	616.9644	746.0485		358	760.4888	914.2901
302	619.4444	748.9621		359	763.0439	917.2789
303	621.9258	751.8771		360	765.6002	920.2689
304	624.4087	754.7936		361	768.1577	923.2601
305	626.8930	757.7115		362	770.7164	926.2525
306	629.3787	760.6308		363	773.2764	929.2461
307	631.8659	763.5515		364	775.8375	932.2409
308	634.3544	766.4736		365	778.3997	935.2369
309	636.8444	769.3972		366	780.9632	938.2341
310	639.3357	772.3221		367	783.5279	941.2324
311	641.8285	775.2485		368	786.0937	944.2320

Table 1.   (*Continued*)

n	log n!	n log n	n	log n!	n log n
369	788.6608	947.2327	426	936.8329	1120.1285
370	791.2290	950.2346	427	939.4633	1123.1927
371	793.7983	953.2377	428	942.0948	1126.2579
372	796.3689	956.2420	429	944.7272	1129.3242
373	798.9406	959.2474	430	947.3607	1132.3914
374	801.5135	962.2540	431	949.9952	1135.4597
375	804.0875	965.2617	432	952.6307	1138.5290
376	806.6627	968.2706	433	955.2672	1141.5993
377	809.2390	971.2807	434	957.9047	1144.6705
378	811.8165	974.2919	435	960.5431	1147.7428
379	814.3952	977.3043	436	963.1826	1150.8161
380	816.9749	980.3178	437	965.8231	1153.8904
381	819.5559	983.3324	438	968.4646	1156.9657
382	822.1379	986.3482	439	971.1071	1160.0419
383	824.7211	989.3651	440	973.7505	1163.1192
384	827.3055	992.3832	441	976.3949	1166.1974
385	829.8909	995.4024	442	979.0404	1169.2766
386	832.4775	998.4227	443	981.6868	1172.3568
387	835.0652	1001.4441	444	984.3342	1175.4380
388	837.6540	1004.4667	445	986.9825	1178.5202
389	840.2440	1007.4904	446	989.6318	1181.6033
390	842.8351	1010.5152	447	992.2822	1184.6875
391	845.4272	1013.5411	448	994.9334	1187.7725
392	848.0205	1016.5681	449	997.5857	1190.8586
393	850.6149	1019.5963	450	1000.2389	1193.9456
394	853.2104	1022.6255	451	1002.8931	1197.0336
395	855.8070	1025.6558	452	1005.5482	1200.1226
396	858.4047	1028.6873	453	1008.2043	1203.2125
397	861.0035	1031.7198	454	1010.8614	1206.3033
398	863.6034	1034.7534	455	1013.5194	1209.3952
399	866.2044	1037.7882	456	1016.1783	1212.4880
400	868.8064	1040.8240	457	1018.8382	1215.5817
401	871.4096	1043.8609	458	1021.4991	1218.6764
402	874.0138	1046.8989	459	1024.1609	1221.7720
403	876.6191	1049.9379	460	1026.8237	1224.8686
404	879.2255	1052.9781	461	1029.4874	1227.9661
405	881.8329	1056.0193	462	1032.1520	1231.0646
406	884.4415	1059.0616	463	1034.8176	1234.1640
407	887.0510	1062.1049	464	1037.4841	1237.2643
408	889.6617	1065.1493	465	1040.1516	1240.3656
409	892.2734	1068.1948	466	1042.8200	1243.4678
410	894.8862	1071.2414	467	1045.4893	1246.5710
411	897.5001	1074.2890	468	1048.1595	1249.6750
412	900.1150	1077.3376	469	1050.8307	1252.7801
413	902.7309	1080.3874	470	1053.5028	1255.8860
414	905.3479	1083.4381	471	1056.1758	1258.9928
415	907.9660	1086.4900	472	1058.8498	1262.1006
416	910.5850	1089.5428	473	1061.5246	1265.2093
417	913.2052	1092.5967	474	1064.2004	1268.3189
418	915.8264	1095.6517	475	1066.8771	1271.4295
419	918.4486	1098.7077	476	1069.5547	1274.5409
420	921.0718	1101.7647	477	1072.2332	1277.6533
421	923.6961	1104.8228	478	1074.9126	1280.7665
422	926.3214	1107.8819	479	1077.5930	1283.8807
423	928.9478	1110.9420	480	1080.2742	1286.9958
424	931.5751	1114.0031	481	1082.9564	1290.1118
425	934.2035	1117.0653	482	1085.6394	1293.2287

Table 1. *(Continued)*

n	log n!	n log n	n	log n!	n log n
483	1088.3234	1296.3465	540	1242.7390	1475.4926
484	1091.0082	1299.4651	541	1245.4722	1478.6597
485	1093.6940	1302.5847	542	1248.2061	1481.8276
486	1096.3806	1305.7052	543	1250.9409	1484.9963
487	1099.0681	1308.8266	544	1253.6765	1488.1658
488	1101.7565	1311.9489	545	1256.4129	1491.3361
489	1104.4458	1315.0720	546	1259.1501	1494.5072
490	1107.1360	1318.1961	547	1261.8881	1497.6791
491	1109.8271	1321.3210	548	1264.6269	1500.8517
492	1112.5191	1324.4468	549	1267.3665	1504.0252
493	1115.2119	1327.5735	550	1270.1068	1507.1995
494	1117.9057	1330.7011	551	1272.8480	1510.3745
495	1120.6003	1333.8296	552	1275.5899	1513.5504
496	1123.2957	1336.9589	553	1278.3327	1516.7270
497	1125.9921	1340.0891	554	1281.0762	1519.9044
498	1128.6893	1343.2202	555	1283.8205	1523.0826
499	1131.3874	1346.3522	556	1286.5655	1526.2616
500	1134.0864	1349.4850	557	1289.3114	1529.4413
501	1136.7862	1352.6187	558	1292.0580	1532.6219
502	1139.4869	1355.7533	559	1294.8054	1535.8032
503	1142.1885	1358.8887	560	1297.5536	1538.9853
504	1144.8909	1362.0250	561	1300.3026	1542.1682
505	1147.5942	1365.1621	562	1303.0523	1545.3518
506	1150.2984	1368.3002	563	1305.8028	1548.5362
507	1153.0034	1371.4390	564	1308.5541	1551.7214
508	1155.7093	1374.5788	565	1311.3062	1554.9074
509	1158.4160	1377.7193	566	1314.0590	1558.0941
510	1161.1235	1380.8608	567	1316.8126	1561.2816
511	1163.8320	1384.0031	568	1319.5669	1564.4698
512	1166.5412	1387.1462	569	1322.3220	1567.6589
513	1169.2514	1390.2902	570	1325.0779	1570.8487
514	1171.9623	1393.4350	571	1327.8345	1574.0392
515	1174.6741	1396.5807	572	1330.5919	1577.2305
516	1177.3868	1399.7272	573	1333.3501	1580.4226
517	1180.1003	1402.8746	574	1336.1090	1583.6154
518	1182.8146	1406.0228	575	1338.8687	1586.8090
519	1185.5298	1409.1718	576	1341.6291	1590.0033
520	1188.2458	1412.3217	577	1344.3903	1593.1984
521	1190.9626	1415.4724	578	1347.1522	1596.3943
522	1193.6803	1418.6240	579	1349.9149	1599.5909
523	1196.3988	1421.7764	580	1352.6783	1602.7882
524	1199.1181	1424.9296	581	1355.4425	1605.9863
525	1201.8383	1428.0836	582	1358.2074	1609.1852
526	1204.5592	1431.2385	583	1360.9731	1612.3848
527	1207.2811	1434.3942	584	1363.7395	1615.5851
528	1210.0037	1437.5507	585	1366.5066	1618.7862
529	1212.7271	1440.7080	586	1369.2745	1621.9880
530	1215.4514	1443.8662	587	1372.0432	1625.1906
531	1218.1765	1447.0252	588	1374.8125	1628.3939
532	1220.9024	1450.1850	589	1377.5827	1631.5979
533	1223.6292	1453.3456	590	1380.3535	1634.8027
534	1226.3567	1456.5070	591	1383.1251	1638.0082
535	1229.0851	1459.6693	592	1385.8974	1641.2144
536	1231.8142	1462.8323	593	1388.6705	1644.4214
537	1234.5442	1465.9962	594	1391.4443	1647.6291
538	1237.2750	1469.1609	595	1394.2188	1650.8376
539	1240.0066	1472.3263	596	1396.9940	1654.0468

Table 1. (*Continued*)

n	log n!	n log n	n	log n!	n log n
597	1399.7700	1657.2567	654	1559.1662	1841.3878
598	1402.5467	1660.4673	655	1561.9824	1844.6380
599	1405.3241	1663.6787	656	1564.7993	1847.8889
600	1408.1023	1666.8907	657	1567.6169	1851.1404
601	1410.8811	1670.1035	658	1570.4351	1854.3926
602	1413.6608	1673.3171	659	1573.2540	1857.6455
603	1416.4411	1676.5313	660	1576.0735	1860.8990
604	1419.2221	1679.7463	661	1578.8938	1864.1532
605	1422.0039	1682.9620	662	1581.7146	1867.4080
606	1424.7863	1686.1784	663	1584.5361	1870.6635
607	1427.5695	1689.3955	664	1587.3583	1873.9196
608	1430.3534	1692.6134	665	1590.1811	1877.1764
609	1433.1380	1695.8319	666	1593.0046	1880.4338
610	1435.9234	1699.0512	667	1595.8287	1883.6919
611	1438.7094	1702.2712	668	1598.6535	1886.9507
612	1441.4962	1705.4919	669	1601.4789	1890.2101
613	1444.2836	1708.7133	670	1604.3050	1893.4701
614	1447.0718	1711.9354	671	1607.1317	1896.7308
615	1449.8607	1715.1582	672	1609.9591	1899.9921
616	1452.6503	1718.3817	673	1612.7871	1903.2541
617	1455.4405	1721.6059	674	1615.6158	1906.5168
618	1458.2315	1724.8309	675	1618.4451	1909.7800
619	1461.0232	1728.0565	676	1621.2750	1913.0440
620	1463.8156	1731.2828	677	1624.1056	1916.3085
621	1466.6087	1734.5099	678	1626.9368	1919.5737
622	1469.4025	1737.7376	679	1629.7687	1922.8396
623	1472.1970	1740.9660	680	1632.6012	1926.1060
624	1474.9922	1744.1952	681	1635.4344	1929.3732
625	1477.7880	1747.4250	682	1638.2681	1932.6409
626	1480.5846	1750.6555	683	1641.1026	1935.9093
627	1483.3819	1753.8867	684	1643.9376	1939.1784
628	1486.1798	1757.1187	685	1646.7733	1942.4480
629	1488.9785	1760.3512	686	1649.6096	1945.7183
630	1491.7778	1763.5845	687	1652.4466	1948.9893
631	1494.5779	1766.8185	688	1655.2842	1952.2608
632	1497.3786	1770.0532	689	1658.1224	1955.5330
633	1500.1800	1773.2885	690	1660.9612	1958.8059
634	1502.9821	1776.5246	691	1663.8007	1962.0793
635	1505.7849	1779.7613	692	1666.6408	1965.3534
636	1508.5883	1782.9987	693	1669.4816	1968.6281
637	1511.3924	1786.2368	694	1672.3229	1971.9035
638	1514.1973	1789.4756	695	1675.1649	1975.1794
639	1517.0028	1792.7150	696	1678.0075	1978.4560
640	1519.8089	1795.9552	697	1680.8507	1981.7332
641	1522.6158	1799.1960	698	1683.6946	1985.0111
642	1525.4233	1802.4375	699	1686.5391	1988.2895
643	1528.2316	1805.6796	700	1689.3842	1991.5686
644	1531.0404	1808.9225	701	1692.2299	1994.8483
645	1533.8500	1812.1660	702	1695.0762	1998.1286
646	1536.6602	1815.4102	703	1697.9232	2001.4096
647	1539.4711	1818.6551	704	1700.7708	2004.6911
648	1542.2827	1821.9006	705	1703.6189	2007.9733
649	1545.0950	1825.1468	706	1706.4678	2011.2561
650	1547.9079	1828.3937	707	1709.3172	2014.5395
651	1550.7215	1831.6412	708	1712.1672	2017.8235
652	1553.5357	1834.8894	709	1715.0179	2021.1082
653	1556.3506	1838.1383	710	1717.8691	2024.3934

Table 1. (*Continued*)

n	log n!	n log n	n	log n!	n log n
711	1720.7210	2027.6793	768	1884.2611	2215.9574
712	1723.5735	2030.9657	769	1887.1470	2219.2773
713	1726.4265	2034.2528	770	1890.0335	2222.5978
714	1729.2802	2037.5405	771	1892.9205	2225.9189
715	1732.1346	2040.8288	772	1895.8082	2229.2405
716	1734.9895	2044.1177	773	1898.6963	2232.5627
717	1737.8450	2047.4072	774	1901.5851	2235.8855
718	1740.7011	2050.6973	775	1904.4744	2239.2088
719	1743.5578	2053.9881	776	1907.3642	2242.5327
720	1746.4152	2057.2794	777	1910.2547	2245.8571
721	1749.2731	2060.5713	778	1913.1456	2249.1821
722	1752.1316	2063.8638	779	1916.0372	2252.5077
723	1754.9908	2067.1570	780	1918.9293	2255.8338
724	1757.8505	2070.4507	781	1921.8219	2259.1604
725	1760.7109	2073.7450	782	1924.7151	2262.4877
726	1763.5718	2077.0400	783	1927.6089	2265.8154
727	1766.4333	2080.3355	784	1930.5032	2269.1438
728	1769.2955	2083.6316	785	1933.3981	2272.4727
729	1772.1582	2086.9283	786	1936.2935	2275.8021
730	1775.0215	2090.2257	787	1939.1895	2279.1321
731	1777.8854	2093.5236	788	1942.0860	2282.4626
732	1780.7499	2096.8221	789	1944.9831	2285.7937
733	1783.6150	2100.1212	790	1947.8807	2289.1254
734	1786.4807	2103.4209	791	1950.7789	2292.4576
735	1789.3470	2106.7212	792	1953.6776	2295.7903
736	1792.2139	2110.0221	793	1956.5769	2299.1236
737	1795.0814	2113.3235	794	1959.4767	2302.4575
738	1797.9494	2116.6256	795	1962.3771	2305.7918
739	1800.8181	2119.9282	796	1965.2780	2309.1268
740	1803.6873	2123.2314	797	1968.1794	2312.4623
741	1806.5571	2126.5353	798	1971.0814	2315.7983
742	1809.4275	2129.8397	799	1973.9840	2319.1349
743	1812.2985	2133.1447	800	1976.8871	2322.4720
744	1815.1701	2136.4503	801	1979.7907	2325.8096
745	1818.0422	2139.7564	802	1982.6949	2329.1478
746	1820.9150	2143.0631	803	1985.5996	2332.4866
747	1823.7883	2146.3705	804	1988.5049	2335.8258
748	1826.6622	2149.6784	805	1991.4106	2339.1657
749	1829.5367	2152.9869	806	1994.3170	2342.5060
750	1832.4117	2156.2959	807	1997.2239	2345.8469
751	1835.2874	2159.6056	808	2000.1313	2349.1884
752	1838.1636	2162.9158	809	2003.0392	2352.5303
753	1841.0404	2166.2266	810	2005.9477	2355.8729
754	1843.9178	2169.5380	811	2008.8567	2359.2159
755	1846.7957	2172.8499	812	2011.7663	2362.5595
756	1849.6742	2176.1625	813	2014.6764	2365.9036
757	1852.5533	2179.4756	814	2017.5870	2369.2483
758	1855.4330	2182.7892	815	2020.4982	2372.5934
759	1858.3132	2186.1035	816	2023.4099	2375.9391
760	1861.1941	2189.4183	817	2026.3221	2379.2854
761	1864.0754	2192.7337	818	2029.2348	2382.6322
762	1866.9574	2196.0497	819	2032.1481	2385.9795
763	1869.8399	2199.3662	820	2035.0619	2389.3273
764	1872.7230	2202.6833	821	2037.9763	2392.6757
765	1875.6067	2206.0010	822	2040.8911	2396.0246
766	1878.4909	2209.3192	823	2043.8065	2399.3741
767	1881.3757	2212.6380	824	2046.7225	2402.7240

Table 1. (*Continued*)

n	log n!	n log n	n	log n!	n log n
825	2049.6389	2406.0745	882	2216.7274	2597.9033
826	2052.5559	2409.4255	883	2219.6734	2601.2833
827	2055.4734	2412.7770	884	2222.6198	2604.6638
828	2058.3914	2416.1291	885	2225.5668	2608.0448
829	2061.3100	2419.4817	886	2228.5142	2611.4263
830	2064.2291	2422.8348	887	2231.4621	2614.8082
831	2067.1487	2426.1884	888	2234.4105	2618.1907
832	2070.0688	2429.5426	889	2237.3594	2621.5736
833	2072.9894	2432.8973	890	2240.3088	2624.9571
834	2075.9106	2436.2525	891	2243.2587	2628.3410
835	2078.8323	2439.6082	892	2246.2091	2631.7254
836	2081.7545	2442.9644	893	2249.1599	2635.1104
837	2084.6772	2446.3212	894	2252.1113	2638.4957
838	2087.6005	2449.6785	895	2255.0631	2641.8816
839	2090.5242	2453.0363	896	2258.0154	2645.2680
840	2093.4485	2456.3946	897	2260.9682	2648.6548
841	2096.3733	2459.7534	898	2263.9214	2652.0421
842	2099.2986	2463.1128	899	2266.8752	2655.4300
843	2102.2244	2466.4726	900	2269.8295	2658.8182
844	2105.1508	2469.8330	901	2272.7842	2662.2070
845	2108.0776	2473.1939	902	2275.7394	2665.5963
846	2111.0050	2476.5553	903	2278.6951	2668.9860
847	2113.9329	2479.9172	904	2281.6512	2672.3763
848	2116.8613	2483.2797	905	2284.6079	2675.7669
849	2119.7902	2486.6426	906	2287.5650	2679.1581
850	2122.7196	2490.0061	907	2290.5226	2682.5498
851	2125.6495	2493.3700	908	2293.4807	2685.9419
852	2128.5800	2496.7345	909	2296.4393	2689.3346
853	2131.5109	2500.0995	910	2299.3983	2692.7277
854	2134.4424	2503.4650	911	2302.3578	2696.1212
855	2137.3744	2506.8310	912	2305.3178	2699.5153
856	2140.3068	2510.1975	913	2308.2783	2702.9098
857	2143.2398	2513.5645	914	2311.2393	2706.3048
858	2146.1733	2516.9321	915	2314.2007	2709.7003
859	2149.1073	2520.3001	916	2317.1626	2713.0963
860	2152.0418	2523.6686	917	2320.1249	2716.4927
861	2154.9768	2527.0377	918	2323.0878	2719.8896
862	2157.9123	2530.4073	919	2326.0511	2723.2869
863	2160.8483	2533.7773	920	2329.0149	2726.6848
864	2163.7848	2537.1479	921	2331.9792	2730.0831
865	2166.7218	2540.5189	922	2334.9439	2733.4819
866	2169.6594	2543.8905	923	2337.9091	2736.8812
867	2172.5974	2547.2625	924	2340.8748	2740.2809
868	2175.5359	2550.6351	925	2343.8409	2743.6811
869	2178.4749	2554.0082	926	2346.8075	2747.0818
870	2181.4144	2557.3817	927	2349.7746	2750.4829
871	2184.3544	2560.7558	928	2352.7421	2753.8845
872	2187.2950	2564.1304	929	2355.7101	2757.2866
873	2190.2360	2567.5054	930	2358.6786	2760.6891
874	2193.1775	2570.8810	931	2361.6476	2764.0921
875	2196.1195	2574.2570	932	2364.6170	2767.4956
876	2199.0620	2577.6336	933	2367.5869	2770.8996
877	2202.0050	2581.0106	934	2370.5572	2774.3040
878	2204.9485	2584.3882	935	2373.5280	2777.7088
879	2207.8925	2587.7662	936	2376.4993	2781.1142
880	2210.8370	2591.1447	937	2379.4710	2784.5200
881	2213.7820	2594.5238	938	2382.4433	2787.9262

Table 1. (*Continued*)

n	log n!	n log n	n	log n!	n log n
939	2385.4159	2791.3330	995	2552.6090	2982.8340
940	2388.3890	2794.7402	996	2555.6072	2986.2663
941	2391.3626	2798.1478	997	2558.6059	2989.6991
942	2394.3367	2801.5559	998	2561.6051	2993.1323
943	2397.3112	2804.9645	999	2564.6046	2996.5659
944	2400.2862	2808.3735	1000	2567.6046	3000.0000
945	2403.2616	2811.7831	1001	2570.6051	3003.4345
946	2406.2375	2815.1930	1002	2573.6059	3006.8694
947	2409.2138	2818.6034	1003	2576.6072	3010.3048
948	2412.1906	2822.0143	1004	2579.6090	3013.7406
949	2415.1679	2825.4256	1005	2582.6111	3017.1769
950	2418.1456	2828.8374	1006	2585.6137	3020.6136
951	2421.1238	2832.2497	1007	2588.6168	3024.0507
952	2424.1024	2835.6624	1008	2591.6202	3027.4882
953	2427.0815	2839.0755	1009	2594.6241	3030.9262
954	2430.0611	2842.4891	1010	2597.6284	3034.3646
955	2433.0411	2845.9032	1011	2600.6332	3037.8034
956	2436.0216	2849.3177	1012	2603.6384	3041.2427
957	2439.0025	2852.7327	1013	2606.6440	3044.6823
958	2441.9838	2856.1481	1014	2609.6500	3048.1225
959	2444.9657	2859.5640	1015	2612.6565	3051.5630
960	2447.9479	2862.9804	1016	2615.6634	3055.0040
961	2450.9307	2866.3972	1017	2618.6707	3058.4454
962	2453.9138	2869.8144	1018	2621.6784	3061.8872
963	2456.8975	2873.2321	1019	2624.6866	3065.3295
964	2459.8815	2876.6502	1020	2627.6952	3068.7722
965	2462.8661	2880.0688	1021	2630.7043	3072.2153
966	2465.8510	2883.4879	1022	2633.7137	3075.6588
967	2468.8365	2886.9074	1023	2636.7236	3079.1028
968	2471.8223	2890.3273	1024	2639.7339	3082.5471
969	2474.8087	2893.7477	1025	2642.7446	3085.9919
970	2477.7954	2897.1686	1026	2645.7557	3089.4372
971	2480.7827	2900.5898	1027	2648.7673	3092.8828
972	2483.7703	2904.0116	1028	2651.7793	3096.3289
973	2486.7584	2907.4338	1029	2654.7917	3099.7754
974	2489.7470	2910.8564	1030	2657.8046	3103.2223
975	2492.7360	2914.2795	1031	2660.8178	3106.6697
976	2495.7254	2917.7030	1032	2663.8315	3110.1174
977	2498.7153	2921.1270	1033	2666.8456	3113.5656
978	2501.7057	2924.5514	1034	2669.8601	3117.0142
979	2504.6965	2927.9762	1035	2672.8751	3120.4633
980	2507.6877	2931.4016	1036	2675.8904	3123.9127
981	2510.6794	2934.8273	1037	2678.9062	3127.3625
982	2513.6715	2938.2535	1038	2681.9224	3130.8128
983	2516.6640	2941.6801	1039	2684.9390	3134.2635
984	2519.6570	2945.1071	1040	2687.9560	3137.7147
985	2522.6505	2948.5347	1041	2690.9735	3141.1662
986	2525.6443	2951.9626	1042	2693.9914	3144.6181
987	2528.6386	2955.3910	1043	2697.0096	3148.0705
988	2531.6334	2958.8199	1044	2700.0284	3151.5233
989	2534.6286	2962.2491	1045	2703.0475	3154.9765
990	2537.6242	2965.6788	1046	2706.0670	3158.4301
991	2540.6203	2969.1090	1047	2709.0869	3161.8842
992	2543.6168	2972.5396	1048	2712.1073	3165.3386
993	2546.6138	2975.9706	1049	2715.1281	3168.7935
994	2549.6111	2979.4020	1050	2718.1493	3172.2487

Table 2. The diversity of species $M(S')$ characteristic of MacArthur's model for various numbers of hypothetical species S'^{a}

S'	$M(S')$	S'	$M(S')$	S'	$M(S')$	S'	$M(S')$
1	0.0000	51	5.0941	102	6.0792	205	7.0783
2	0.8113	52	5.1215	104	6.1069	210	7.1128
3	1.2997	53	5.1485	106	6.1341	215	7.1466
4	1.6556	54	5.1749	108	6.1608	220	7.1796
5	1.9374	55	5.2009	110	6.1870	225	7.2118
6	2.1712	56	5.2264	112	6.2128	230	7.2434
7	2.3714	57	5.2515	114	6.2380	235	7.2743
8	2.5465	58	5.2761	116	6.2629	240	7.3045
9	2.7022	59	5.3004	118	6.2873	245	7.3341
10	2.8425	60	5.3242	120	6.3113	250	7.3631
11	2.9701	61	5.3476	122	6.3350	255	7.3915
12	3.0872	62	5.3707	124	6.3582	260	7.4194
13	3.1954	63	5.3934	126	6.3811	265	7.4468
14	3.2960	64	5.4157	128	6.4036	270	7.4736
15	3.3899	65	5.4378	130	6.4258	275	7.5000
16	3.4780	66	5.4594	132	6.4476	280	7.5259
17	3.5611	67	5.4808	134	6.4691	285	7.5513
18	3.6395	68	5.5018	136	6.4903	290	7.5763
19	3.7139	69	5.5226	138	6.5112	295	7.6008
20	3.7846	70	5.5430	140	6.5318	300	7.6250
21	3.8520	71	5.5632	142	6.5521	310	7.6721
22	3.9163	72	5.5830	144	6.5721	320	7.7177
23	3.9779	73	5.6027	146	6.5919	330	7.7620
24	4.0369	74	5.6220	148	6.6114	340	7.8049
25	4.0937	75	5.6411	150	6.6306	350	7.8465
26	4.1482	76	5.6599	152	6.6495	360	7.8870
27	4.2008	77	5.6785	154	6.6683	370	7.9264
28	4.2515	78	5.6969	156	6.6867	380	7.9648
29	4.3004	79	5.7150	158	6.7050	390	8.0022
30	4.3478	80	5.7329	160	6.7230	400	8.0386
31	4.3936	81	5.7506	162	6.7408	410	8.0741
32	4.4381	82	5.7681	164	6.7584	420	8.1087
33	4.4812	83	5.7853	166	6.7757	430	8.1426
34	4.5230	84	5.8024	168	6.7929	440	8.1757
35	4.5637	85	5.8192	170	6.8099	450	8.2080
36	4.6032	86	5.8359	172	6.8266	460	8.2396
37	4.6417	87	5.8524	174	6.8432	470	8.2706
38	4.6792	88	5.8687	176	6.8596	480	8.3009
39	4.7157	89	5.8848	178	6.8758	490	8.3305
40	4.7513	90	5.9007	180	6.8918	500	8.3596
41	4.7861	91	5.9164	182	6.9076	550	8.4968
42	4.8200	92	5.9320	184	6.9233	600	8.6220
43	4.8532	93	5.9474	186	6.9388	650	8.7373
44	4.8856	94	5.9627	188	6.9541	700	8.8440
45	4.9173	95	5.9778	190	6.9693	750	8.9434
46	4.9483	96	5.9927	192	6.9843	800	9.0363
47	4.9787	97	6.0075	194	6.9992	850	9.1236
48	5.0084	98	6.0221	196	7.0139	900	9.2060
49	5.0375	99	6.0366	198	7.0284	950	9.2839
50	5.0661	100	6.0510	200	7.0429	1000	9.3578

^{a} After Lloyd and Ghelardi, 1964.

Index_____